HIGHER CHEMISTRY

HIGHER CHEMISTRY

John MacGregor

Oliver & Boyd O&B

Oliver & Boyd
Longman House, Burnt Mill, Harlow, Essex CM20 2JE, England and Associated Companies throughout the world.

An imprint of the Longman Group UK Ltd

First published 1992
Second impression 1993
ISBN 0 050 05080 X

Designed by Martin Adams

Set in 10/12 pt Times Roman
Printed in Singapore

Produced by Longman Singapore Publishers Pte Ltd

The Publisher's policy is to use paper manufactured from sustainable forests.

Acknowledgements

We are grateful to the following for permission to reproduce photographs: Bryan and Cherry Alexander, 5.48 and cover, and 4.11; Allsport UK, 7.24 (photo Chris Raphael); Aviation Picture Library, 8.13; Barnaby's Picture Library, 5.5 (photo Dick Cramp); B & H (Leics) Ltd, Loughborough, manufacturers of RigiDrain, 2.9; BP Chemicals, 2.40; British Alcan Primary and Recycling, 5.55; British Nuclear Fuels, 8.22; Cadburys 1.35; Calor Gas, 3.12; Camera Press, 1.1 (photo J. Paul); John Cleare Mountain Camera, 8.16; Dust Control (Scotland) 1.11; Dr. Alex Eliot, Department of Clinical Physics and Bioengineering, Glasgow, 8.28; English Adhesives and Chemicals, 5.38; Esab Group (UK) for Murex, 6.4; Exxon Chemical Olefins Inc, 2.15; Fyffes, 2.7; William Grant and Sons, 2.25; Holt Studios International, 4.19 (photo Nigel Cattlin); ICCE Photo Library, 7.25 (photo Philip Steele); ICI Chemicals and Polymers, 7.15; ICI Films, Dumfries, 4.25; Industrial and Mining Bit Manufacturing, Moudon, 5.14; M.W. Kellogg, Houston, 2.49; Johnson Matthey Catalytic Systems, 1.30; Oxford Scientific Films, 4.16 (photo M. Gibson/Animals Animals) and 7.23 (photo David Fox); Panos Pictures, 2.51 (photo Paul Harrison), 4.2 and 4.7 (both Ron Giling); Science Photo Library, 4.28 (photo Adam Hart-Davis), 4.32 (photo St. Bartholomew's Hospital), 4.41 (photo Alex Bartel) and 8.17 (photo Novosti); Brian Simpson Photography/Associated Octel, 5.59; Steetley Magnesia Products, 7.33 (photo Turners Photography); Welding Institute, Cambridge, 6.21.

Figure 2.24 was taken by John Birdsall; 5.24 by Gareth Boden; 8.6 by Longman Photographic Unit. All other photographs by Graeme Fleming.

Cover photo: B. & C. Alexander Icebergs in North-west Greenland (also see section 5.15.2)

CONTENTS

PREFACE

This book has been written specially for the Higher Grade Chemistry syllabus in Scotland. The aim throughout is to help you, the student, develop a clear understanding of the many chemical concepts and ideas which underpin the learning outcomes. These outcomes are set out in the Scottish Examination Board *Arrangements* document (see overleaf) and are a guide to the level of knowledge and understanding required.

The text is divided into the eight units of the syllabus. Learning outcomes for each unit are developed and put into context. Placed before each unit is a short summary of 'assumed knowledge and understanding'. This is content, relevant to the Higher unit, which is covered in the Standard Grade syllabus: it is a base upon which further knowledge and understanding can be constructed rather than starting from scratch.

In the syllabus, the learning outcomes for the course have been divided into fundamental (F) and non-fundamental (NF) topics. This differentiation was introduced to help students distinguish 'core' content from 'extension' content. Approximately two-thirds of examination marks are allocated to assessment of fundamental outcomes and the remaining one-third to non-fundamental outcomes. A broad outline of core and extension content is given before each unit to let you see the overall framework. More specific guidance is given at the end of a unit; learning outcomes, condensed or amalgamated where appropriate in the interests of economy, are listed with F outcomes coming first. NF outcomes are identified by a label (▲) and placed towards the end of the list.

Higher Chemistry students are required to demonstrate their familiarity with the practical side of certain experiments; called 'prescribed practical activities' (PPAs) by the Scottish Examination Board. Theoretical support for each of the PPAs is fitted into the relevant part of the text so that experiment and theory are combined naturally and put into proper context. Also dispersed throughout the text, in appropriate places, are 'sample exercises'. These worked examples serve to consolidate 'numerical' ideas and develop your answering technique.

A selection of original problem solving exercises, developed to be in line with current Higher type questions and reflecting the philosophy of the Higher syllabus, is provided at the end of each unit. Such exercises give you the opportunity to apply your chemical knowledge and understanding to situations and contexts which may be unfamiliar. The problem solving abilities or skills to be assessed are listed in the *Arrangements* and are as follows:

1 selecting information from different and/or multiple sources

2 constructing tables, charts, graphs and diagrams in order to present data

3 selecting an appropriate format for the presentation of data

4 selecting and/or combining numerical procedures in order to carry out a calculation

5 selecting or suggesting an experimental procedure

6 differentiating between functioning and non-functioning variables in an experimental procedure

7 identifying and controlling functioning variables

8 drawing conclusions from chemical information including experimental results and equations

9 explaining results and conclusions

10 using hypotheses or generalisations to make predictions for related situations.

The exercises offered span the range of problem solving skills and also vary in their level of difficulty. The key ability being tested is indicated at the end of the question (for example, ability 1 is coded PS skill 1). It is assumed that you have access to the official databook (see below). At the end of the book outline answers are provided; these are for the purpose of allowing you to check your responses and not to short-circuit the answering process!

The Higher Chemistry course is a very demanding exercise: the more thorough the understanding you can achieve, the more satisfaction you will gain. You will come to regard facts and figures less as a burden and more as a form of helpful supporting evidence or a challenge to your intellect. A study of chemistry should, when all is said and done, be an enjoyable experience: I sincerely hope you find it to be so.

JMG (1992)

Scottish Examination Board publications

Scottish Certificate of Education: Higher Grade and Certificate of Sixth Year Studies. Revised Arrangements in Chemistry (1990)

Chemistry (Revised) – Higher Grade and Certificate of Sixth Year Studies. Data Booklet (1990)

Unit 1
Controlling Reaction Rates

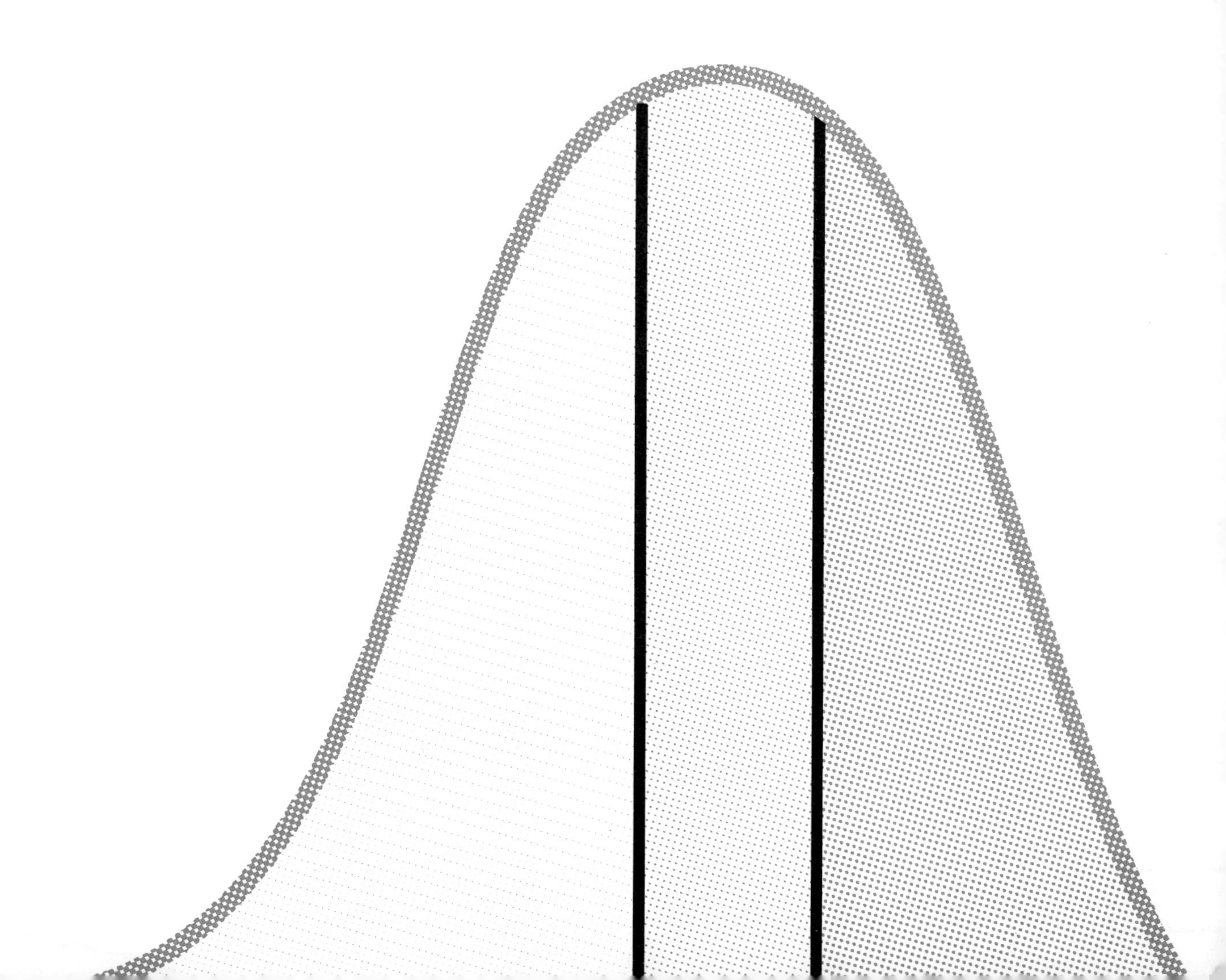

Assumed Knowledge and Understanding

Before starting on Unit 1 you should know and understand:
the effect of concentration, particle size, temperature and catalysts on the speed of a reaction.

Outline of the Core and Extension Material

As you progress through Unit 1 you should, at least, try the fundamental content listed under **core** and, if possible, pick up the extra and/or more difficult content listed under **extension**.

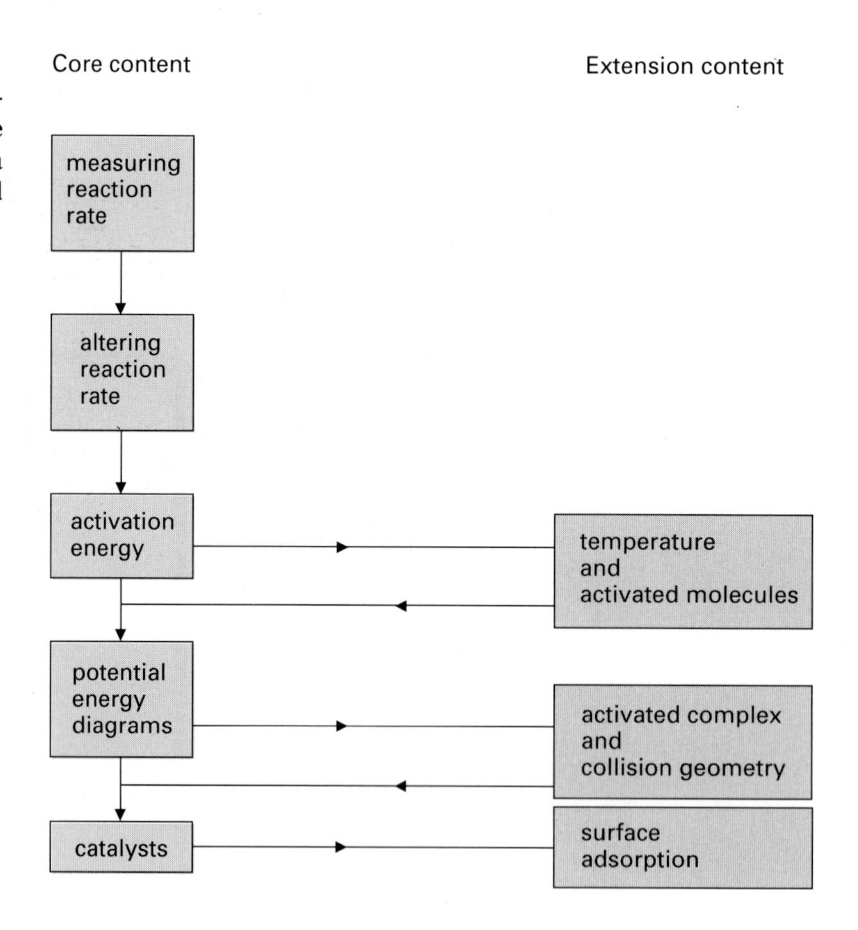

1 CONTROLLING REACTION RATES

1.1 Reaction rates

Chemical kinetics is the study of rates of reaction and how the rate depends on certain factors: these factors include concentration, particle size, temperature and catalysts.

Industrial chemists are concerned not merely with converting one substance into another but with bringing about the chemical change as quickly, easily and cheaply as possible. To achieve this economic goal they need to understand how the reaction takes place, i.e. the reaction mechanism. Such understanding is gained from studying reaction rates.

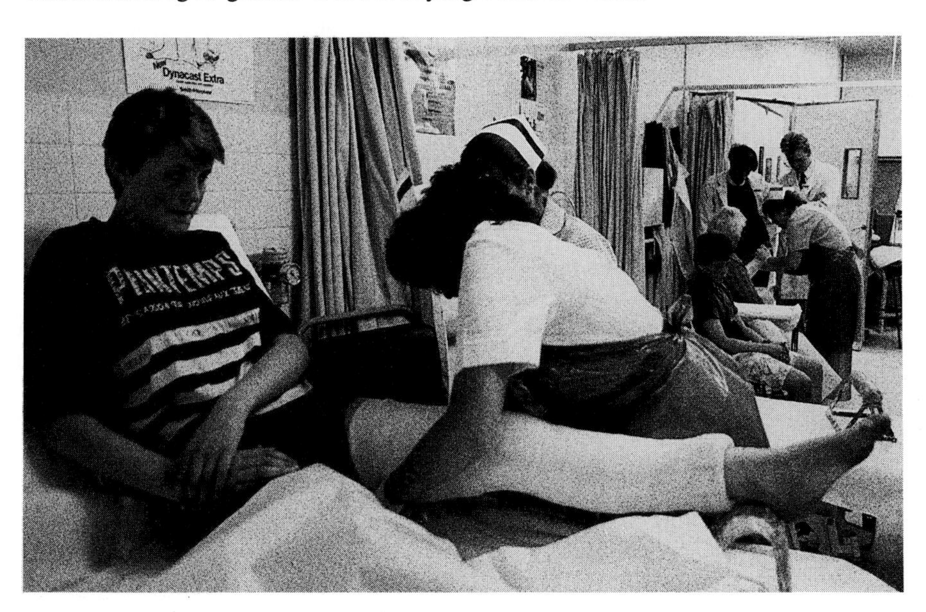

Figure 1.1 Applying a plaster cast; the rate at which the plaster sets is important – it must be neither too fast or too slow

1.2 Measuring reaction rate

During a chemical reaction reactants are being used up and products are forming. As the concentration of reactants falls so the concentration of products rises.

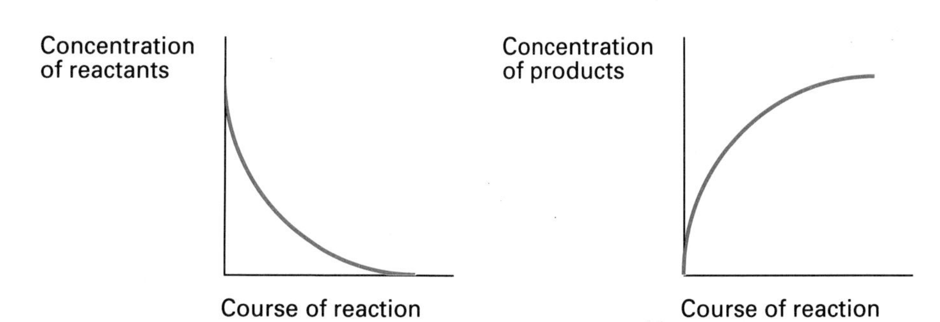

Figure 1.2 Change of concentration with time

To measure the rate of a reaction we have to measure how much of a reactant is being used up, or how much of a product forms, in a given time interval.

$$\text{Rate} = \frac{\text{change in concentration of a reactant or product}}{\text{time taken for the change to occur}}$$

$$= \frac{\Delta\,(\text{concentration})}{\Delta\,(\text{time})} \quad \text{where } \Delta = \text{change in value of} \ldots$$

Much practical ingenuity has been used to determine the rates of reactions. Any property which changes in step with concentration can be measured. For average-paced reactions, properties such as changes in mass, volume, pressure, conductivity and colour intensity have been observed in order to measure rate. Reactions which are extremely slow or tremendously fast present extra challenges in the estimation of rates. New techniques have been devised to allow us to study reactions which are finished in a matter of microseconds.

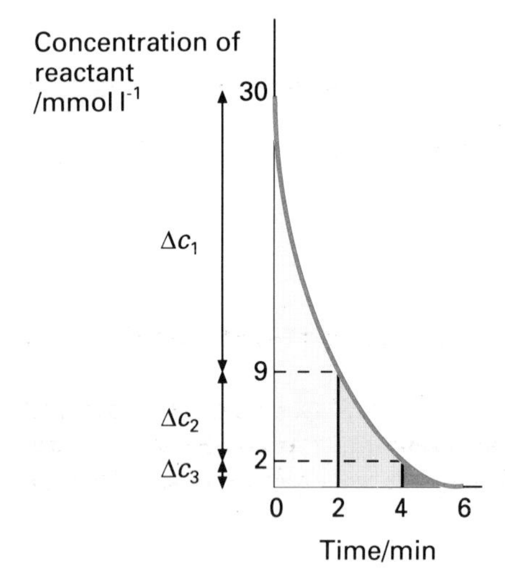

1.2.1 Average rate of reaction

The rate of reaction is defined as change in concentration over a given time interval. Change of concentration may refer to either reactant or product. Which time interval is chosen decides the **type of rate** measured. Figure 1.3 represents the rates of disappearance of reactant and appearance of product for a reaction. (For simplicity, concentration is given in millimoles per litre.) The rate of reaction changes as it is being measured. We can measure, therefore, only the **average** rate of reaction over a period of time.

If the total time taken for the reaction to run its full course is divided into three equal time intervals, Δt, then the **average** rate of reaction over each of the three stages is as follows.

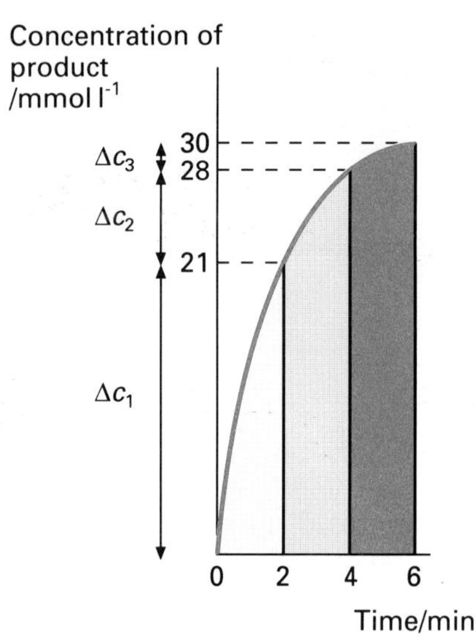

Figure 1.3 Three stages of reaction

$$\text{Average rate for stage 1} = \frac{\Delta c_1}{\Delta t} = \frac{21\ \text{mmol l}^{-1}}{2\ \text{min}} = 10.5\ \text{mmol l}^{-1}\ \text{min}^{-1}$$

$$\text{Average rate for stage 2} = \frac{\Delta c_2}{\Delta t} = \frac{7\ \text{mmol l}^{-1}}{2\ \text{min}} = 3.5\ \text{mmol l}^{-1}\ \text{min}^{-1}$$

$$\text{Average rate for stage 3} = \frac{\Delta c_3}{\Delta t} = \frac{2\ \text{mmol l}^{-1}}{2\ \text{min}} = 1.0\ \text{mmol l}^{-1}\ \text{min}^{-1}$$

Clearly the rate of reaction is decreasing all the time, being relatively fast over the initial stage and very much slower in the final stage when the reaction is approaching completion. The average rate for the whole reaction could be calculated by considering the **total** reaction time.

$$\text{Average rate for whole reaction} = \frac{\text{total } \Delta\,(\text{concentration})}{\text{total } \Delta\,(\text{time})} = \frac{30\ \text{mmol l}^{-1}}{6\ \text{min}}$$

$$= 5\ \text{mmol l}^{-1}\ \text{min}^{-1}$$

1.2.2 Instantaneous rate of reaction

If the time interval over which changes in concentration, etc., are observed is made smaller and smaller, the final result would be to obtain an **instantaneous** rate of reaction at any selected time t: a rate $\Delta c/\Delta t$ given by the **slope** (gradient) of the tangent to the curve at time t.

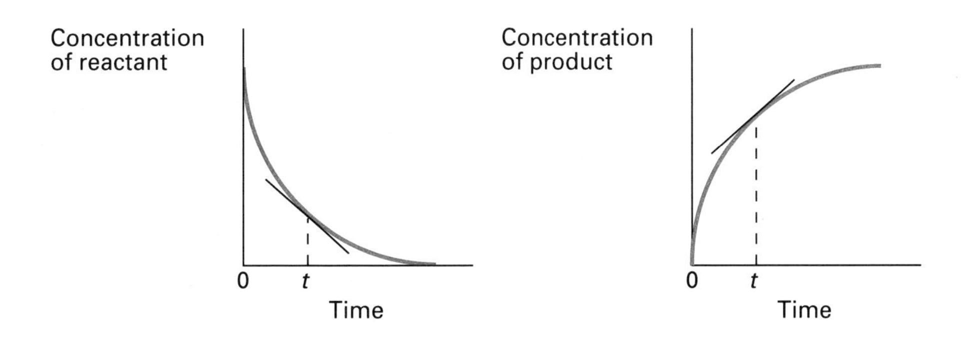

Figure 1.4 Instantaneous rate at time t

The steepness of the slope at any point gives an indication of the size of the rate.

1.2.3 Reciprocals of times taken as a measure of rate

The rates of the same reaction under different conditions can be compared by measuring the times taken for the **same amount** of reaction to occur. If the rate is high the time taken will be small, and conversely if the rate is low the time taken will be large.

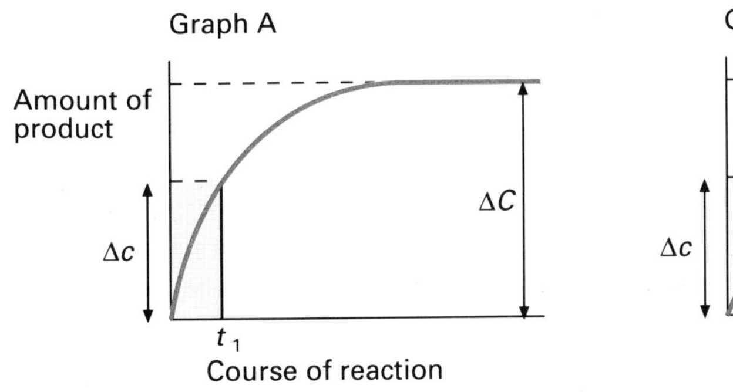

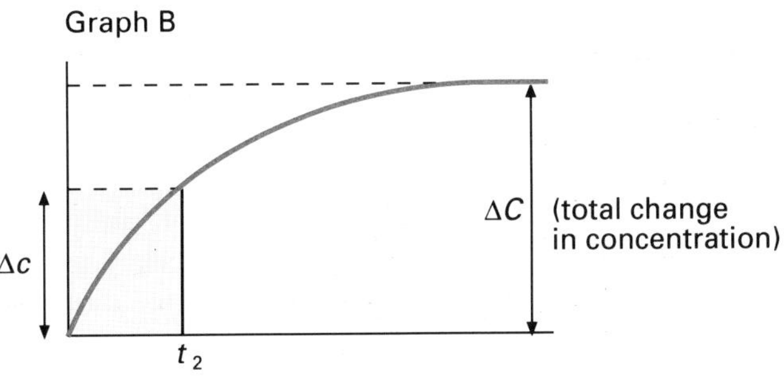

Figure 1.5 Comparing reaction times

Graph B represents a slower reaction than graph A. The average rates of reaction for the **initial** stages in A and B clearly differ.

$$\text{In A, average initial rate} = \frac{\Delta c}{t_1}$$

$$\text{In B, average initial rate} = \frac{\Delta c}{t_2}$$

Since the change in concentration, Δc, is the same in each experiment, the rates in A and B are proportional to **reciprocals** of times taken, i.e. rate A $\propto$ $1/t_1$ and rate B $\propto$ $1/t_2$ where $\propto$ = 'is proportional to'. It is important to make measurements of rate over periods of time which are short compared to the total reaction time.

$\Delta c/\Delta C$ must represent a small fraction of the total reaction for $1/t$ values to be valid indicators of the initial rate of reaction. The smaller the value of Δc taken, the more the average initial rate ($\Delta c/\Delta t$) approximates to the instantaneous rate of reaction at the very beginning.

Sample exercise

$10\ cm^3$ of CO_2 are produced in 18 seconds by a chalk–acid reaction (reaction A). A second chalk–acid reaction (reaction B) produces the same volume of CO_2 ($10\ cm^3$) in 6 seconds. Compare the rates of the reactions.

Method

If rate $\propto \dfrac{1}{\text{time}}$, $\dfrac{\text{rate of reaction A}}{\text{rate of reaction B}} = \dfrac{\text{time taken in B}}{\text{time taken in A}}$

$$= \frac{6\ \text{seconds}}{18\ \text{seconds}}$$

$$= \frac{1}{3}$$

Rate of reaction B is therefore three times as large as the rate of reaction A.

1.3 Effect of concentration on rate

Reactions generally become slower with the passage of time. For a reaction involving a single reactant such as

$$A \longrightarrow \text{products}$$

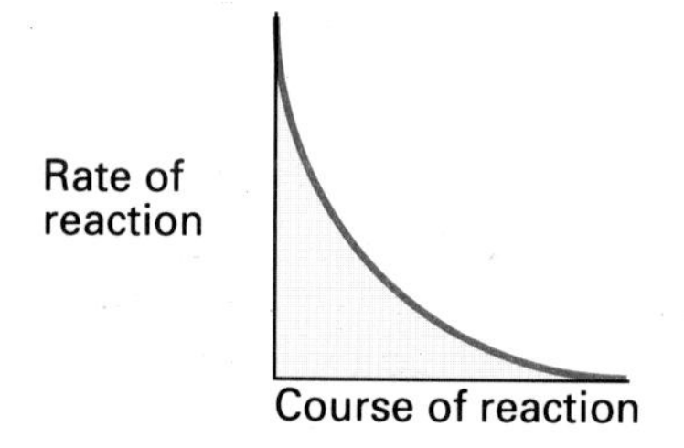

Figure 1.6 Change of reaction rate with time

this gradual reduction in rate can be attributed to the gradual lowering of the concentration of A, as A is used up. To investigate the effect of concentration more fully, it is necessary to start with different **initial** concentrations and measure **average** rates of reaction over a selected time. This time may be the whole duration of the reaction or merely the time taken for the reaction to reach a certain point. In a reaction involving **two** reactants such as

$$A+B \longrightarrow \text{products}$$

the concentration of B must be fixed while that of A is deliberately altered and vice versa.

1.3.1 Thiosulphate–acid reaction

A suitable case for study is the reaction between solutions of thiosulphate ions ($S_2O_3^{2-}$) and acid (H^+ ions) which causes the formation of pale yellow sulphur:

$$\text{thiosulphate} + \text{acid} \longrightarrow \text{sulphur} + \text{other products}$$

A convenient 'end-point' for this reaction is when the emerging sulphur finally obscures a cross marked on paper placed beneath the reaction flask. The effect of changing the concentration of either the thiosulphate or acid can

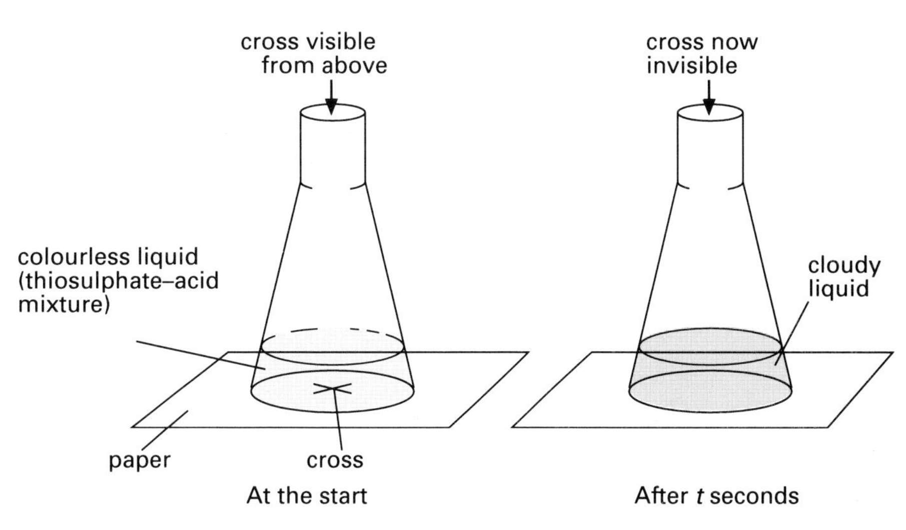

Figure 1.7 Thiosulphate–acid reaction

be investigated by measuring the rate at different concentrations. In each experiment,

$$\text{rate} = \frac{\text{amount of sulphur required to obscure the cross}}{\text{time taken for that amount of sulphur to be formed}}$$

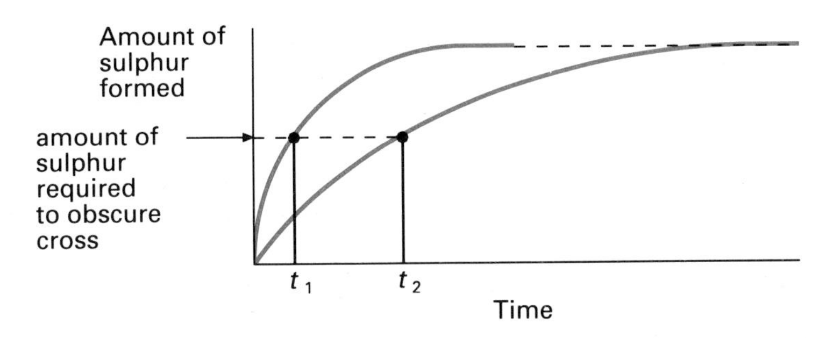

Figure 1.8 Formation of sulphur

The quantity 'amount of sulphur required to obscure cross' is assumed to be the same in each experiment and so the rate of reaction is proportional to ($\propto$) the **reciprocal of the time taken** for the cross to be obscured (t).

$$\textit{Therefore rate of reaction} \propto \frac{1}{t}$$

A shorter time taken (t_1) means a faster reaction, and a longer time (t_2) means a slower reaction.

If the reciprocals of the measured times are plotted against the various concentrations taken, the following graphs are obtained.

Note $[x]$ = 'concentration of x'.

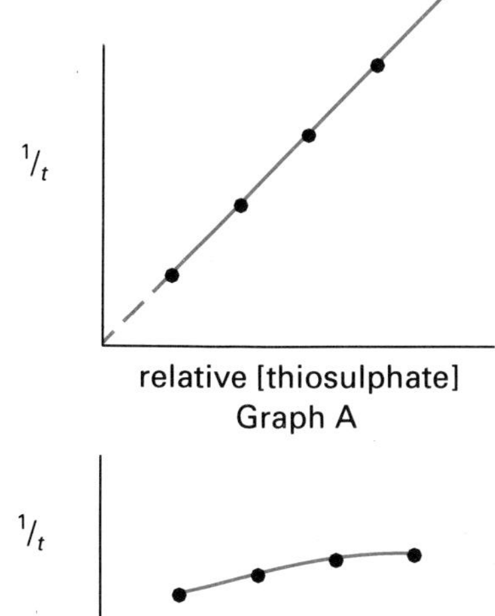

Figure 1.9 Effect of change of concentration on rate

From graph A (a straight line passing through the origin) we can conclude that the rate of this reaction is **directly proportional** to the concentration of thiosulphate ions, e.g. doubling the concentration of thiosulphate doubles the rate, and so on.

Graph B, on the other hand, indicates that there is only a slight increase in rate for a twofold, threefold and fourfold increase in the acid concentration. Doubling the concentration of acid slightly raises the rate of reaction but obviously not by a factor of two.

The relationships between reaction rate and concentration of reactants are often complicated. There is no simple way of predicting the relationship in advance. Each concentration/rate relationship must be investigated in carefully designed experiments and even then experimental results are often difficult to interpret. One can only say that **for many reactions** increasing the concentration of a reactant increases the reaction rate.

Frequently, as concentration increases, rate increases.

1.4 Effect of particle size on reaction rate

We can easily alter the concentration of a solution, but how can we increase the concentration of particles of solid exposed to another reactant? The answer is to increase the surface area of the solid by crushing it into smaller granules or even powder.

In the chalk–acid reaction (where carbon dioxide is evolved) the effect of particle size can be investigated by carrying out experiments with chalk in different particle sizes. The graph obtained is of the type shown in Figure 1.10.

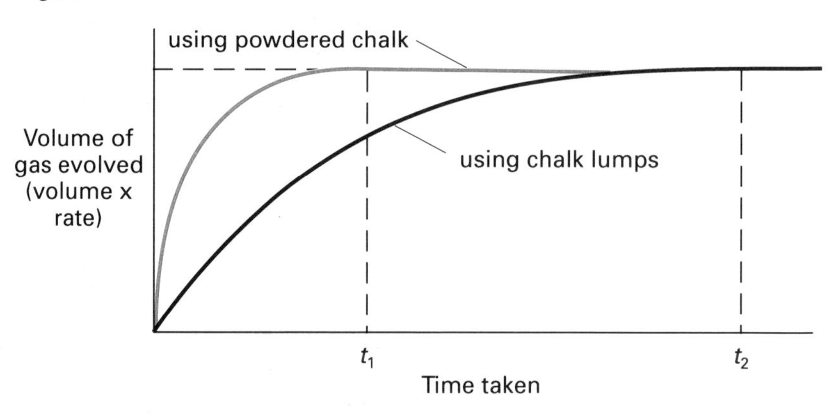

Figure 1.10 Effect of altering particle size

As the graph shows, the reaction of acid with powdered chalk takes place at a faster rate and finishes sooner (around t_1) than the reaction of acid with lump chalk which finishes around time t_2. Generally, increasing the surface area of any solid reactant raises the rate.

As fine division (surface area) increases, rate increases.

Figure 1.11 Because it is so finely divided and can therefore react very fast indeed, dust can present a high explosion risk and must be removed; this is a dust extraction system at a grains recovery plant in Glasgow

1.5 The collision theory of reaction rate

The collision theory suggests that, for a chemical reaction to occur, **particles must collide**. The particles may be atoms, molecules or ions and the collisions may take place in solution or in the gas phase. As a result of collisions, there can be a rearrangement of atoms, electrons and chemical bonds, leading to the formation of new particles.

The collision model adequately accounts for the effects of concentration and particle size on rate.

(a) Increasing the concentration of a reactant increases the number of particles in a given space and this leads to more frequent collisions. The higher frequency of collisions results in a higher rate of reaction.

As concentration increases, the number of collisions per second increases: therefore rate increases.

(b) In the chalk–acid reaction the reactants are in different phases: the chalk is solid and the acid is liquid. Here the rate of reaction depends upon the area of contact between the phases, in this case between the solid chalk and thc acid solution.

Powdering the chalk enlarges the contact area and creates the opportunity for more collisions in the same period of time.

The more finely divided a solid reactant is, the faster the rate of reaction.

As surface area increases, the number of collisions per second increases: therefore rate increases.

20 particles on a surfaceof the solid

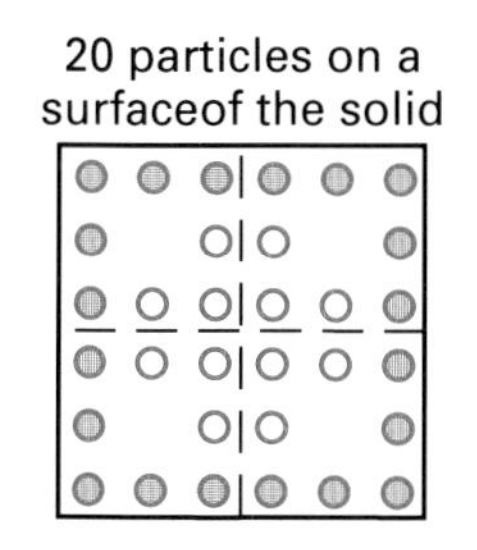

32 particles on surfaces of the quarters

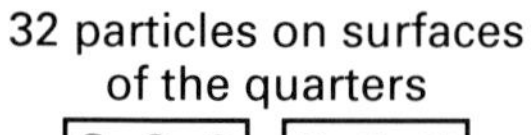

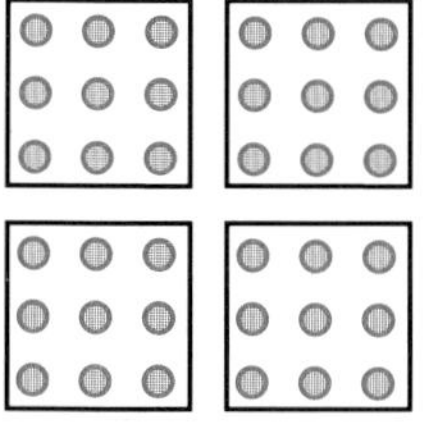

Figure 1.12 Increasing surface area

1.6 Effect of temperature on rate of reaction

The thiosulphate–acid reaction can also be used to investigate the effect of temperature on reaction rate. Typical results are shown in Figure 1.13. Temperature is shown in kelvins (0°C = 273 K). The reaction rate increases with a rise in temperature. On the basis of a simple collision theory this is what we would expect. With rising temperature, particles move more quickly and collide more frequently, thus boosting the rate of reaction.

What remains to be explained, however, is an experimentally observed dramatic rise in rate arising from a small increase in temperature. For a rise in temperature of approximately 10 K the rate of the thiosulphate–acid reaction roughly doubles. This effect is repeated in many other reactions.

It may be calculated however that, for a rise of 10 K, the actual number of collisions per second increases by only one to two per cent. Clearly rate of reaction depends on more than simply the number of collisions per second.

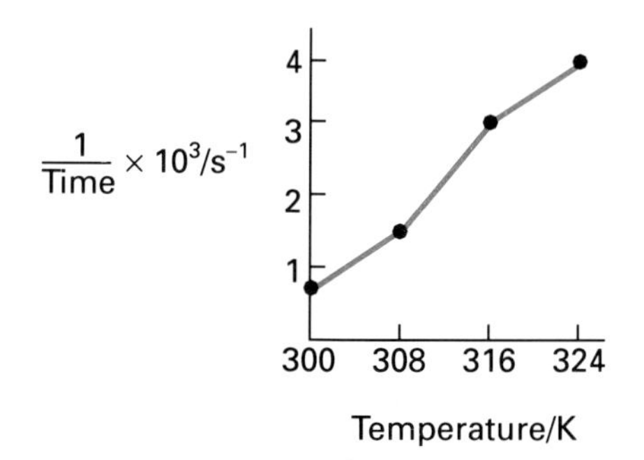

Figure 1.13 Rate of reaction at different temperatures

1.6.1 Activation energy

The simple collision concept must be altered to take account of this temperature effect. The idea that molecules need merely collide in order to react ignores one important fact: that the formation of new products requires bonds to be broken, other bonds to be formed and energy to be redistributed, and a considerable restructuring may be necessary.

Not only must particles collide in order to react, they must have **enough energy** to reorganise their bonds. This energy is described as **activation energy** and can be thought of as the minimum energy required for **effective** collisions, i.e. collisions which lead to reaction.

Collisions	energy $\geqslant$ activation energy →	products
	energy $<$ activation energy →	no products

High energy collisions are much more likely to lead to the breaking and making of chemical bonds than less forceful collisions. An insufficiently energetic collision will be unsuccessful in that the colliding molecules simply separate again without reacting.

Collisions between particles in solution, however, are complicated by the fact that the particles have to make their way through the solvent. The effect of the solvent is therefore very important in determining the movement of the particles and the nature of their collisions. Activation energy is a much more complex quantity in solution than it is in the gas phase.

Where does this activation energy come from?

1.6.2 Kinetic energy and temperature

The simplest situation for considering particles and their energies is that of the gas phase. Molecules in a gas are neutral and in rapid motion. The **temperature** of the gas is a measure of the **average kinetic (movement) energy** of the molecules. Some molecules will have high kinetic energies, some will have low kinetic energies. The energy of any molecule can change as it bumps into other molecules and either loses or gains energy as a result of these collisions. For the whole gas however, at a fixed temperature, the overall kinetic energy remains the same.

Figure 1.14 shows the distribution of molecular kinetic energies at two different temperatures: T_2 is higher than T_1.

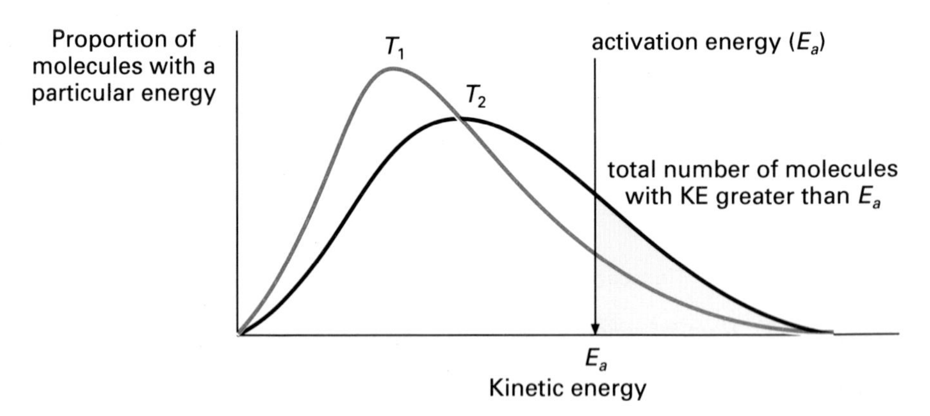

Figure 1.14 Effect of temperature on the proportion of molecules with higher energy

At the higher temperature there are not only fewer molecules with low energies but there are many more with high energies.

The activation energy for reaction is given the symbol E_a. The number of molecules with energies greater than the activation energy (indicated by the shaded portions of the graph) increases rapidly as the temperature rises.

At the temperature T_1 a **proportion** of molecules has more than the activation energy E_a and the reaction will proceed at a certain rate. At the higher temperature of T_2 a **greater proportion** of molecules possess energies above the activation energy threshold and the reaction rate will be higher.

An increase in temperature has therefore two effects on the movement of molecules.

(a) It increases the frequency of all collisions.

(b) It increases the energy of each collision.

However, it is the increase in the frequency of **successful** (energy $> E_a$) collisions that mainly causes an increase in the reaction rate.

1.7 The activated complex theory

The collision theory is a useful model for reactions in the gas phase. A more sophisticated theory and one which can be applied to all types of reaction is the **activated complex** theory. In this theory particular attention is paid to what happens just before reaction.

Take, for example, a reaction between an atom A and a diatomic molecule BC.

$$\underset{\text{reactants}}{A + BC} \longrightarrow \underset{\text{products}}{AB + C}$$

The process may be visualised in terms of **simultaneous** bond breaking and bond forming, with a half-way stage between reactants and products referred to as the **activated complex**.

$$A + B\text{—}C \longrightarrow \underset{\text{activated complex}}{A\text{---}B\text{---}C} \longrightarrow A\text{—}B + C$$

Full lines represent full covalent bonds while broken lines represent partially formed or partially broken covalent bonds.

The sequence of events is as follows. A and BC approach one another. If they get close enough bonding electrons are redistributed, resulting in partial formation of a bond between A and B and partial weakening of the bond between B and C. An activated complex ABC is now created but it is highly unstable as it can readily decompose one way or the other.

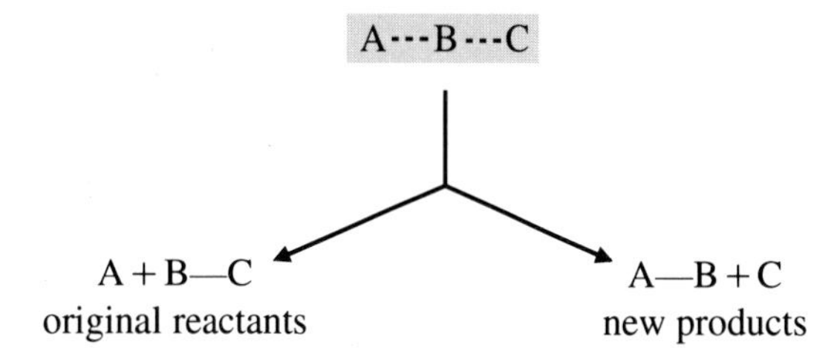

If the activated complex decomposes to form AB and C, a reaction has occurred.

1.7.1 Energy and the activated complex

The three stages of approach, reorganisation of bonding and separation may be viewed from an **energy** standpoint.

In any reaction, some chemical bonds are broken and some new bonds are formed. During the reaction the **total energy** of the participating particles **remains constant** but this total energy can be interconverted between kinetic energy (KE) and potential energy (PE) of the particles.

Compressed or coiled springs possess potential energy by virtue of the tensions created through squeezing or stretching them. In much the same way atoms, ions and molecules have potential energy by virtue of the attractive and repulsive forces which are in delicate balance within them.

An activated complex is a highly unstable arrangement of atoms and represents **a state of high potential energy**. The reaction sequence can be described in three stages and shown diagrammatically by means of an energy profile (Figure 1.15).

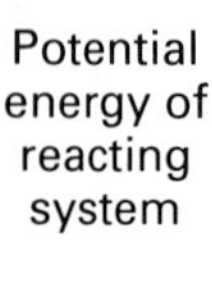

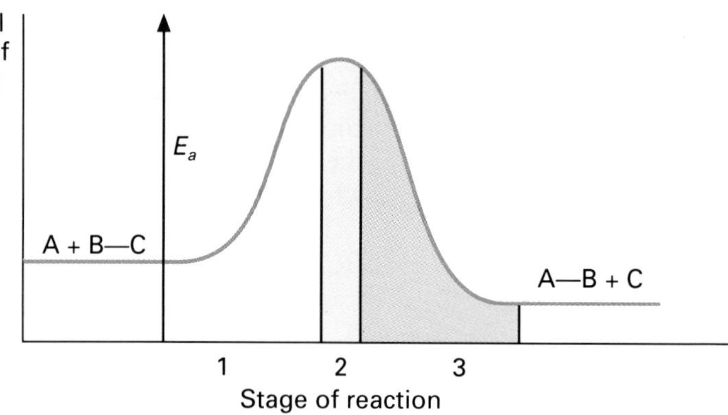

Figure 1.15 Energy profile for reaction

Stage 1 As the reacting particles A and BC approach each other their electron clouds repel (PE rises) and they slow down (KE falls).

If the particles are not moving fast enough there will be insufficient kinetic energy to drive them into the highly strung arrangement of the activated complex. They will rebound without reaction.

However, if A and BC carry kinetic energy equal to or greater than the activation energy (E_a), their KE can be converted into an equivalent amount of PE: the potential energy of the activated complex.

In a successful collision, fast-moving reactant molecules have sufficient KE to provide the energy of activation and produce the high energy activated complex.

Stage 2 The activated complex is a highly strained and stressed arrangement which is short-lived. The slightest displacement of A or C away from B causes it to decompose spontaneously into either reactants (therefore no reaction) or products (reaction). Its potential energy is partly reconverted into kinetic energy.

Stage 3 If the activated complex disintegrates to give new products AB and C, these products will have a different total potential energy from that of the reactants as the new bonding contains different attractive forces from the previous set-up.

The top of the curve (Figure 1.15) represents the potential energy barrier which must be surmounted if reaction is to occur.

The activation energy for the forward reaction is E_a: the difference between the potential energy of the reactants and the potential energy of the activated complex.

1.7.2 Activation energy: the effect of orientation

A simple example of an activated complex may be found in some single-step reactions between diatomic molecules, e.g. the decomposition of AB.

$$2AB \longrightarrow \underset{\text{activated complex}}{(A_2B_2)} \longrightarrow A + B$$

Not only must the two molecules of AB be brought together in order to collide, they must be held in precisely the right orientation to ensure that reaction can take place.

Figure 1.16 outlines two possible orientations. One is unfavourable towards reaction; in the other situation the geometrical arrangement of the molecules is favourable to a reaction.

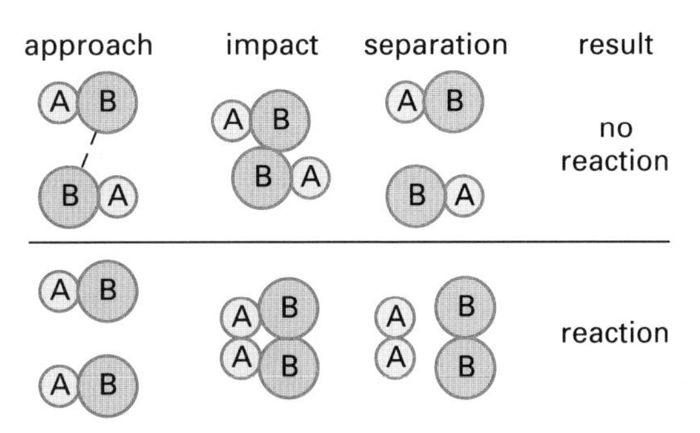

Figure 1.16 Favourable and unfavourable orientations

The energy profile for any reaction corresponds to some particular orientation or collision geometry. The experimentally observed activation energy will correspond to the most favourable orientation. Less favourable orientations of reacting particles will mean higher activation energies.

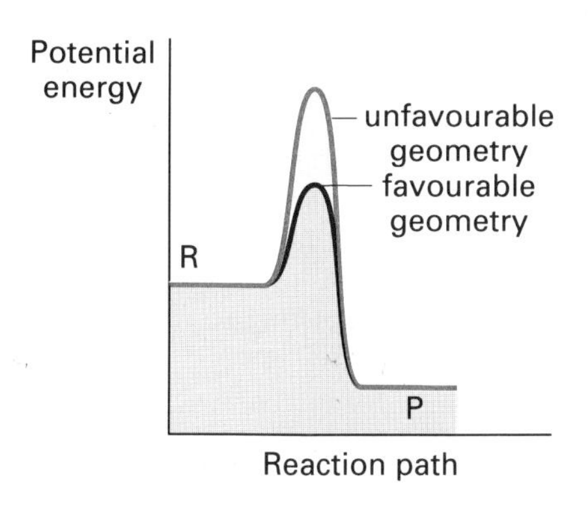

Figure 1.17 Effect of geometrical orientation on activation energy

1.8 Exothermic and endothermic reactions

Fast reactant particles give up their kinetic energy when they form the activated complex (AC): the KE is changed into PE within the complex. One might think that after a short time there would be no fast particles left. However, much if not all of the energy absorbed by the activated complex is converted back into kinetic energy when the complex breaks up. How much of that potential energy is regained by the products as kinetic energy? This depends on the natural potential energy level of the products.

If the potential energy of the products is less than the energy of the reactants, then surplus kinetic energy is carried off by the product particles and released to the surroundings as heat. The reaction is said to be **exothermic**.

The decrease in PE ⟶ surplus of KE among product particles ⟶ heat **given out** to surroundings

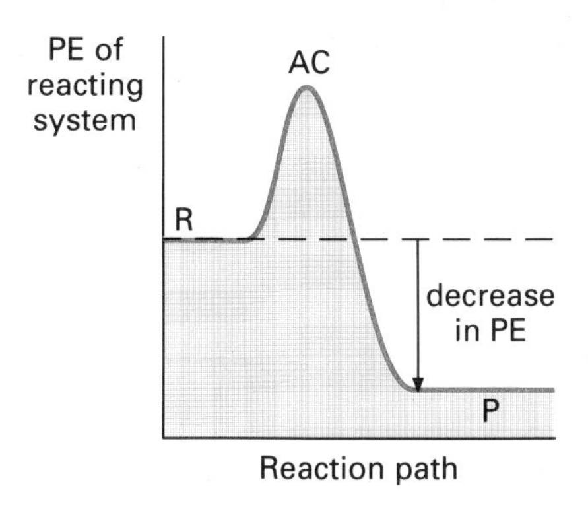

Figure 1.18 An exothermic reaction

If the potential energy of the products is more than the energy of the reactants, the kinetic energy returned when the activated complex breaks up is not able to compensate for the kinetic energy used up in forming the complex.

Consequently, extra kinetic energy must be supplied, from external sources, to maintain the supply of fast particles. This extra energy takes the form of heat absorbed from the surroundings. Such a reaction is said to be **endothermic**.

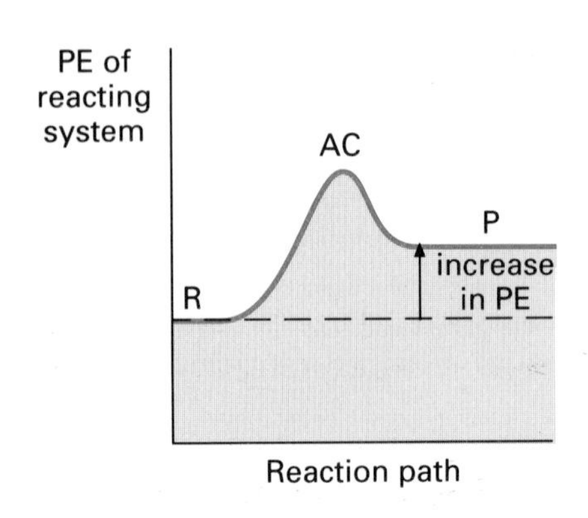

Figure 1.19 An endothermic reaction

The increase in PE ⟶ shortfall in KE ⟶ heat **taken in** from surroundings

The **net** energy change in a reaction is the difference between the energies of reactants and products and is **independent** of activation energies.

The rate at which the reaction occurs depends on the size of the activation energy and not the size of the net energy change. For example, a very exothermic reaction capable of releasing lots of energy may take place at a barely observable rate, if the activation energy barrier happens to be high.

The heat released in an exothermic reaction can be used to supply activation energy for the remainder of the reactants. Thus, once the exothermic reaction has been set in motion by a supply of heat at the start, the heat given out, as products form, can maintain a supply of activated molecules until reaction is complete. Combustion of candle wax is exothermic but it still requires heat, supplied by a lighted match, to activate a small proportion of hydrocarbon molecules into combination with oxygen. Thereafter the reaction is self-sustaining.

Figure 1.20 Once reaction has been initiated, this combustion is self-sustaining

1.9 Catalysis

Catalysts have been used since ancient times in such activities as making wine, bread and cheese. In 1894, Wilhelm Ostwald offered a definition of a catalyst which is still applicable today: 'catalysts are substances that speed up the rate of chemical reactions without themselves being consumed during the reactions'.

Catalysts are widely used in the manufacture of fuels, foods, chemicals and medicines. They also play a vital role in natural processes such as photosynthesis, cell metabolism, nitrogen fixation and so on.

The majority of industrial catalysts are metals which belong to the middle group of the periodic table (described as 'transition metals'), or oxides of these metals.

Catalysts are highly specific in what they do. One substance may catalyse a particular type of reaction and produce one type of product from a given reactant. On the other hand, another substance may catalyse a different type of reaction on the same reactant and thus produce a completely different product.

One important starting point for chemical synthesis is the mixture of carbon monoxide and hydrogen called 'syngas'. Syngas can be converted into a variety of useful chemicals through the agency of different catalysts.

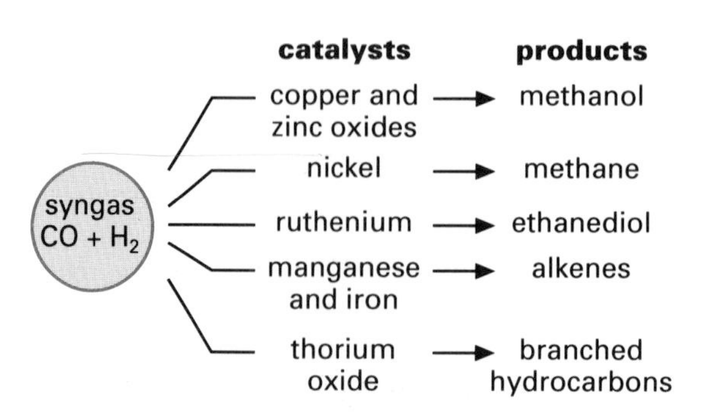

Figure 1.21 Variety of catalysts for converting syngas

1.9.1 Classification of catalysts

Catalysis is of two main types:

(a) heterogeneous catalysis

(b) homogeneous catalysis.

In **heterogeneous** catalysis, the catalyst is usually a solid; the reactants are liquids or, more frequently, gases. **The catalyst and reactants are in different phases or states.**

In **homogeneous** catalysis, the catalyst and reactants are usually present in a common liquid phase, i.e. in solution. **The catalyst and reactants are in the same state.**

Any catalyst, heterogeneous or homogeneous, catalyses a reaction by offering reactants an alternative lower energy pathway to products.

1.10 Heterogeneous catalysis

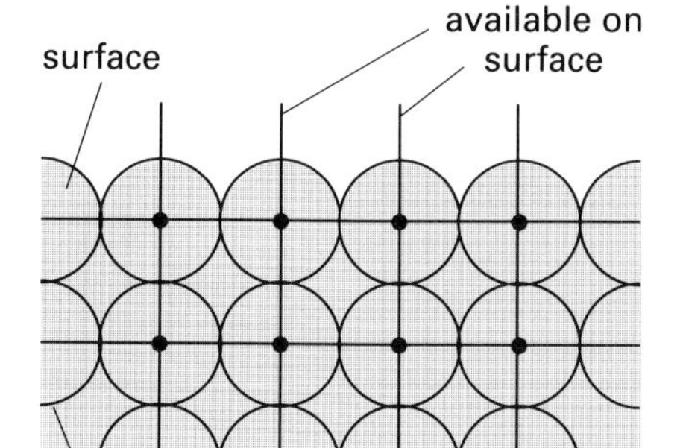

Figure 1.22 Surface bonding

The surfaces of metals and ionic solids are chemically reactive. In the inside of a metal each atom is fully bonded to its neighbours. At the outside surface, however, atoms have 'spare' bonds because there are no neighbouring atoms on one side. The surface therefore has the ability to form weak covalent bonds with atoms of another substance. The process of surface bonding is called **chemical adsorption** or **chemisorption**. Inevitably the structure of the adsorbed molecules will be different after chemisorption compared to before. Some bonds of the molecules may be weakened or even broken. (**Ad**sorption should not be confused with **ab**sorption which involves penetration of the substance into the inside of the solid.)

When two substances are adsorbed on the same surface and beside one another, reaction between them is more easily brought about. Reaction is being catalysed by the solid surface.

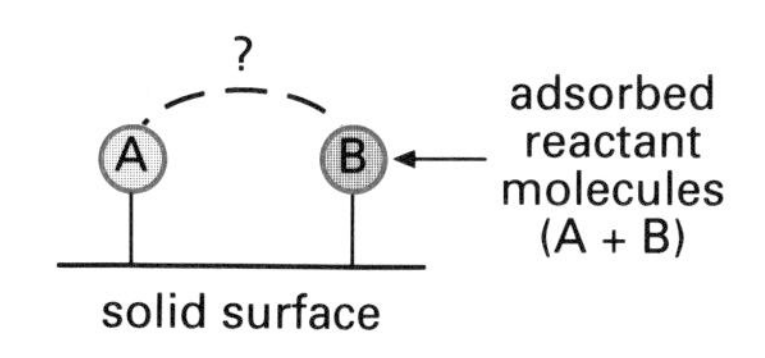

Figure 1.23 Reaction catalysed by a solid surface

A heterogeneous catalyst is a solid material prepared with a large surface area on which a chemical reaction can take place easily and speedily. Heterogeneous catalysts have been used for the production of large scale chemicals such as methanol and ammonia and in the conversion of oil fractions into petrol.

The effectiveness of a heterogeneous catalyst depends on how rapidly each surface site can adsorb reactants, allow them to rearrange chemically and then release the newly formed species (the product), so that the process can be repeated at that site.

One simple example of a heterogeneously catalysed reaction is the combination of hydrogen and iodine in the gas phase.

1.10.1 Catalysed hydrogen–iodine reaction

The combination of hydrogen and iodine can be brought about with the help of a heterogeneous catalyst such as platinum.

$$H_2(g) + I_2(g) \xrightarrow{Pt} 2HI(g)$$

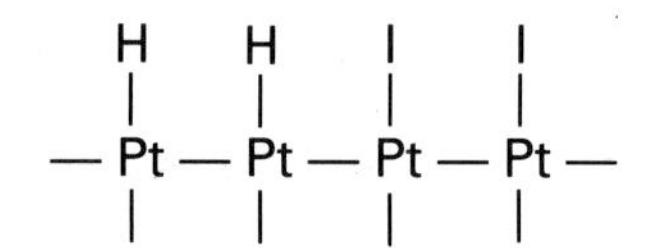

Figure 1.24 Co-adsorbed H and I atoms on Pt (the reaction intermediate)

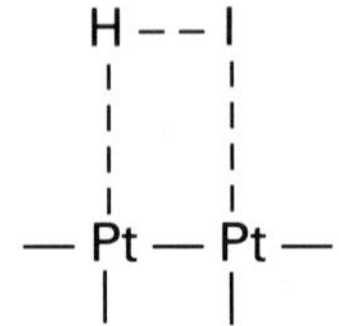

Figure 1.25 The second activated complex

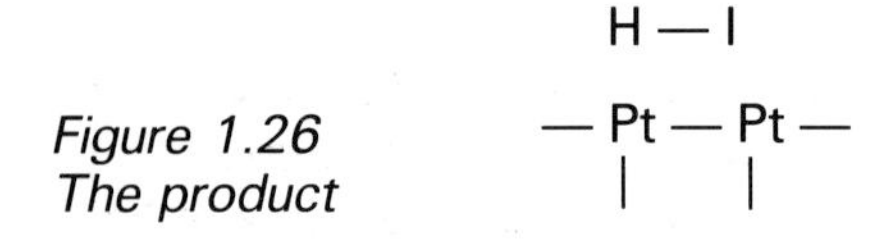

Figure 1.26 The product

The following mechanism has been suggested to explain the catalytic action.

(a) When H_2 and I_2 are adsorbed on platinum the molecules dissociate into atoms and the **atoms** are separately bonded to metal atoms.

(b) A second short-lived activated complex forms. It involves the platinum and adjacent hydrogen and iodine atoms.

(c) As the bond between H and I shortens and strengthens, the bonds to the platinum simultaneously stretch and weaken.

(d) A molecule of HI forms and departs, leaving the surface site vacant for adsorption of further atoms.

The **uncatalysed** reaction between hydrogen and iodine molecules depends on **random** collisions in the gas phase.

The presence of a platinum surface alters the situation. The hydrogen and iodine can now react through a stepwise process which might be condensed into two stages.

Stage 1 The formation of an intermediate species, H—Pt—I (Figure 1.25).

Stage 2 The simultaneous decomposition of this intermediate and the formation of hydrogen iodide.

There is an activated complex for each stage (A_1 and A_2) and two energy barriers to surmount, but each is much lower than that for the uncatalysed reaction.

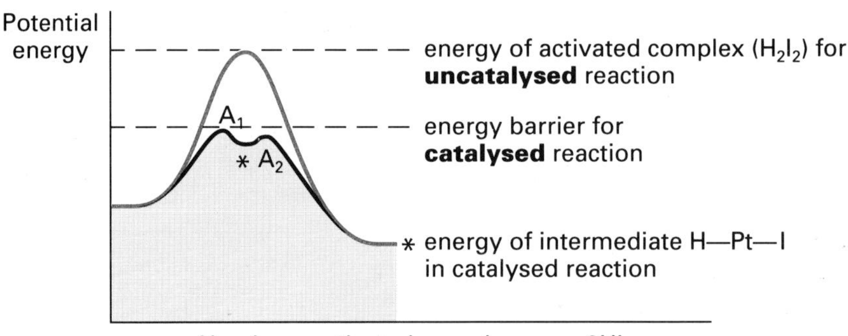

Figure 1.27 Energy profile diagram for uncatalysed and Pt-catalysed combination of H_2 and I_2

The catalyst provides a different mechanism or pathway for the reactant molecules to join together. There may be more steps than in the uncatalysed reaction but each step has a low energy barrier.

For the hydrogen–iodine reaction the relative values of activation energies are shown in Figure 1.28.

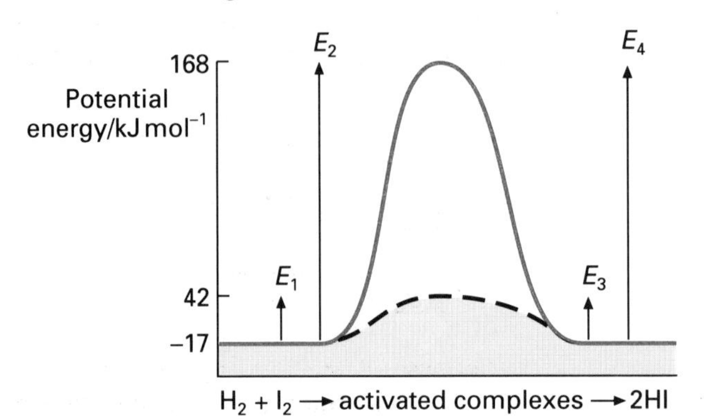

Figure 1.28 Energy profile for uncatalysed and Pt-catalysed reaction $H_2 + I_2 \rightleftharpoons 2HI$

The formation of hydrogen iodide is slightly exothermic and easily reversible. A catalyst lowers the activation energy for both forward and reverse reactions as shown by the figures for the respective activation energies given in Table 1.1.

Code	Reaction	Activation energy/ kJ mol^{-1}
E_1	Pt-catalysed $H_2 + I_2 \longrightarrow 2HI$	42
E_2	uncatalysed $H_2 + I_2 \longrightarrow 2HI$	168
E_3	Pt-catalysed $2HI \longrightarrow H_2 + I_2$	59
E_4	uncatalysed $2HI \longrightarrow H_2 + I_2$	185

Table 1.1 Activation energies for catalysed and uncatalysed reactions

A catalysed reaction proceeds through a lower activation energy route, each step involving a new activated complex. Many more reactant particles can now clear the energy barrier and so the rate of reaction is considerably increased without the need for a rise in temperature.

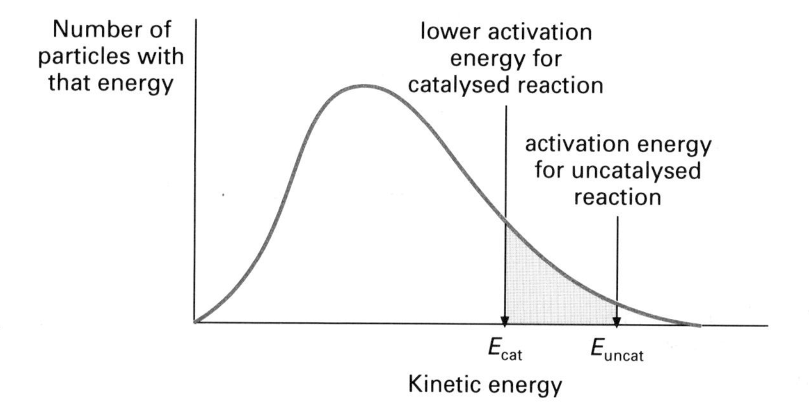

Figure 1.29 Effect of catalysis on proportion of particles which have the energy needed for reaction

1.10.2 Improving and destroying catalytic activity

Catalytic converters are fitted to car exhaust systems in order to reduce air pollution. So-called 'three-way' converters have been developed to simultaneously convert carbon monoxide, hydrocarbons and oxides of nitrogen in the exhaust gas to carbon dioxide, water and nitrogen.

$$CO \xrightarrow{\text{oxidation}} CO_2$$

$$C_xH_y \xrightarrow{\text{oxidation}} CO_2 + H_2O$$

$$NO_x \xrightarrow{\text{reduction}} N_2$$

The task of accelerating the oxidation of two pollutants and reduction of a third, all at the same time, has been given to a mixture of three expensive

Figure 1.30 Automobile catalysts of various forms, together with a catalytic converter sectioned to show its construction; the catalysts convert harmful pollutants to less harmful gases

metals. The metals are rhodium, platinum and palladium. To increase the catalysts' surface area the three precious metals are supported on a material which is largely alumina (aluminium oxide).

This supported catalyst (Rh, Pt and Pd metals on alumina) is coated on to honeycomb blocks made of ceramic or steel. This creates an overall surface area of a single converter unit equivalent to that of several football pitches. Traces of impurities may reduce the performance of a heterogeneous catalyst or even render it useless. It is for this reason that such catalysts operate only with cars using 'lead-free' petrol. The presence of lead compounds in the exhaust gases would interfere with the surface activity of the catalyst – a process known as **poisoning**. The 'poison' is adsorbed on the catalyst surface in preference to the desired reactant(s) and this prevents the catalyst operating properly.

Catalyst poisoning is also the reason why it is not possible to replace the very costly rhodium, platinum and palladium metals with cheaper base metals such as copper and nickel. These base metals are vulnerable to poisoning by the trace amounts of sulphur dioxide always present in car exhaust gases.

1.11 Homogeneous catalysis

Catalysts act by offering new reaction paths with lower activation energy 'hurdles'. In heterogeneous catalysis the new path results from the reactants being activated through adsorption on a solid surface.

With homogeneous catalysis the catalyst and reactants are closely mixed but the principle of offering a new reaction pathway still applies.

Take, for example, a reaction such as $A+B \rightarrow C$. A homogeneous catalyst may act by first forming an activated complex with A. This decomposes to give an intermediate compound. (Intermediate compounds differ from short-lived highly unstable activated complexes: they are sufficiently stable to be capable of isolation.) The intermediate then reacts with B to form a second activated complex which decomposes to give C and regenerate the catalyst. This catalytic cycle is outlined in Figure 1.31.

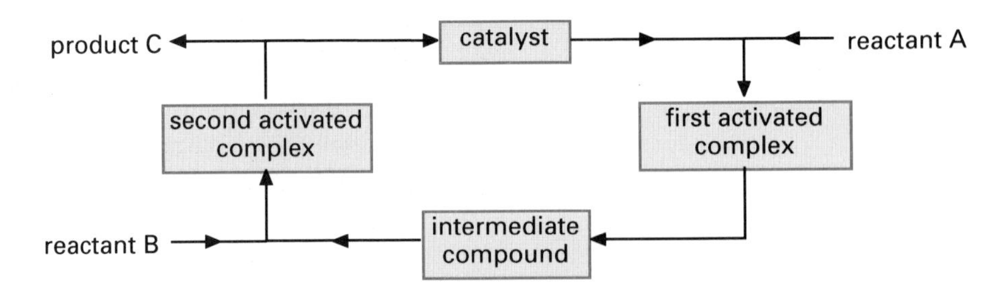

Figure 1.31 Catalytic cycle

An example of such homogeneous catalysis in the gas phase is the decomposition of ozone, O_3, in the stratosphere. Ozone molecules react with oxygen atoms to form oxygen molecules.

$$O_3 + O \longrightarrow 2O_2$$

This decomposition is catalysed by chlorine atoms which form an intermediate compound ClO.

$$Cl + O_3 \longrightarrow O_2 + ClO$$
$$ClO + O \longrightarrow Cl + O_2$$

Overall change: $O_3 + O \longrightarrow 2O_2$

The catalysis follows a circular path (see Figure 1.31), with chlorine atoms used up in the first step and then regenerated in the last step. The potential energy diagram for the reaction is similar to that described in Figure 1.27 for heterogeneous catalysis.

The layer of ozone gas in the stratosphere shields us from over-exposure to ultraviolet radiation from the sun. It is therefore very important that we restrict release, into the atmosphere, of any gases which might produce catalysts for decomposing the ozone.

Although industrial processes involving heterogeneous catalysts far exceed (in terms of volumes of chemicals produced) those using homogeneous catalysts, the latter, nevertheless, make a valuable contribution. Homogeneous catalysts may be more difficult to separate from the reaction mixture but they are more specific and controllable than heterogeneous catalysts.

A good example of a homogeneous catalyst helping to produce a single desired product is the manufacture of ethanoic acid from methanol and carbon monoxide.

$$\underset{\text{methanol}}{CH_3OH} + \underset{\text{carbon monoxide}}{CO} \longrightarrow \underset{\text{ethanoic acid}}{CH_3COOH}$$

The conversion is catalysed by a rhodium compound, and 99 per cent of the product obtained is ethanoic acid; this is a big improvement on other industrial processes for making the acid in which there is a smaller proportion of ethanoic acid in the mixture of products obtained.

1.12 Enzyme catalysis

The many complicated chemical changes which take place in the living cells of plants and animals are catalysed by large protein molecules called **enzymes**. Enzyme catalysis does not fit neatly into either category of catalysis, i.e. heterogeneous or homogeneous.

Enzymes function by joining, temporarily, with a **substrate** (reactant). The substrate molecule fits into a cavity on the large enzyme molecule in much the same way as a key fits into a lock. Here the substrate is held in a **precise** orientation which is favourable to the decomposition of that substrate or reaction with a second different substrate. This **enzyme–substrate** intermediate then decomposes to form products which move away from the site and allow the process to be repeated with other substrate molecules.

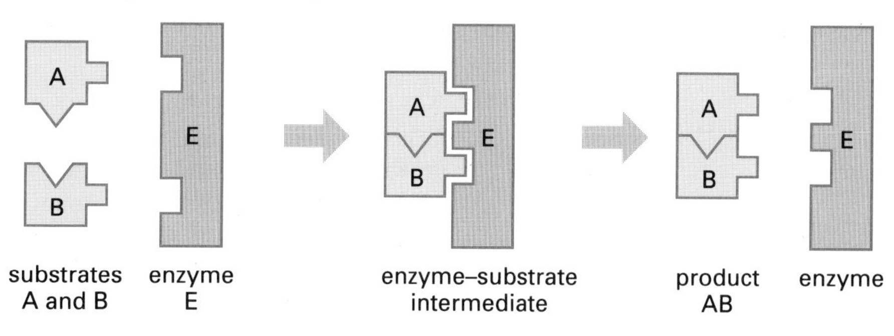

Figure 1.32 A model of enzyme action

Enzymes 'process' molecules for reaction at a phenomenal rate. **Catalase** is an enzyme which is present in most animal cells. Cell reactions involving dissolved oxygen often produce **hydrogen peroxide**, H_2O_2, as a by-product. This compound is toxic and has to be removed immediately, a task carried out by catalase. The catalase speeds up the decomposition of the peroxide into water and oxygen.

$$2H_2O_2 \longrightarrow 2H_2O + O_2$$

Such is catalase's ability to 'turn over' H_2O_2 molecules that up to five million molecules of the peroxide are decomposed by one catalase molecule every minute! This rate compares very favourably with that achieved by other non-enzyme catalysts and, of course, with the rate for the uncatalysed decomposition of H_2O_2. The **differences** in rate are caused by the different activation energies involved.

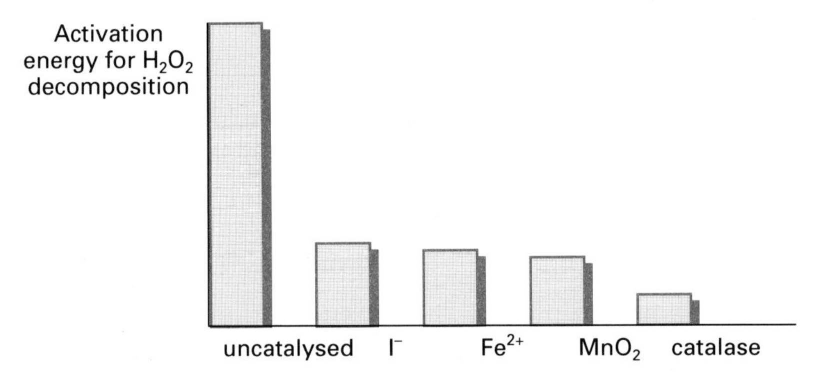

Figure 1.33 Catalysts for the decomposition of H_2O_2

Apart from their tremendous catalytic activity, enzymes are generally quite specific and selective in their action. Catalase's role in animal cells is simply to pick out hydrogen peroxide from amongst a crowd of other molecules and speed up its decomposition to prescribed products (water and oxygen). It has no other function to perform.

Other enzymes may be more versatile and operate on different substrates, but in such cases these enzymes are usually catalysing the same type of reaction, such as splitting a particular kind of bond, etc.

1.12.1 Enzymes in industry

Enzymes are natural biological catalysts and they are becoming more and more important in industry. They are 'clean' chemicals on at least two counts. Firstly, because enzymes are catalysts, unchanged by chemical reaction, they produce **few pollutants** when put to work in a chemical plant. Enzymes, however, have to be produced. A natural source has to be found and this is usually bacteria or fungi. The micro-organism is then grown in fermenters before the key enzyme is isolated. Any waste produced is in the form of a **fertile** slurry rather than harmful or toxic by-products and this provides a second good reason for using enzymes where possible.

The use of enzymes for commercial purposes is not a recent development; the enzyme **rennin** (or rennet) has been used for thousands of years in the making of cheese. It breaks down the protein (casein) in milk to produce curds.

Figure 1.34 'Biological' washing powders contain enzymes that decompose organic stains

Perhaps the most familiar industrial use of enzymes has been in washing powders. The first biological detergent marketed was a German product, 'Burnus', in 1913. Dirt is a mixture of various inorganic compounds and biological material which may be protein, fat (oil) or carbohydrate. These 'biological' stains include blood, urine, perspiration, food and drink. Protein stains are normally very difficult to remove and require a very hot wash (85–95 °C). The addition of protease (protein decomposing) enzymes to detergent powders greatly improves stain removal. The protease enzymes maintain their catalytic activity up to 65 °C. Enzymes called lipases are now added to washing powders to assist the breakdown of fat and oily stains.

A rather different reason for washing textiles is found in one area of fashion. Many people like jeans to look faded and worn. Manufacturers have therefore been obliged to put new denim through an expensive pre-washing exercise, using pumice stone. This 'stone-washing' process can now be more cheaply carried out using an enzyme called **cellulase**.

The food and drink industry is the major user of enzymes. Most of the enzymes used belong to the class called **hydrolases**. These enzymes catalyse the hydrolysis (breaking down) of large carbohydrate, protein and fat molecules into smaller molecules. Only a few examples can be selected here.

Baking

Amylase, a starch-decomposing enzyme, is naturally present in wheat flour but in variable amounts. The enzyme is now added to wheat flour to maintain a standard level and allow bakers to produce loaves of consistent quality. Amylase is also used extensively in the manufacture of gums and sugar syrups.

Brewing

When beer is chilled to temperatures below 283 K a fine precipitate known as 'chill haze' appears. The precipitate is an insoluble protein complex and can be removed by a protease enzyme. **Papain**, an enzyme extracted from papaya fruit, is commonly used for this purpose.

Tenderising

Meat can be made more tender by adding **protease** enzymes which break down connective tissue and soften the meat. Similar enzymes are used to help process skins and hides into leather. They increase the pliability of the hide, a process known as 'leather-bating'.

Sugar production

Fruit processing involves decomposing larger, less soluble carbohydrate molecules, such as starch, into smaller, more soluble and sweeter molecules, such as sugars. Originally acids were used to decompose starchy components but now enzymes do the job more cheaply and efficiently.

Pectinase is a pectin (carbohydrate) decomposing enzyme used to speed up the separation of fruit juice, e.g. apple juice, from the insoluble residue and produce a clear liquid.

Figure 1.35 Newly coated chocolates leaving the enrobing machine

Invertase splits the sugar, sucrose, into glucose and the highly sweet fructose.

$$\text{Sucrose} \xrightarrow{\text{invertase}} \text{glucose} + \text{fructose}$$

This enzyme is used to produce soft-centred confectionery. A soft fluid material cannot be effectively covered by chocolate on a production line. Instead firm sucrose, containing a small amount of invertase enzyme, is covered (enrobed) with chocolate. The sucrose gradually breaks down to form the more soluble fructose and glucose sugars; the interior of the chocolate becomes more liquid and a 'soft centre' is achieved.

Enzymes, like other homogeneous catalysts, are dispersed in solution and difficult to separate from the products. It is possible, however, to 'immobilise' enzymes by encapsulating them in plastic gels. This allows their recovery from solution and repeated use.

Enzymes tend to work within certain limits of temperature and pH; they are also easily destroyed by strong chemicals. Nevertheless they are so efficient and so selective in the changes they bring about that their potential for use in all areas of industry is enormous.

Summary of Unit 1

Having read and understood the information and ideas given in this unit, you should now be able to:

calculate average rates from experimental values of concentration (or volume or mass) and time

indicate the relationship between rate and the reciprocal of time

outline the collision theory and use it to explain the effects of concentration and particle size on reaction rate

explain the effect of temperature on rate with reference to activation energy

use potential energy diagrams to explain exothermic and endothermic reactions and to calculate activation energies

compare the effect of activation energy and net energy change on reaction rate

explain, with reference to activation energy, how catalysts increase rate

contrast heterogeneous and homogeneous catalysis and describe the main features of enzyme catalysis

give examples of key industrial catalysts and catalyst poisons

▲ use energy distribution diagrams to explain the effect of temperature change on molecular kinetic energy and reaction rate

▲ explain the formation of an activated complex and its relationship to activation energy

▲ show the importance of favourable orientation or collision geometry to the formation of an activated complex

▲ explain the process of heterogeneous catalysis in terms of adsorption and how a substance may effectively poison a surface catalyst.

PROBLEM SOLVING EXERCISES

1. The reaction between ozone and nitrogen monoxide gases is important in the formation of smog in the atmosphere.

$$O_3 + NO \longrightarrow O_2 + NO_2$$

The activation energy, E_a, for the reaction is 10 kJ mol^{-1}. The energy change (E products $-$ E reactants) is -200 kJ mol^{-1}.

Sketch an energy diagram for the reaction.

(PS skill 2)

2. Ethane reacts with steam in the presence of a heterogeneous catalyst, nickel. The rate of reaction for ethane–water mixtures containing different amounts of sulphur impurity was measured and results graphed as shown.

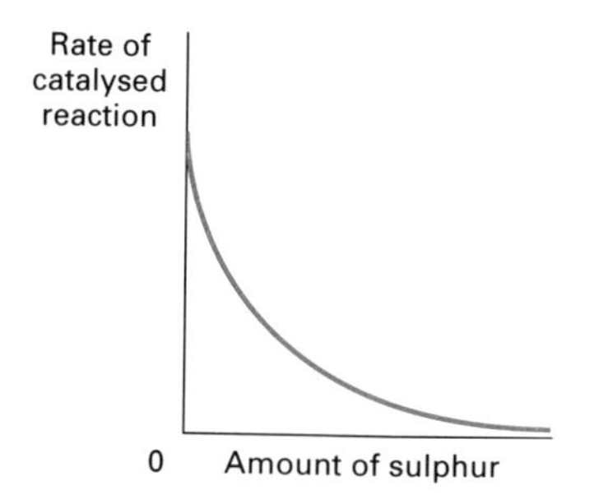

Figure 1.36

(PS skill 9)

Put forward an explanation for the reduction in rate with increasing concentration of sulphur. Why did the rate fall to zero?

(PS skill 2)

3. The concentration of sucrose, as it decomposed in acid solution, was measured periodically over a period of 210 minutes.
The average rate of decomposition was calculated to be 0.00081 mol l^{-1} min^{-1}.
If the starting concentration of sucrose was 0.32 mol l^{-1}, what would be its concentration after 210 minutes?

(PS skill 4)

4. Ammonia was decomposed at 773 K over nine nickel catalysts which differed in surface area. The results are shown in the spike chart below.

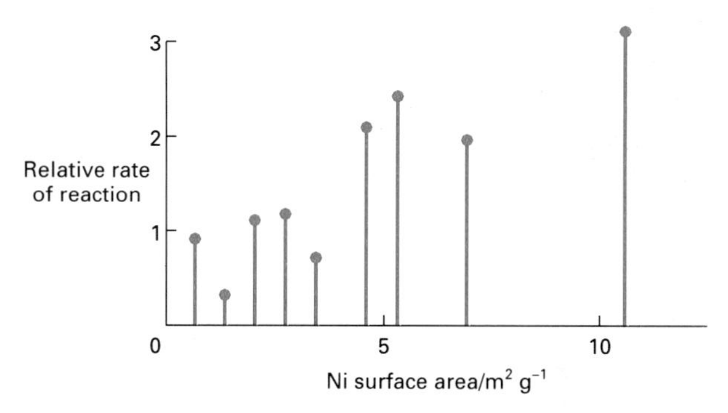

Figure 1.37

What conclusion do you draw from these results?

(PS skill 8)

5. The times taken for a cement–water mixture to complete reaction (i.e. harden or set) were recorded at different temperatures. The graph is shown below.

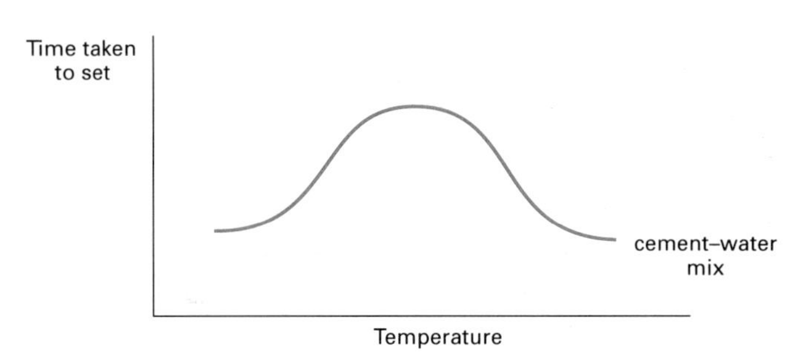

Figure 1.38

The trials were repeated with a small amount of 'retarder' added to the cement–water mixture.
A retarder slows down the rate of reaction between the cement and water.

Sketch the graph you would expect to obtain for a cement–water–retarder mixture.

(PS skill 10)

PROBLEM SOLVING EXERCISES

6. An experiment involves the catalysed decomposition of hydrogen peroxide solution.

$$H_2O_2(aq) \longrightarrow H_2O(l) + \frac{1}{2}O_2(g)$$

The concentration of $H_2O_2(aq)$ is 1 mol l^{-1}.
The grid below contains possible effects of altering the conditions of an experiment.

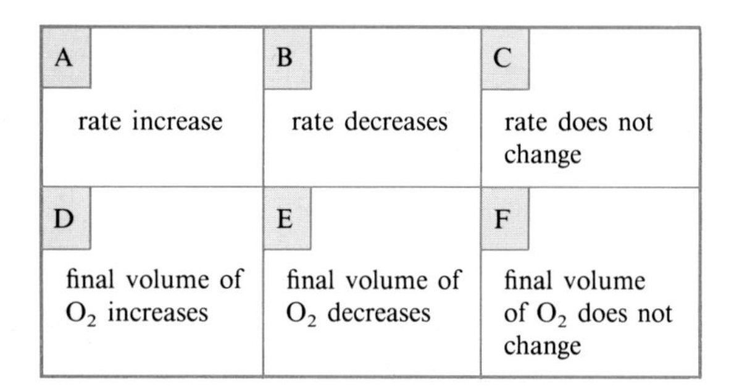

A	B	C
rate increase	rate decreases	rate does not change
D	**E**	**F**
final volume of O_2 increases	final volume of O_2 decreases	final volume of O_2 does not change

Which **pair** of effects would you expect in each case if the experimenter added, to the original $H_2O_2(aq)$, an equal volume of
(a) 4 mol l^{-1} $H_2O_2(aq)$?
(b) 0.1 mol l^{-1} $H_2O_2(aq)$?
(c) water?

(PS skill 6)

7. The graph below shows the rate of a reaction (measured as the reciprocal of the time taken), plotted against the concentration of a reactant.

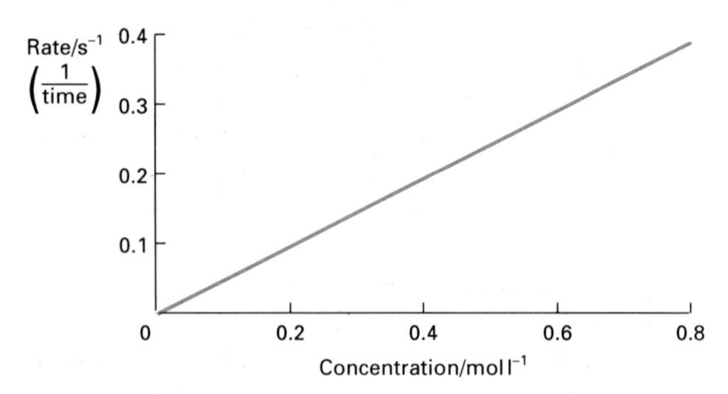

Figure 1.39

How long would it take to complete the reaction when the concentration of the reactant is 0.5 mol l^{-1}?

(PS skill 8)

8. The addition of yeast to a solution of glucose brings about an enzyme catalysed reaction in which glucose is converted into ethanol and carbon dioxide.

$$C_6H_{12}O_6 \longrightarrow 2C_2H_5OH + 2CO_2$$

With the help of a labelled diagram, describe how you would measure the rate of this reaction.

(PS skill 5)

9. A homogeneously catalysed reaction in solution takes place in three steps as follows.

(a) $Ce^{4+} + Mn^{2+} \longrightarrow Ce^{3+} + Mn^{3+}$

(b) $Mn^{3+} + Ce^{4+} \longrightarrow Ce^{3+} + Mn^{4+}$

(c) $Mn^{4+} + Tl^{+} \longrightarrow Mn^{2+} + Tl^{3+}$

Which of the ions in the reaction scheme above was added, as catalyst, to the original mixture of reactants?

(PS skill 8)

10. The grid below contains energy diagrams drawn to the same scale for four different reactions.

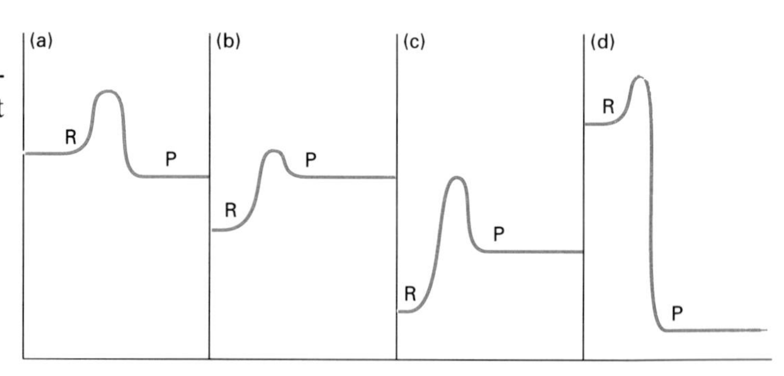

Figure 1.40

Which energy diagram shows the largest activation energy for the reaction R $\longrightarrow$ P?

(PS skill 7)

Unit 2
Feedstocks and fuels

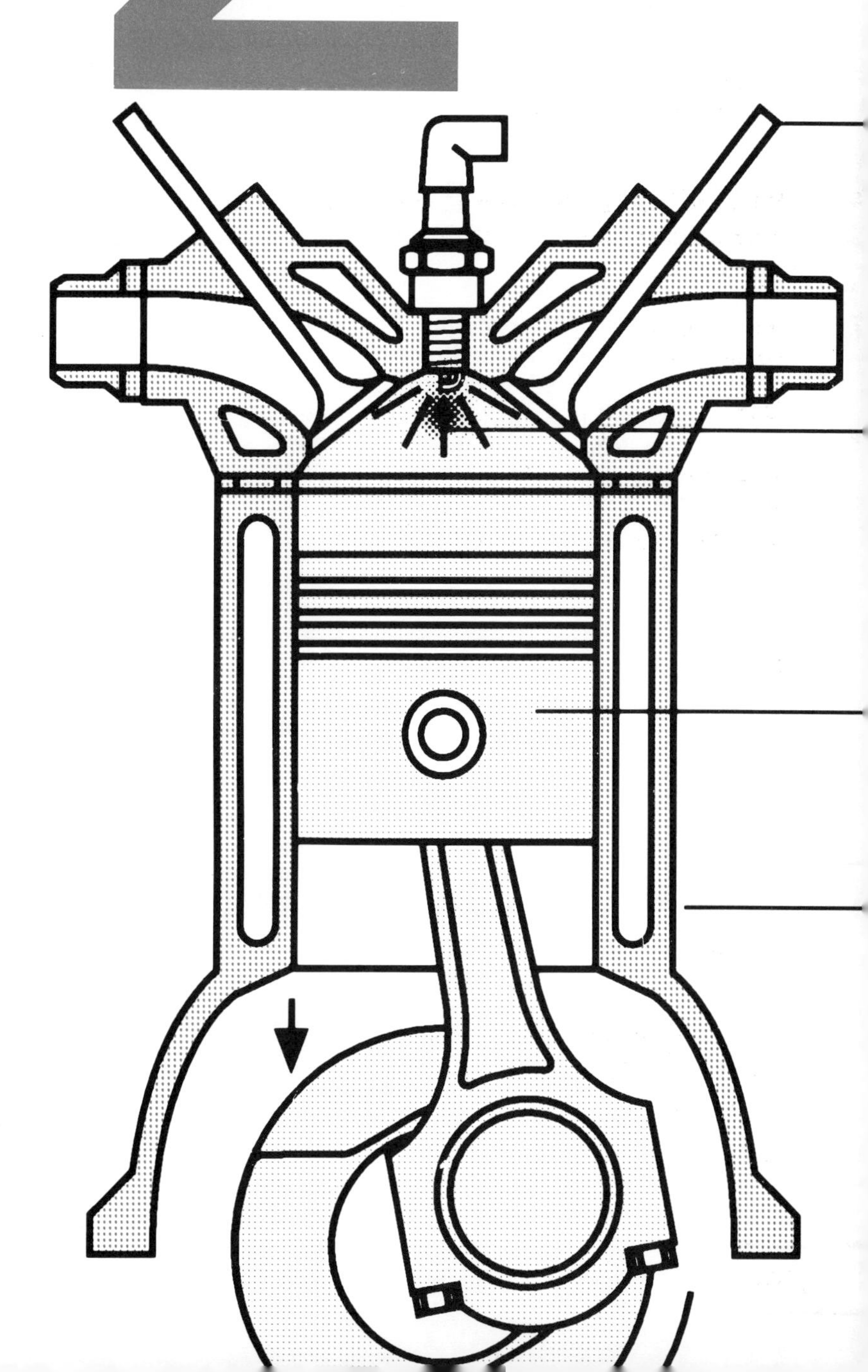

ASSUMED KNOWLEDGE AND UNDERSTANDING

Before starting on Unit 2 you should know and understand:

the nature of fuels and combustion

how coal, oil and natural gas are formed

how crude oil is distilled into fractions

the relationship between the physical properties of oil fractions and their molecular structures

the pollutants formed during combustion of fuels

the action of catalytic converters and lean-burn engines

alkanes, cycloalkanes and alkenes/saturated and unsaturated hydrocarbons

molecular and structural formulae/isomerism/homologous series

addition reactions/cracking

addition polymerisation

thermoplastic and thermosetting polymers

how sugars are fermented to produce alcohol

how distillation is used as a means of increasing the proportion of alcohol

rates of reaction and catalysts.

OUTLINE OF THE CORE AND EXTENSION MATERIAL

As you progress through Unit 2 you should, at least, try the fundamental content listed under **core** and, if possible, pick up the extra and/or more difficult content listed under **extension**.

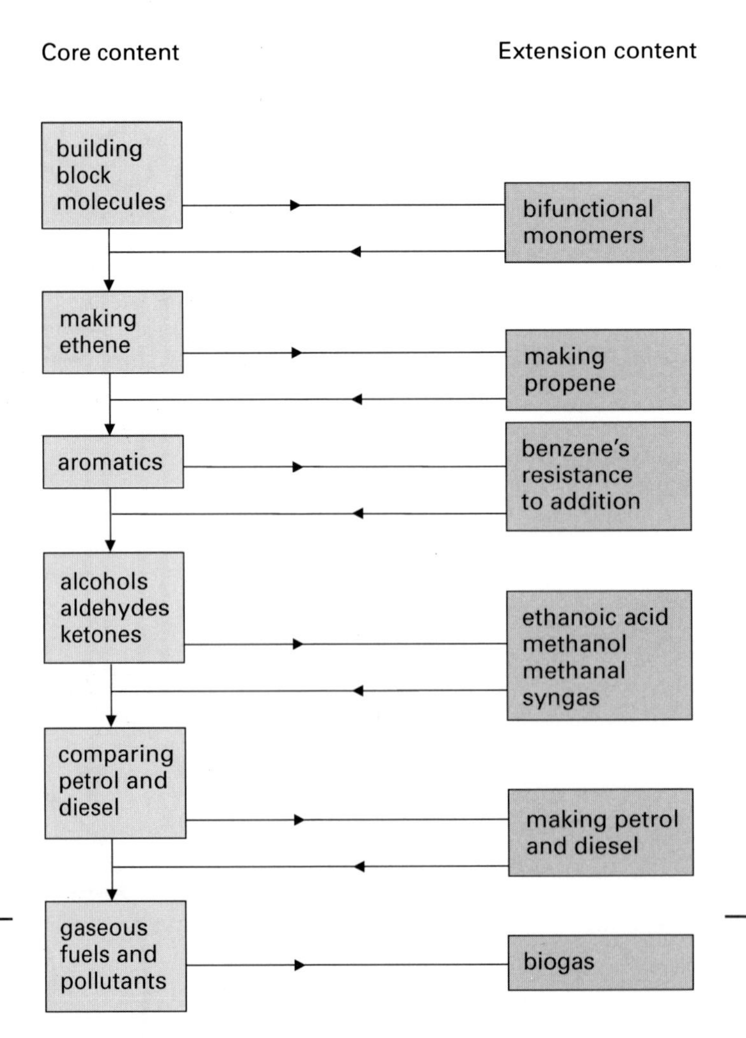

2 FEEDSTOCKS AND FUELS

2.1 Organic chemicals

Chemicals are used in almost all areas of human activity. There is, particularly in developed countries, a huge demand for the plastics, textiles, pharmaceuticals, agrochemicals, dyes, detergents, paints, etc., which improve, at least in a materialistic way, the quality of our lives.

These consumer products are made, by and large, from organic chemicals. 'Organic' was originally used to distinguish compounds extracted from living material (organisms) from those obtained from minerals. However, chemists were gradually able to synthesise, in the lab, similar compounds to those obtained naturally. As a result the name 'organic' was extended to cover all carbon compounds except carbonates and the oxides of carbon.

Organic chemicals are obtained mainly from four natural sources: **coal**, crude **oil** (petroleum), natural **gas** and **biomass** (plant material, chiefly carbohydrate).

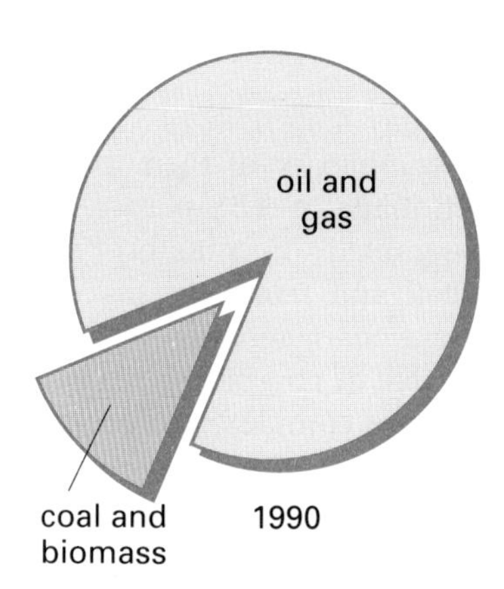

Figure 2.1 Sources of organic chemicals in Britain

The relative consumption of these four natural sources varies from country to country and from time to time. In 1950, for example, 60 per cent of organic chemicals produced in Britain came from coal. In 1990 the position was quite different, with oil and gas supplying 90 per cent of organic chemicals.

2.1.1 Renewable and non-renewable resources

Biomass, being plant material, can be replaced by harvesting the trees, crops, etc., clearing the ground and replanting with new seeds. Biomass is therefore described as a **renewable** resource.

Coal, oil and natural gas, on the other hand, are **fossil fuels** which have taken millions of years to form. Once used they cannot be replaced and they are consequently referred to as **non-renewable** resources.

There is a limited amount of these fossil fuels and this makes their continued exploitation a matter of serious concern. The position with regard to oil and gas is especially worrying; their economically recoverable reserves are small in comparison to those of coal.

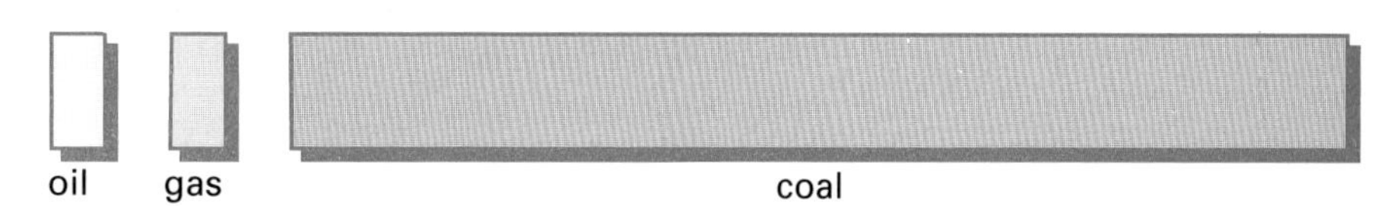

Figure 2.2 Relative reserves of fossil fuels: Western Europe

Oil and gas supplies could possibly run out, given current rates of consumption, sometime in the second half of the twenty-first century. Coal reserves, being so much larger, should last for another two or three hundred years.

2.1.2 Feedstocks or fuels

Organic chemicals may be used in two ways. They may be burned to release much needed energy for heating, power generation and transport. Alternatively, they may be converted into other useful organic chemicals.

Chemicals from which other chemicals can be extracted or synthesised are described as **feedstocks**. The choice is therefore simple: do we use coal, oil and gas as fuels or do we use them as feedstocks?

There are sources of **energy** other than the combustion of fossil fuels. Much interest is focussed on nuclear and solar energy, and increasing attention is being paid to geothermal energy (the heat within the earth) and to wind, wave and tidal power.

What are the alternative **chemical** resources to coal, oil and gas? At the present time much research is going into the production of chemicals from **biomass** rather than from fossil fuels.

2.2 Building molecules

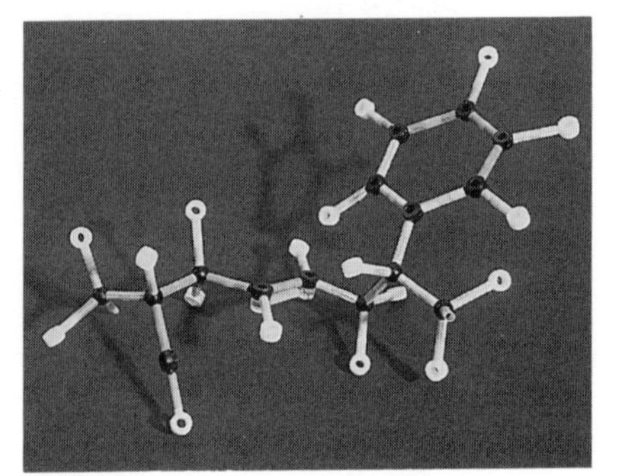

Figure 2.3 A model of the repeating unit in the ABS molecule

If we examine the chemical structures of everyday materials, plastics, fibres, dyes, drugs, etc., what may strike us is the complexity of their molecules. Plastic toy assembly building bricks are often made of ABS plastic.

How are such complicated molecules assembled? Fortunately (as with toys!) the 'building bricks' are relatively small and relatively simple. The small molecular bricks must have one important characteristic: they must be chemically reactive. That is, they must be capable of combining readily with other molecules in order to form the larger, more complex, molecules.

2.2.1 Simple reactive molecules

Reactive organic molecules have reactive 'parts' attached to the carbon skeleton of the molecule. These reactive parts are groups of atoms called **functional groups**. Four common functional groups are listed in Table 2.1.

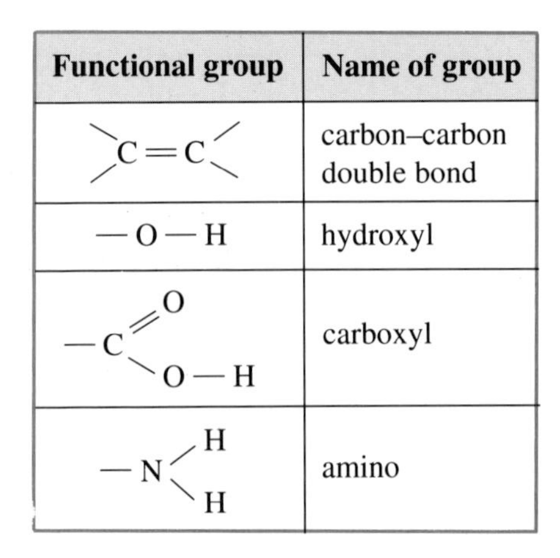

Functional group	Name of group
$>C=C<$	carbon–carbon double bond
$-O-H$	hydroxyl
$-C(=O)-O-H$	carboxyl
$-N(H)H$	amino

Table 2.1 Common functional groups

The $-NH_2$ group is found in many biological molecules, especially proteins. If one (or more) of the hydrogen atoms of ammonia, NH_3, is replaced by a hydrocarbon group, the resulting compound is called an **amine**. It is named from the alkyl group followed by the ending -amine. The four simplest amines are as follows.

CH_3NH_2	methylamine
$CH_3CH_2NH_2$	ethylamine
$CH_3CH_2CH_2NH_2$	propylamine
$CH_3CH_2CH_2CH_2NH_2$	butylamine

Many of the smaller amines have unpleasant 'fishy' smells and are produced, along with ammonia, by the decomposition of dead plant or animal matter.

The properties of an organic compound depend on which functional group is present in the molecule. Four compounds which each have two carbons per molecule (C_2) but carry different functional groups are ethene, ethanol, ethanoic acid and ethylamine.

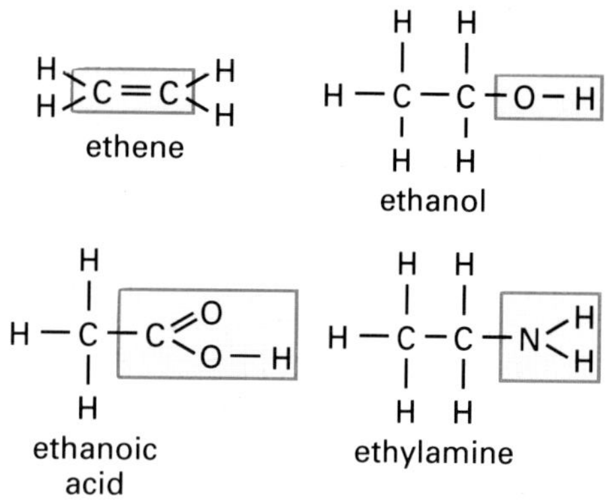

Figure 2.4 C_2 compounds with different functional groups

The different reactions of the C_2 compounds with water reflect the different characters of the functional groups.

Compound	Formula	Soluble in water	pH of solution
ethene	C_2H_4	no	–
ethanol	C_2H_5OH	yes	7
ethanoic acid	CH_3COOH	yes	less than 7
ethylamine	$CH_3CH_2NH_2$	yes	more than 7

Table 2.2 Reactions of C_2 compounds with water

2.2.2 Molecules with two functional groups

Molecules which contain **two** functional groups may be particularly reactive and easy to synthesise into very large molecules (macromolecules). Four such 'bifunctional' molecules are shown below. For simplicity the remainder of each molecule is represented by a circle.

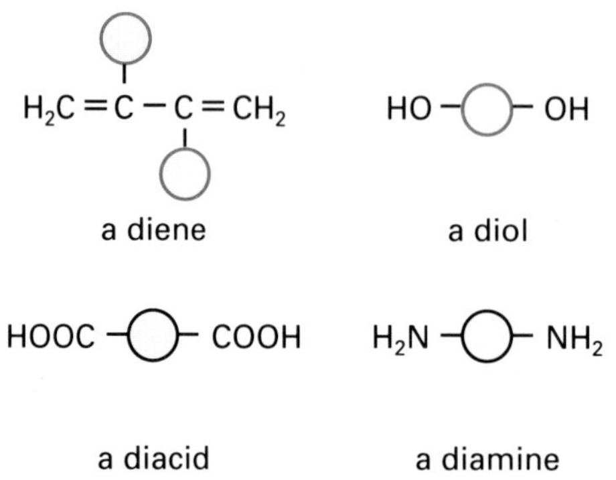

$H_2C=C-C=CH_2$ (with a circle attached to each central C)

a diene

HO–○–OH

a diol

HOOC–○–COOH

a diacid

H_2N–○–NH_2

a diamine

Figure 2.5 Four 'bifunctional' molecules

Synthetic rubbers are manufactured by polymerising dienes. Diols, diacids and diamines can condense indefinitely with one another to form polymers such as polyesters and polyamides (nylons).

$$\text{diol} + \text{diacid} \xrightarrow{\text{polymerisation}} \text{polyester}$$

$$\text{diamine} + \text{diacid} \xrightarrow{\text{polymerisation}} \text{polyamide (nylon)}$$

2.3 Building block chemicals

The chemical industry is in the business of making chemicals which it can sell either to suppliers of fuels or to manufacturers of consumer goods.

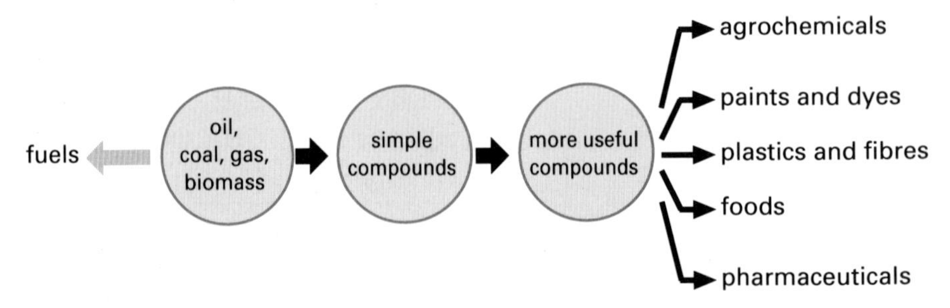

Figure 2.6 The chemical industry

Over one million different organic chemicals have been synthesised, and of this number around thirty thousand are regarded as being of commercial importance. The majority of these chemicals come from a handful of simple 'building block' chemicals, all of them hydrocarbons.

These hydrocarbons are listed below (in order of size).

one C_2 compound	ethene
one C_3 compound	propene
four C_4 compounds	butenes (3) and butadiene
one C_6 compound	benzene
one C_7 compound	toluene
two C_8 compounds	xylenes

2.4 Ethene

Figure 2.7 In this store, bananas are artificially ripened by exposure to ethene

The most popular and versatile building block chemical is unquestionably the simple alkene, ethene. Its manufacturing output is greater than any other organic chemical. A minor but interesting use of ethene is in stimulating the ripening of fruit. Fruit suppliers often transport their produce unripe and then expose it to ethene gas at its destination.

2.4.1 Poly(ethene)

Poly(ethene), first made in 1933 by ICI, is formed through the addition polymerisation of ethene molecules. In Scotland, poly(ethene) is made by BP at Grangemouth.

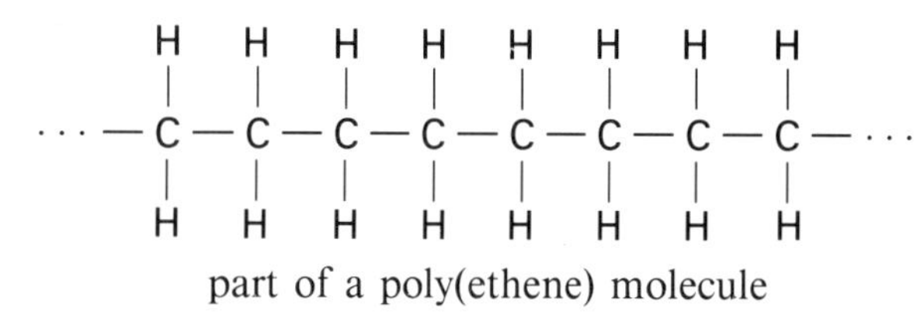

part of a poly(ethene) molecule

The molecular chain may be thousands of $—CH_2—$ units long and is not perfectly linear as shown, but has a zig-zag form and contains side branches of varying lengths, attached at irregular intervals. The polymerisation process can be controlled through varying temperature, pressure and catalysts to produce two forms of poly(ethene):

(a) low density poly(ethene), LDPE
(b) high density poly(ethene), HDPE.

The difference in physical properties such as density is due to a difference in molecular structure. The molecular chains for LDPE are more **highly branched** than the relatively linear chains of HDPE. This branching prevents LDPE molecules from packing so closely together and its density is therefore less. LDPE chains are more compact, less entangled and more easily separated from one another. LDPE is therefore more flexible and has a lower softening point, as shown in Table 2.3.

Figure 2.8

Table 2.3

	Density/g cm^{-3}	Softening point/K
LDPE	0.92	358
HDPE	0.96	400

Figure 2.9 HDPE ducting for underground electric cables obviously has to be tough!

Ethene can react with other building block chemicals to form substituted ethene monomers which may then be polymerised to form a range of important plastics. Three such monomers are **chloroethene** (vinyl chloride), **phenylethene** (styrene) and **ethenyl ethanoate** (vinyl acetate) which polymerise to form poly(vinyl chloride) (PVC), polystyrene (PS) and poly(vinyl acetate) (PVA) respectively.

$$CH_2{=}CH_2 \rightarrow CH_2{=}CHCl \rightarrow \text{PVC polymer}$$
$$CH_2{=}CH_2 \rightarrow C_6H_5CH{=}CH_2 \rightarrow \text{PS polymer}$$
$$CH_2{=}CH_2 \rightarrow CH_2{=}CHOOCCH_3 \rightarrow \text{PVA polymer}$$

2.5 Propene

Propene is a widely used building block chemical. Its major use is in the manufacture of **poly(propene)** [polypropylene]. As with poly(ethene) the properties of the polymer depend on its molecular structure.

Early efforts to polymerise propene produced soft waxy materials which were too soft to be useful. This was due to the methyl groups being arranged **randomly** round the axis of the molecular chain as shown in Figure 2.10.

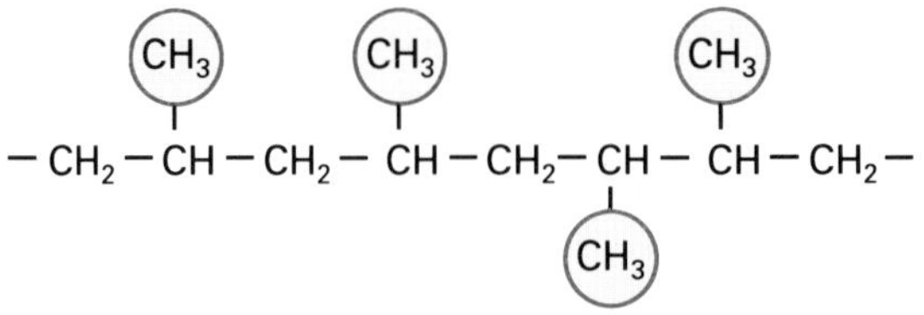

$$-CH_2-\underset{}{\overset{CH_3}{CH}}-CH_2-\overset{CH_3}{CH}-CH_2-\underset{CH_3}{CH}-\overset{CH_3}{CH}-CH_2-$$

Figure 2.10 Random arrangement of methyl groups

An Italian chemist, Giulio Natta, however, produced a commercially useful poly(propene) by employing a special catalyst. This allowed polymerisation to proceed in a highly regular fashion, producing an ordered structure for poly(propene), with all the methyl groups arranged on the same side of the polymer chain, as shown in Figure 2.11.

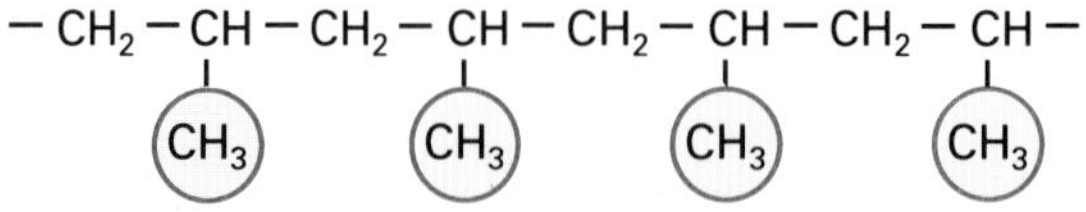

$$-CH_2-\underset{CH_3}{CH}-CH_2-\underset{CH_3}{CH}-CH_2-\underset{CH_3}{CH}-CH_2-\underset{CH_3}{CH}-$$

Figure 2.11 Methyl groups arranged on the same side of the polymer chain

More than one third of polypropene is extruded into fibre for carpets, ropes and clothing. Another very important clothing fibre is poly(propenonitrile) [polyacrylonitrile], which is better known as 'acrylic' and is made by polymerising a monomer, $CH_2{=}CHCN$, derived from propene.

Between them, ethene and propene 'generate' the world's five top-selling plastics and three top-selling synthetic fibres.

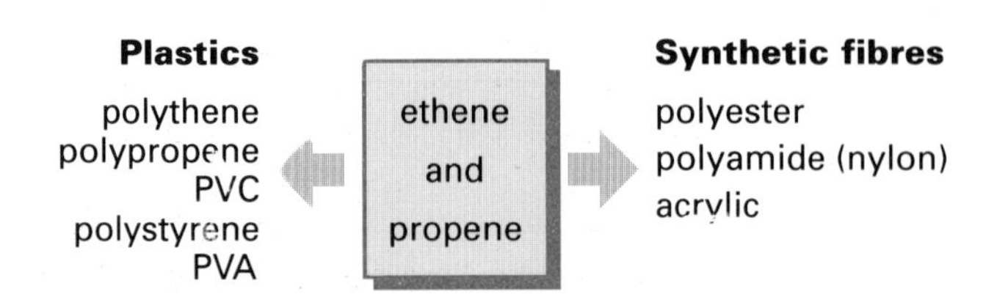

Figure 2.12 Major plastics and fibres derived from ethene and propene

2.6 C_4 alkenes

The IUPAC (International Union of Pure and Applied Chemistry) system is a means of naming, systematically, even the most complicated compounds. There are three key points to note.

(a) In naming alkenes we select, as the parent structure, the longest continuous chain that contains the double bond.

(b) The position of the double bond in the parent chain is indicated by a number and we always choose to number the carbons from the end nearest the double bond.

(c) The positions of alkyl groups attached to the parent chain are also shown by numbers.

Butene, for example, has three isomers which are named as shown.

```
                                                          H
                                                          |
                                                      H — C — H
    H   H   H                  H   H       H          H   |
    |   |   |      H           |   |       |          |   |      H
H — C — C — C = C /       H — C — C = C — C — H   H — C — C = C /
    |   |         \ H          |       |   |          |         \ H
    H   H                      H       H   H          H
```

but-1-ene — but-2-ene — methylpropene

A very important C_4 alkene is the diene, buta-1,3-diene, C_4H_6, which is copolymerised with styrene to produce the synthetic rubber SBR (styrene butadiene rubber).

```
H           H                      H   H   H   H       H
 \          |      H               |   |   |   |       |
  C = C — C = C /        ··· — C — C — C — C = C — C — ···
 /    |         \                  |   |   |       |   |
H     H           H                |   H   H       H   H
                                 C6H5
```

buta-1,3-diene — part of SBR polymer chain

2.7 Alkenes derived from natural gas

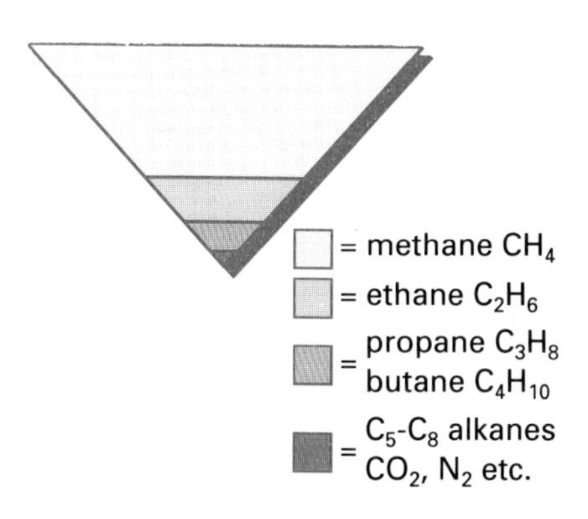

Figure 2.13 Composition of natural gas

Natural gas may be found with oil or it may occur alone. The composition varies from source to source but it is chiefly methane. Also present are some non-hydrocarbon gases such as nitrogen and carbon dioxide along with higher alkanes in amounts which decrease with increasing size. In Scotland, North Sea gas (about 80 per cent methane) is brought ashore at St Fergus, north of Peterhead. Here the methane is removed and sold to British Gas for direct entry into the national gas grid.

The remaining components are called **natural gas liquids** since, with a small change in pressure, they can be either gases or liquids. These natural gas liquids (NGL) are carried by pipeline down to the Shell Expro separation plant at Mossmorran in Fife. Here the liquids are separated by fractional distillation. The ethane, sometimes with some propane and butane, is piped to steam cracking furnaces nearby (Exxon) or at Grangemouth (BP).

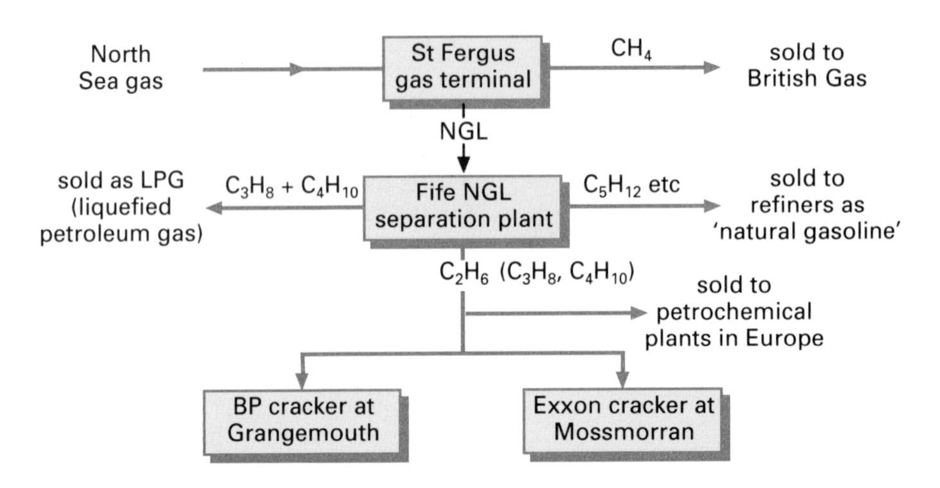

Figure 2.14 Separation of North Sea gas components

Crackers now have to be sufficiently flexible to handle different feedstocks.

In the cracking process alkane molecules can be broken down into smaller molecules of hydrogen, lower alkanes and alkenes.

In a steam cracking furnace, the preheated ethane feedstock is mixed with steam and further heated to over 1070 K. Under these conditions, ethane molecules break up into ethene (around 80 per cent efficiency) and by-products. The by-products, mainly methane (CH_4) and hydrogen, are used as fuel gases for the furnaces. Uncracked ethane is recycled.

Propane, C_3H_8, and butane, C_4H_{10} can be used to supplement ethane as sources of lower alkenes when they are cracked.

Table 2.4 Products from cracking ethane, propane and butane

Feedstock	Products from cracking				
C_2H_6	H_2	CH_4	C_2H_4		
C_3H_8	H_2	CH_4	C_2H_4	C_3H_6	
C_4H_{10}	H_2	CH_4	C_2H_4	C_3H_6	C_4H_8 C_4H_6

Figure 2.15 The Exxon ethene plant, Mossmorran, Fife

2.8 Alkenes from crude oil

Alkenes are not present in natural gas and are only present to a very slight extent in crude oil. They must therefore be made from suitable alkanes, found in oil or gas. Distillation of crude oil separates the complex mixture of hydrocarbons into several less complex mixtures called fractions. The separation is based on boiling range and therefore relates largely to the number of carbon atoms in the molecule.

Table 2.5 lists typical fractions with an indication of the size and type of molecules present. 'Aromatics' are unsaturated cyclic compounds and will be dealt with later in this unit.

Table 2.5 Crude oil fractions

Fraction	Carbon atoms	Type of molecule
gas	C_1–C_4	short alkanes
gasoline	C_5–C_6	straight and branched alkanes
naphtha	C_6–C_{10}	straight and branched alkanes, cycloalkanes and simple aromatics
kerosine	C_{10}–C_{14}	complex alkanes, cycloalkanes and aromatics
gas oil	C_{14}–C_{19}	complex alkanes, cycloalkanes, aromatics, mixed types
residue	C_{19+}	complex alkanes, cycloalkanes, complex aromatics, mixed types

2.8.1 Cracking of naphtha

Naphtha, unlike its neighbouring fractions gasoline and kerosine, is not suitable for direct inclusion in a fuel. The fraction is, however, a valued feedstock for other chemicals. The straight chain alkanes within naphtha are a good source of lower alkenes when naphtha is subjected to steam cracking.

The naphtha is vaporised, mixed with steam and heated to above 1000 K in a cracking furnace. The alkenes produced are prevented from recombining by rapidly cooling the reaction products as they leave the reactor.

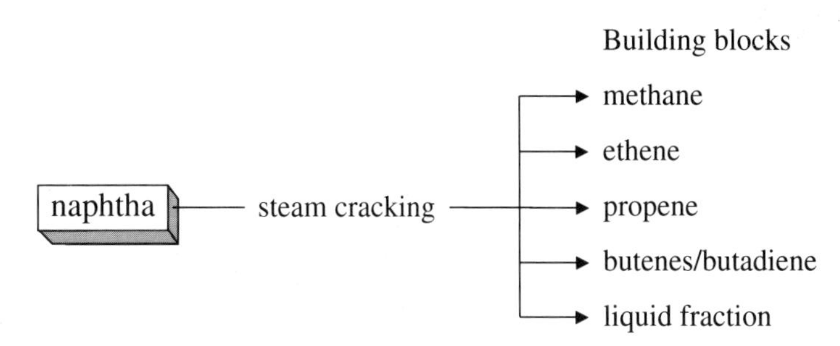

The alkenes are used as reactive building block chemicals to make polymers and other chemicals.

The liquid fraction from cracked naphtha is rich in simple aromatic hydrocarbons. The simplest aromatic hydrocarbon is a colourless liquid called benzene.

2.9 Benzene

In the early 1800s certain natural substances were discovered which contained at least six carbon atoms per molecule and could not be readily converted to compounds with less than six carbon atoms. Such compounds often had characteristic smells and were called **aromatic** compounds.

The fundamental aromatic compound is the hydrocarbon benzene, which has a molecular formula C_6H_6. Despite this high degree of unsaturation (eight hydrogens 'short') benzene is unreactive, compared to alkenes, and reacts by substitution rather than by addition, e.g.

$$C_6H_6 + Br_2 \xrightarrow{\text{catalyst}} C_6H_5Br + HBr$$

Benzene has been known since 1825 but the nature of its chemical structure was a problem which proved hard to solve. In 1865 Friedrich Kekulé (University of Bonn) proposed the idea of two cyclic structures between which the benzene would alternate.

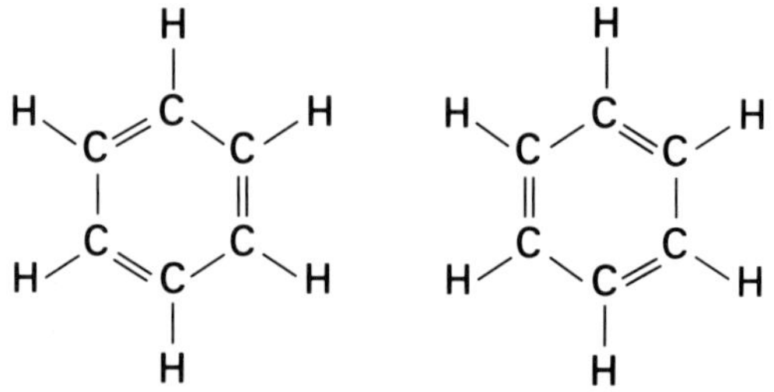

These Kekulé structures did not explain the extraordinary stability and **resistance to addition** shown by benzene. Clearly the benzene molecule did not have three single bonds and three double bonds.

The revised and now accepted structure is as follows. Benzene is a flat molecule with the six carbon atoms forming the corners of a regular hexagon. The carbon–carbon bonds are of **equal** length and strength: **intermediate** between single and double bonds.

Six electrons (one per carbon atom) are not fixed to particular carbon atoms. These electrons are said to be **'delocalised'**.

The delocalised electrons form a continuous bond which is represented by a circle drawn within the hexagon. The presence of the circle distinguishes the benzene ring, C_6H_6, from the cyclohexane ring, C_6H_{12}, which is often represented by a plain hexagon. Unlike benzene, cyclohexene, C_6H_{10}, does have a typical C=C bond.

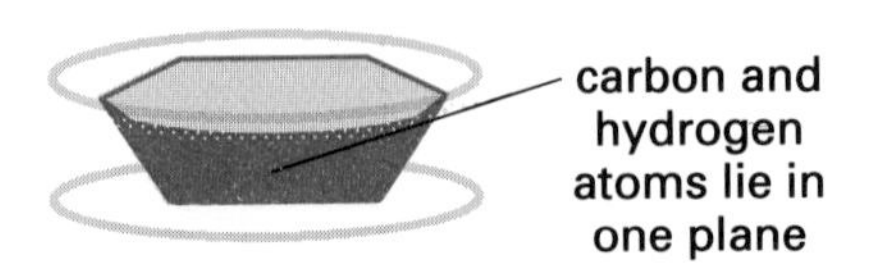

Figure 2.16 Model for structure of benzene

Figure 2.17 C_6 ring compounds

Of these three C_6 ring compounds, only cyclohexene will undergo an addition reaction with bromine.

$$\underset{\text{cyclohexene}}{C_6H_{10}} + Br_2 \longrightarrow C_6H_{10}Br_2$$

In drawing benzene rings, it is understood that a hydrogen atom is attached to each corner of the hexagon unless another atom or group is indicated. The group C_6H_5, derived from benzene by removing a hydrogen atom, is called the **phenyl group**. For example styrene, $C_6H_5CH{=}CH_2$, is systematically named 'phenylethene'.

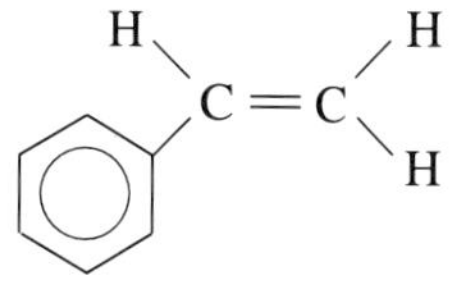

phenylethene

The name 'phenyl' may appear a strange choice but it is derived from the Greek word *phaino* which means 'I give light'. Benzene was extracted from coal tar, which is formed in the production of coal gas ('illuminating' gas) for lighting and heating homes.

2.9.1 Aromatic compounds

The benzene ring, being particularly stable, survives reactions and conditions which decompose less stable molecular arrangements. The benzene or 'aromatic' ring is therefore able to stay intact during chemical synthesis. Benzene rings appear regularly in the molecular structure of everyday chemicals including dyes and drugs (Figure 2.18). The reasons for the stability of the benzene ring are complex, but they include the lowering of potential energy that results when bonding electrons are able to 'delocalise' and spread around the ring of carbon atoms.

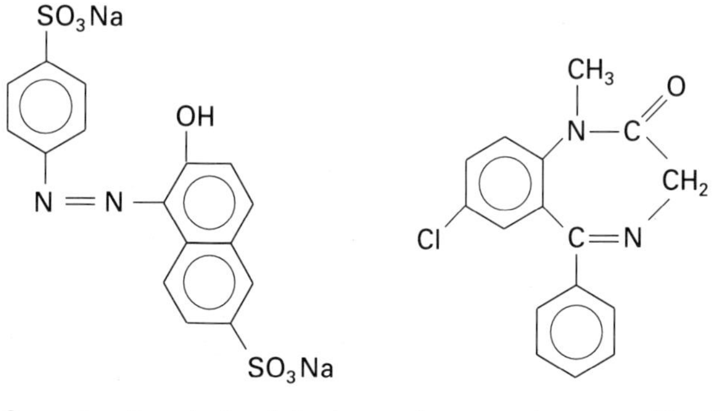

Sunset yellow (a food dye) diazepam (a tranquilliser)

Figure 2.18 Benzene rings in dye and drug molecules

2.10 Aromatics from oil

In systematically naming benzene derivatives, i.e. **compounds in which the dominant feature is the benzene ring**, the name of the group replacing a hydrogen atom is prefixed to the word 'benzene'. If several groups are attached to a benzene ring their relative positions as well as their identities must be indicated.

The four main aromatic building block chemicals are benzene itself and those of its derivates, namely **toluene** and two isomers of **xylene**.

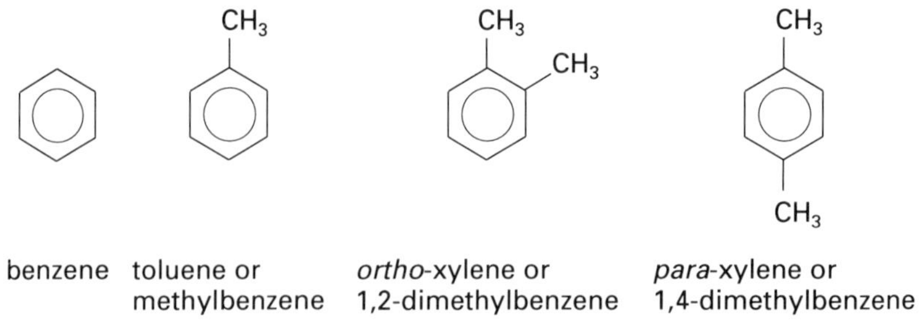

Figure 2.19 The four main aromatic building block chemicals

The liquid fraction from cracked naphtha is rich in simple aromatic hydrocarbons such as benzene, toluene and xylenes. Some of these aromatics may have been present in the original naphtha and have survived breakdown, thanks to the stability of the benzene ring. The conditions under which cracking occurs, however, bring about the conversion of cycloalkanes into aromatics, and this accounts for the increased concentration of aromatics produced.

The conversion of cycloalkanes into aromatics is described as **reforming**. Reforming is the process of changing the structures of molecules without necessarily altering their sizes.

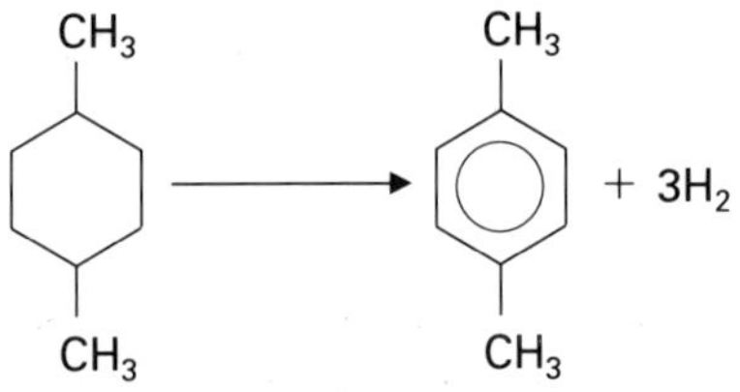

In oil refineries both cracking and reforming are major processes. Steam cracking is used to convert small alkanes into alkenes for use as feedstocks. **Catalytic** cracking is used to split the large molecules of heavy oil fractions into smaller ones, by the action of heat and the aid of a catalyst (oxides of silicon and aluminium).

Catalytic reforming is a set of processes, the aim of which differs slightly from the aims of steam or catalytic cracking. The purpose of reforming in an oil refinery is to rearrange the structures of molecules to produce more branched and aromatic molecules.

Naphtha feedstock is reformed by being passed over a platinum catalyst at raised temperatures and pressures. Hydrogen gas, produced during reforming, is used in other processes within the refinery.

The reformed naphtha, now richer in branched and aromatic hydrocarbons, may be used to make high octane petrol. Alternatively the aromatics, mainly benzene, toluene and xylenes, can be extracted from the mixture using a selective solvent, i.e. one which dissolves aromatics but not non-aromatics.

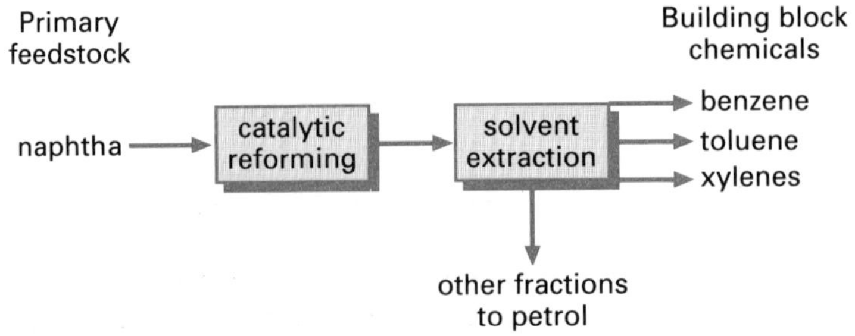

Figure 2.20 An example of reforming

Figure 2.21 Aromatics from reformed naphtha

2.11 Aromatics as building block chemicals

Benzene, toluene and xylene are not only important components of high grade petrol but essential building block chemicals (secondary feedstocks) for the manufacture of other chemicals.

Benzene

The chief outlet for benzene is in the manufacture of the plastic **polystyrene** [systematic name poly(phenylethene)]. Benzene reacts with ethene to form ethylbenzene which is then dehydrogenated to form styrene. Polymerisation to polystyrene can then be brought about.

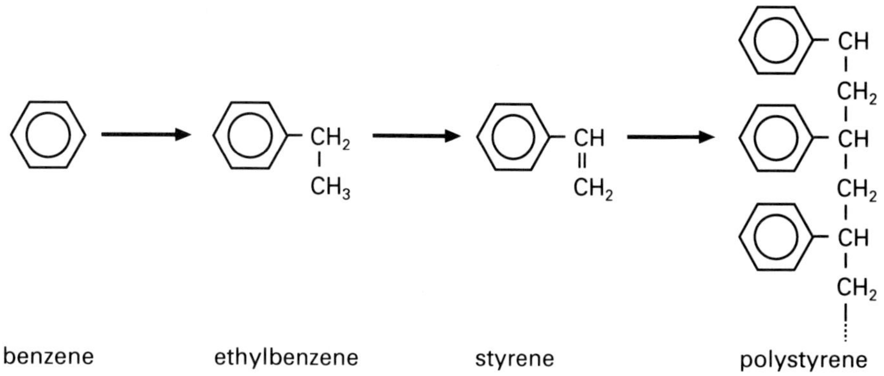

benzene → ethylbenzene ($C_6H_5-CH_2-CH_3$) → styrene ($C_6H_5-CH=CH_2$) → polystyrene

Figure 2.22 Synthesis of polystyrene

Expanded polystyrene is made, initially, as polystyrene beads containing pentane. When heated in steam, the pentane vaporises to form expanded, foamed polymer particles which can be blown into moulds.

Toluene

More than half the toulene extracted from reformed naphtha or cracked fractions is converted to benzene, which is less available but in more demand as a feedstock. The remaining toluene is used as a solvent or as a building block chemical in the manufacture of products such as **polyurethane** resins and trinitrotoluene (TNT) explosives.

Xylene

Xylene (dimethylbenzene) has three isomers depending on where the two methyl groups are attached to the benzene ring. A simplified representation of the structures is shown in Figure 2.23.

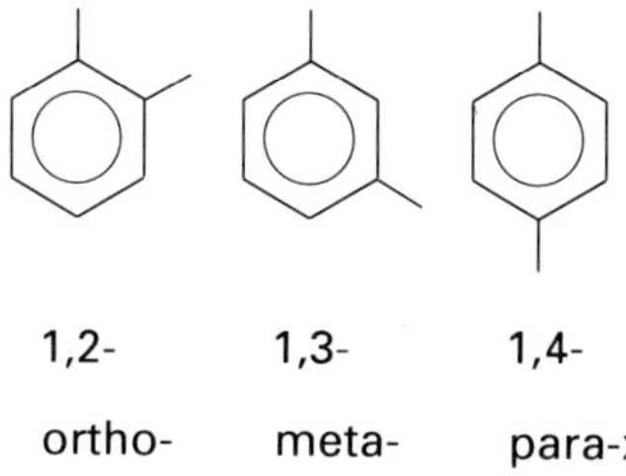

1,2- 1,3- 1,4- dimethylbenzene

ortho- meta- para-xylene

Figure 2.23 Three isomers of xylene

A mixture of the three xylene isomers is called **xylol** and is used as a solvent. *ortho*-Xylene is a feedstock for the production of a **crosslinked polyester** for

Figure 2.24 Creosote kills the micro-organisms that cause timber fencing to rot

paints. *para*-Xylene, on the other hand, is starter for the manufacture of a **linear polyester** for fibres. *meta*-Xylene is unimportant as a feedstock.

2.12 Aromatics from coal

Traditionally, chemicals have been obtained from coal, by a process called coal carbonisation. When coal is heated strongly, in the absence of air, it is broken down into:

(a) coal gas
(b) coal tar
(c) coke.

Coal gas, formerly an important domestic and industrial fuel, has been replaced by natural gas. Coke's major customer is the steel industry. The third product, **coal tar**, can be further distilled into light, medium and heavy oils. These oils contain, amongst other compounds, aromatic hydrocarbons which increase in complexity the heavier the oil. Creosote oil is one fraction from coal tar. It is used to preserve wood fencing.

2.13 Organic chemicals from biomass

The primary feedstocks, coal, oil and natural gas, are fossil fuels and produce compounds of carbon which are generally hydrocarbons. Biomass is the main direct source of organic chemical feedstocks containing the element oxygen. For example, crops with a high starch or sugar content, such as maize, cassava, pineapples, potatoes, sugar beet and sugar cane, can be fermented to produce **ethanol**.

During fermentation, micro-organisms such as fungi, yeast, bacteria and moulds feed on the biomass and produce particular chemicals as by-products. The best known example is the fermentation of sugary residues to produce alcoholic drinks.

$$\underset{\text{sugar}}{C_6H_{12}O_6} \xrightarrow{\text{fermentation}} \underset{\text{ethanol}}{CH_3CH_2OH}$$

In Brazil, sugar cane is grown on a large scale and fermented to produce ethanol. Most of the cars in Brazil run on sugar cane based ethanol – some on pure ethanol and the rest on a mixture of ethanol and petrol.

In the USA the production by fermentation of ethanol for use as a motor fuel is also a major industry. A variety of feedstocks is used, including all forms of carbohydrate waste from potato-processing, paper-making, brewing, etc.

The main advantage of using biomass is that it is a renewable resource. There are, however, some major disadvantages attached to the production of '**bioethanol**', i.e. ethanol from biomass. Firstly, the growing and harvesting of solid plant material is labour intensive. Secondly, the process of fermentation not only takes a matter of days (slow by industrial standards) but is likely to stop if micro-organisms react badly to a change in conditions. An additional problem is the purity of the product. Bioethanol is usually obtained as a dilute aqueous solution. Separating the ethanol from the water is a difficult and costly business.

Fermentation of sugary juices by yeast stops when the ethanol content reaches about 16 per cent. Drinks with a higher alcohol content have to be produced by **distilling** the mixture produced through fermentation.

Figure 2.25 Copper stills in a whisky distillery

2.14 Ethanol from oil/natural gas

Ethanol, a compound of carbon, hydrogen and oxygen, is manufactured **industrially**, i.e. without using biochemical processes, by reacting the most reactive hydrocarbon, ethene, with the simplest oxygen-containing compound, water. The ethene, produced by cracking oil (naphtha) or gas (ethane) fractions, is directly **hydrated** to form ethanol.

$$\begin{array}{ccccccc} & \mathrm{H} & & \mathrm{H} & & & \\ & | & & | & & & \\ \mathrm{H}- & \mathrm{C} & = & \mathrm{C} & -\mathrm{H} \end{array} + \mathrm{H_2O} \xrightarrow{\text{hydration}} \begin{array}{ccccccc} & \mathrm{H} & & \mathrm{H} & & \\ & | & & | & & \\ \mathrm{H}- & \mathrm{C} & - & \mathrm{C} & -\mathrm{H} \\ & | & & | & & \\ & \mathrm{H} & & \mathrm{OH} & & \end{array}$$

ethene ethanol

Steam and ethene, at temperatures around 570 K and pressures around 67 atmospheres, are passed into a reactor containing a catalyst (phosphoric acid absorbed on the surface of silica). The conversion of feedstock to product is low (4–5 per cent) but the unconverted ethene is **recycled**, leading ultimately to a conversion rate approaching 100 per cent. The product, an ethanol–water mixture, is concentrated by fractional distillation to give 'rectified spirit' which is 95 per cent ethanol, 5 per cent water.

Why do we not get 100 per cent ethanol? If a mixture of ethanol and water is distilled, the lowest boiling component is neither ethanol (bp 351.3 K) nor water (bp 373 K), but a mixture of 95 ethanol:5 water (bp 351.15 K). No matter how efficient the fractionating column used, the proportion of ethanol cannot be increased simply by further distillation. To produce 100 per cent ethanol, the 5 per cent water has to be removed by a drying agent.

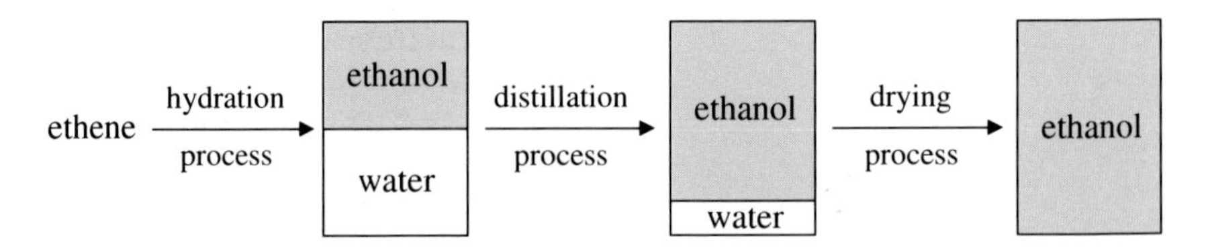

The methylated spirits sold by pharmacists is a mixture of rectified spirit and methanol (poisonous). A purple dye is added to help identification and a trace of foul-tasting chemical included to discourage drinking!

2.14.1 Dehydration of ethanol

The hydration of ethene to give ethanol can be reversed. Ethanol can be **dehydrated** to form ethene. Before 1950 the most popular route to non-aromatic organic chemicals was by dehydration of fermentation ethanol. This 'bioethanol' was dehydrated to give the ethene which acted as a key feedstock in the synthesis of a whole range of compounds such as poly(ethene), etc.

$$\underset{\text{ethanol}}{CH_3CH_2OH} \xrightarrow{\text{dehydration}} \underset{\text{ethene}}{CH_2{=}CH_2}$$

In the school lab the dehydration can be carried out by passing ethanol vapour over heated aluminium oxide.

2.15 Alcohols

A hydroxy molecule in which the OH group is attached to a carbon atom of a benzene ring is called a **phenol**. An example is the TCP molecule.

OH, Cl, Cl, Cl

Figure 2.26 Example of a phenol

If the OH group is attached to any other type of carbon, the molecule is described as an **alcohol.** Three alcohols are shown in Figure 2.27.

CH_3CH_2OH	OH	CH_2CH_2OH
ethanol	cyclohexanol	phenylethanol

Figure 2.27 Examples of alcohols

Phenylethanol is an aromatic alcohol used in 'rose' perfumes. It can be extracted from rose petals.

2.15.1 Alkanols

The simplest alcohols are those whose structures are obtained by replacing the H atom of an alkane by an OH group. This subset of alcohols forms a family of compounds called **alkanols**.

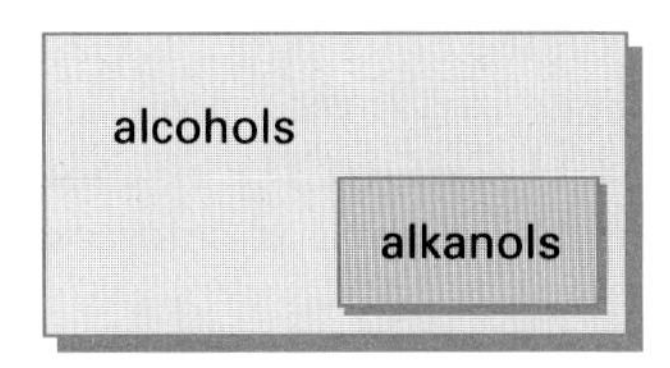

Figure 2.28 Alkanols: a subset of alcohols

The alkanols have the general formula $C_nH_{2n+1}OH$. They are named by replacing the 'e' of the alkane by 'ol'. Table 2.6 lists the first eight members of the series.

Table 2.6 Alkanols

Name	Formula
methanol	CH_3OH
ethanol	C_2H_5OH
propanol	C_3H_7OH
butanol	C_4H_9OH
pentanol	$C_5H_{11}OH$
hexanol	$C_6H_{13}OH$
heptanol	$C_7H_{15}OH$
octanol	$C_8H_{17}OH$
etc.	etc.

2.15.2 Primary, secondary and tertiary alcohols

The classification of alcohols into three types, primary, secondary and tertiary, can be explained by looking at the formation of some simple alkanols.

C_3 alkanols

One of the first alkanol syntheses practised commercially was that based on the hydration of propene. Hydration here means addition of water across the double bond and, in the case of propene, there are two possible products.

```
                                    H   H   H
                                    |   |   |
                                H — C — C — C — H    propan-1-ol
                                    |   |   |
     H   H   H                      H   H   OH
     |   |   |        +H2O    ↗
 H — C — C = C — H   ———————<
     |                        ↘
     H                              H   H   H
                                    |   |   |
     propene                    H — C — C — C — H    propan-2-ol
                                    |   |   |
                                    H   OH  H
```

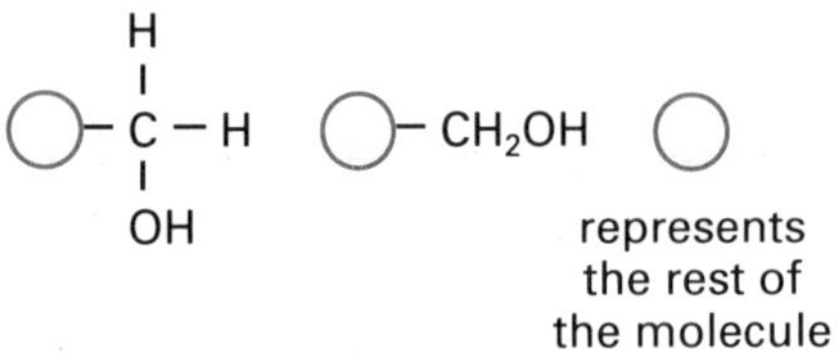

H
|
O– C –H O– CH_2OH O
|
OH represents the rest of the molecule

Figure 2.29 Primary alcohol

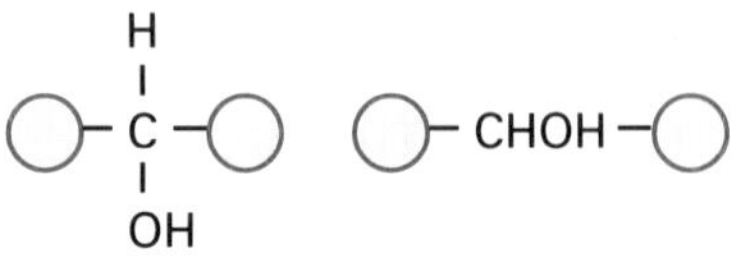

H
|
O– C –O O– CHOH –O
|
OH

Figure 2.30 Secondary alcohol

We indicate the position of the OH group by numbering the carbon atoms in the parent chain from the end which will give the lowest figure, e.g. C^3-C^2-C^1. In practice, hydration of propene yields only **propan-2-ol**.

These isomers of propanol represent different categories of alcohol (alkanol). Propan-1-ol is classed as a **primary** alcohol since the carbon atom holding the OH group is linked to no more than one other carbon atom. Propan-2-ol, on the other hand, is classed as a **secondary** alcohol since the OH group is attached to a carbon atom already linked to two other carbon atoms. Propan-2-ol, $CH_3CHOHCH_3$, is a useful solvent for lacquers, cosmetics and industrial polishes, as well as being used to extract perfume oils from flowers.

C_4 alkanols

C_4 alkanols can be made by hydrating the isomers of butene. Both linear (straight chain) butenes yield the secondary alkanol, butan-2-ol, a good solvent for dyes and inks.

$$\text{but-1-ene} \xrightarrow{+H_2O} \text{butan-2-ol} \xleftarrow{+H_2O} \text{but-2-ene}$$

The branched isomer is 2-methylpropene. Hydration of this branched butene provides the alcohol TBA (tertiary butyl alcohol) which is put into petrols in order to raise their octane ratings. TBA has the systematic name 2-methylpropan-2-ol and is classified as a **tertiary** alcohol since the carbon atom holding the OH group is linked to three other carbon atoms.

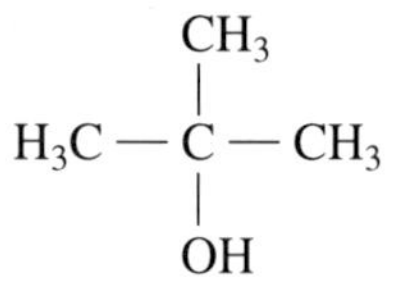

CH_3
|
H_3C — C — CH_3
|
OH

2-methylpropan-2-ol

2.16 Solvents

Solvents are used in a wide range of industries to dissolve other substances: either for removal purposes, as in cleaning and extraction of oils from seeds and flowers, or as carriers to place dissolved substances in position, as in painting and surface coating.

Solvents, other than water, may be divided into two groups:

(a) **hydrocarbon** solvents, such as white spirit, which are by-products of oil refining
(b) **chemical** solvents which contain other elements such as oxygen, nitrogen and chlorine.

Those chemical solvents which contain oxygen include a group of compounds called **ketones**.

2.16.1 Ketones

Figure 2.31 Ketone

Ketones are compounds containing the **carbonyl group**, C=O, placed between other carbon atoms in the molecule. The simplest ketones are the **alkanones**: compounds in which the carbonyl group is flanked on either side by alkyl groups such as methyl, CH_4 ethyl, C_2H_5 etc. In naming alkanones, the longest chain carrying the CO group is considered the parent structure and the 'e' of the alkane replaced by 'one'. We can indicate the position of any group by numbering the carbons in the chain. For example, MIBK is a ketone used for solvent extraction during the production of antibiotics. The **systematic name** is 4-methylpentan-2-one.

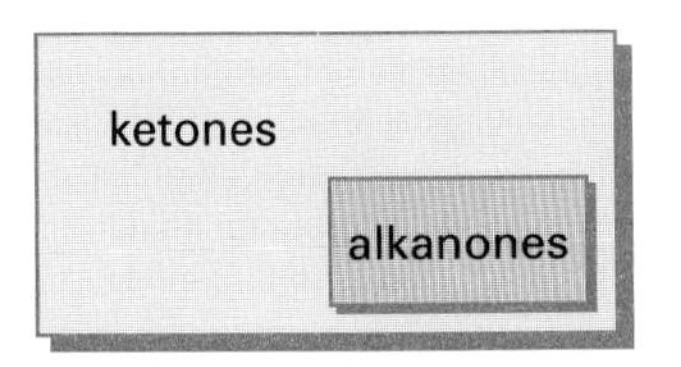

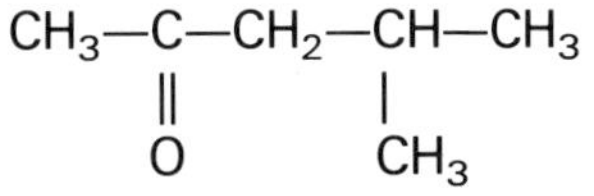

Figure 2.32 Alkanones: a subset of ketones

$$CH_3{-}\underset{\displaystyle \overset{\|}{O}}{C}{-}CH_2{-}\underset{\displaystyle \overset{|}{CH_3}}{CH}{-}CH_3$$

C_3 and C_4 alkanones are excellent solvents widely used in the paints, cosmetics and pharmaceutical industries. They can be made by oxidising the corresponding C_3 and C_4 secondary alkanols.

$$\underset{\text{propan-2-ol}}{CH_3CHOHCH_3} \xrightarrow{\text{oxidation}} \underset{\text{propanone}}{CH_3COCH_3}$$

$$\underset{\text{butan-2-ol}}{CH_3CHOHCH_2CH_3} \xrightarrow{\text{oxidation}} \underset{\text{butanone}}{CH_3COCH_2CH_3}$$

Propanone (acetone), as well as being a popular solvent, is also a building block chemical for the production of **methyl methacrylate**. This compound polymerises to form a plastic of excellent transparency which substitutes for glass and has the common brand name 'Perspex'.

2.17 Aldehydes

If a carbonyl group is linked to only one carbon atom in a molecule, the compound is described as an **aldehyde.**

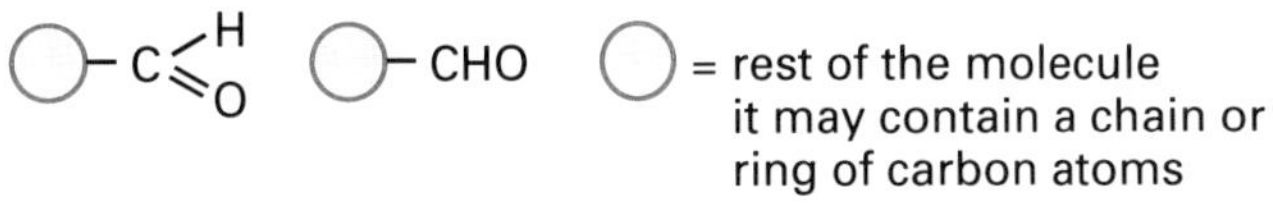

○–C(=O)H ○–CHO ○ = rest of the molecule it may contain a chain or ring of carbon atoms

Figure 2.33 Aldehydes

Aldehydes are formed by the gentle oxidation of **primary** alcohols. Phenylmethanal, for example, is used commercially for scenting soap, adding an 'almond' flavour to foods, and in the manufacture of dyes and antibiotics. It can be produced by oxidising the corresponding primary alcohol.

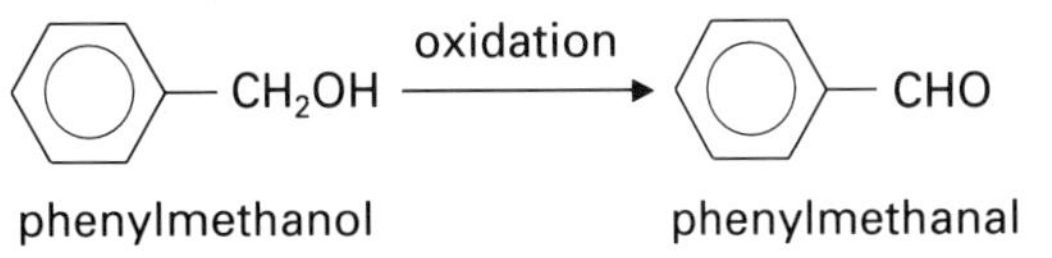

$$\underset{\text{phenylmethanol}}{C_6H_5{-}CH_2OH} \xrightarrow{\text{oxidation}} \underset{\text{phenylmethanal}}{C_6H_5{-}CHO}$$

Figure 2.34 Forming aldehydes

The simplest aldehydes are those derived from primary alkanols and are called **alkanals**. Alkanals make up a **subset** within the larger set of aldehydes (like alkanones within ketones).

The alkanals form a series with the general formula $C_nH_{2n+1}CHO$ and are named by replacing the 'e' of the parent alkane by 'al'.

Name	Formula
methanal	$HCHO$
ethanal	CH_3CHO
propanal	C_2H_5CHO
butanal	C_3H_7CHO
etc.	etc.

Table 2.7 Alkanals

The oxidation of primary alcohol to aldehyde formed the basis of a motorist's 'breathalyser' test for alcohol. The driver was asked to breathe into a bag containing an oxidising agent, acidified dichromate ion. If the orange crystals turned green the message for drivers was certainly not 'go ahead'!

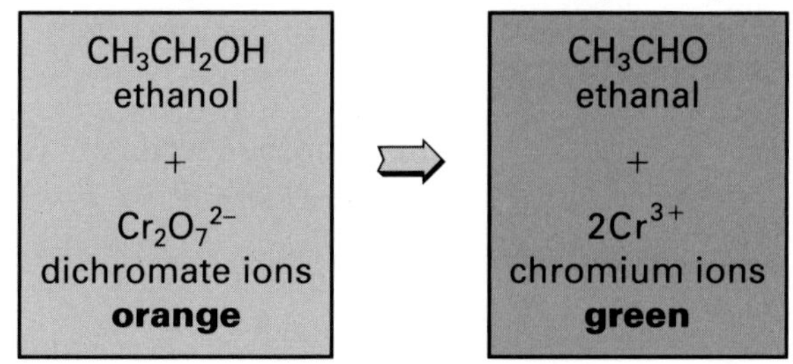

Police now use a different method of measuring the amount of alcohol.

2.17.1 Oxidation of aldehydes

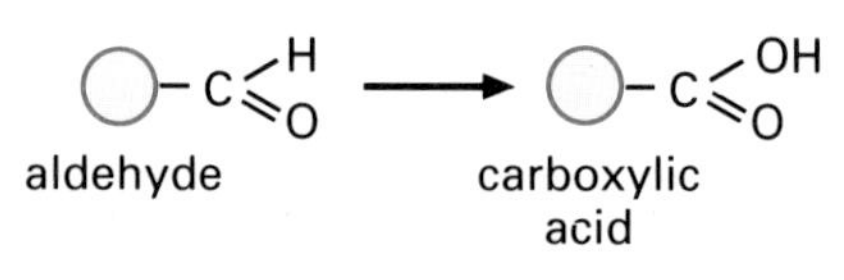

Figure 2.35 Oxidation of aldehydes

Aldehydes occur naturally in plant oils, and are partly responsible for the flavouring and smell of many fruits and plants. A disadvantage of using aldehydes in perfumes is their **inclination to undergo oxidation** – to the corresponding carboxylic acids.

Anisaldehyde has a pleasant smell like hawthorn flowers and is an important ingredient of floral perfumes. In its preparation, care must be taken to stop it oxidising to anisic acid.

CH_3O–⌬–CHO ⟶ CH_3O–⌬–$COOH$

Figure 2.36 Anisaldehyde oxidising to anisic acid

Ketones, unlike aldehydes, are **resistant to oxidation** and this difference provides a means of distinguishing these two subsets of carbonyl compounds. Aldehydes (being readily oxidisable) will reduce, for example, acidified dichromate ions to chromium(III) ions (Figure 2.37). Ketones will not.

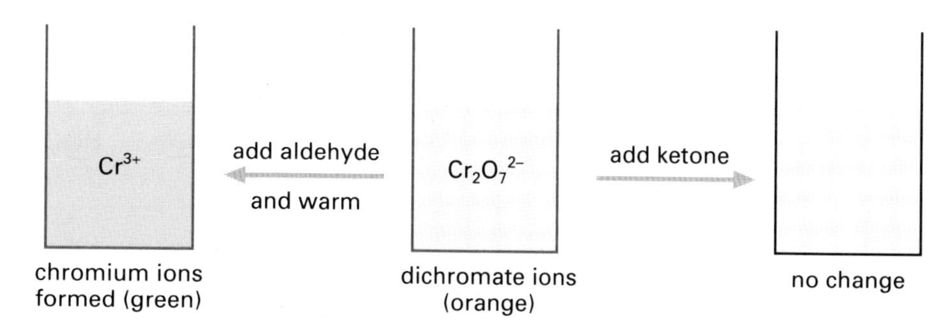

Figure 2.37 Effect of mixing acidified dichromate solution with aldehyde or ketone

2.17.2 Reduction of aldehydes

The ease with which aldehydes can be reduced, as well as oxidised, makes them valuable intermediates in chemical pathways.

Primary alcohols, such as butan-1-ol, are useful industrial solvents. Unfortunately, it cannot be manufactured by hydrating the appropriate alkene (but-1-ene) as this results in the preferential formation of butan-2-ol.

A method of making butan-1-ol is required and a successful one involves an aldehyde at an intermediate stage. In a process known as the Oxo process, a mixture of carbon monoxide and hydrogen (known as synthesis gas) reacts with an alkene in the presence of a catalyst. The product is an aldehyde (with an extra carbon atom in the chain) which can then be **reduced** to the primary alcohol.

$$C_3 \text{ alkene} \xrightarrow{+CO/H_2} C_4 \text{ aldehyde (intermediate compound)} \xrightarrow{\text{reduction}} C_4 \text{ primary alcohol}$$

e.g. $$\underset{\text{propene}}{CH_3CH{=}CH_2} \longrightarrow \underset{\text{butanal}}{CH_3CH_2CH_2CHO} \longrightarrow \underset{\text{butan-1-ol}}{CH_3CH_2CH_2CH_2OH}$$

2.18 Carboxylic acids

Carboxylic acids are produced naturally when aldehydes and primary alcohols undergo oxidation.

$$\bigcirc\!-CH_2OH \longrightarrow \bigcirc\!-CHO \longrightarrow \bigcirc\!-COOH$$

These acids contain the **carboxyl group**, which may be represented as

$$-COOH, \quad -\overset{\overset{\displaystyle O}{\|}}{C}-O-H \quad \text{or} \quad -C\begin{matrix}{/\!\!/}O \\ \backslash O-H\end{matrix}$$

Many carboxylic acids are involved in the food industry.

COOH (attached to benzene ring)

$HOOC-CH_2-C(OH)(COOH)-CH_2-COOH$

Benzoic acid preserves food by inhibiting the growth of bacteria.

Citric acid occurs naturally as the chief acid in citrus fruits.

The simplest carboxylic acids are those structurally derived from alkanes and are called **alkanoic acids**. They are named by replacing the 'e' of the parent alkane with 'oic acid' and have the general formula $C_nH_{2n+1}COOH$. Alkanoic acids form a subset within the whole set of carboxylic acids.

Name	Formula
methanoic	$HCOOH$
ethanoic	CH_3COOH
propanoic	C_2H_5COOH
butanoic	C_3H_7COOH
pentanoic	C_4H_9COOH
hexanoic	$C_5H_{11}COOH$
heptanoic	$C_6H_{13}COOH$
octanoic	$C_7H_{15}COOH$
etc	etc

Table 2.8 Alkanoic acids

Figure 2.38 The smell of rancid butter is due to butanoic acid, formed as the butter goes stale

The smaller acids have strong characteristic smells. Vinegar, for instance, has the smell of ethanoic acid and rancid butter the smell of butanoic acid.

2.18.1 Comparing alkanols and alkanoic acids

The smaller alkanols and alkanoic acids are generally colourless liquids. They contain the OH group within their molecules, as does water. As a result of this similarity in structure, small alkanols and small alkanoic acids are soluble in water. This solubility diminishes as the size of molecules increases and the hydrocarbon chain becomes the dominant feature.

alkanol: $R-CH_2-O-H$ alkanoic acid: $R-C(=O)-O-H$

alkanol alkanoic acid

The presence of the C=O group in an alkanoic acid has an effect on the neighbouring O—H bonds. This O—H bond is now more likely to split (dissociate) and form a hydrogen ion. When CH_3COOH, for example, is dissolved in water, it ionises to form an acid solution (pH less than 7).

$$CH_3C(=O){-}O{-}H \xrightarrow{H_2O} CH_3C(=O){-}O^- + H^+$$

CH_3CH_2OH, however, does not ionise when dissolved in water and its molecules remain undissociated. An aqueous solution of ethanol has a pH of 7.

It requires reaction with an extremely reactive metal, such as sodium, to bring about ionisation of the O—H bond in ethanol. Hydrogen gas is released at a moderate rate.

$$CH_3CH_2{-}O{-}H + Na \longrightarrow CH_3CH_2O^- \; Na^+ + \tfrac{1}{2}H_2$$

2.19 Ethanoic acid

The most important alkanoic acid is ethanoic (acetic) acid, CH_3COOH. It is required for the synthesis of many products but chiefly to make PVA resin (for paints and adhesives), cellulose acetate (for plastic film or textiles) and compounds called esters (for solvents).

Ethanoic acid can be manufactured from four different primary feedstocks, namely wood, coal, biomass and oil. Biomass and oil (or gas) are currently the more important sources.

2.19.1 Ethanoic acid from biomass

The making of **vinegar**, a solution of ethanoic acid in water, is one of the oldest biochemical processes. Vinegar is the product of a double fermentation.

(a) Fermentation of a sugary mash by a suitable yeast to give ethanol.

(b) Oxidation of the ethanol to ethanoic acid in a second fermentation by a suitable bacterial culture.

$$\underset{\text{sugar}}{C_6H_{12}O_6} \xrightarrow{a} \underset{\text{ethanol}}{CH_3CH_2OH} \xrightarrow{b} \underset{\text{ethanoic acid}}{CH_3COOH}$$

The raw materials used to produce the sugary feedstock vary from malted barley to any type of waste fruit. Malted barley leads to malt vinegar, grapes to wine vinegar, apple juice to cider vinegar and so on.

The souring of wine left exposed to the air is due to oxidation of the ethanol in the wine to ethanoic acid.

The fermentation method produces ethanoic acid in a solution with water (concentration up to 10 per cent). The variability of the sugary feedstock (carbohydrate residues) and the expense of separating acid from water makes this method less competitive than wholly artificial, non-biochemical manufacture.

Vinegar is basically a 4–5 per cent solution of ethanoic acid in water. The same solution can be produced by making ethanoic acid in a non-biochemical way and dissolving it in water. The product is called **non-brewed condiment**.

2.19.2 Ethanoic acid from oil or gas feedstocks

Ethanoic acid can be artificially synthesised by three different industrial methods:

(a) the ethene route
(b) the naphtha route
(c) the methanol/carbon monoxide route.

The ethene route is the oldest method and involved catalytic oxidation of ethene to ethanal which is then oxidised to the acid.

$$\underset{\text{ethene}}{CH_2CH_2} \longrightarrow \underset{\text{ethanal}}{CH_3CHO} \longrightarrow \underset{\text{ethanoic acid}}{CH_3COOH}$$

In Britain the other two routes are preferred. BP, the second largest producer of ethanoic acid in the world, manufactures the acid at Hull using different processes in adjacent plants.

The naphtha route

Naphtha is a major fraction (about 20 per cent by mass) of North Sea oil. It is therefore a readily available feedstock in Britain. Light naphtha (the more volatile sub-fraction) is oxidised by compressed air to give ethanoic acid. Smaller amounts of methanoic and propanoic acids are also produced as by-products.

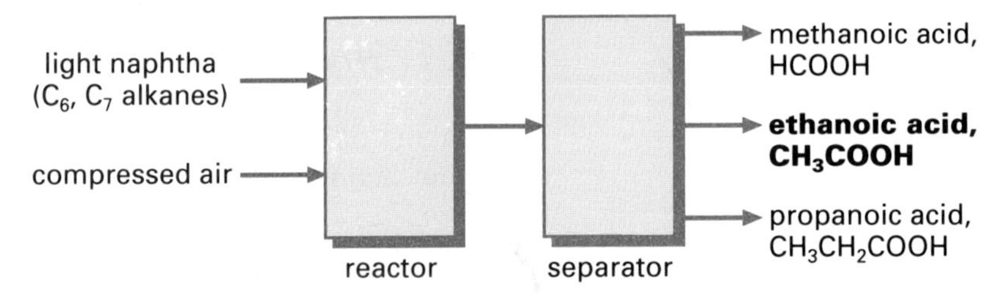

Figure 2.39 Manufacture of ethanoic acid from naphtha

The methanol route

The development of a rhodium compound catalyst, patented by the Monsanto company, enabled methanol and carbon monoxide to combine, under pressure, to form ethanoic acid.

$$CH_3OH + CO \longrightarrow CH_3COOH$$

This method has the big advantage of having a choice of primary feedstocks as both methanol and carbon monoxide can be obtained from either coal, oil or gas. A second attraction of the method is the high yield of ethanoic acid obtained. By-products are almost negligible.

Figure 2.40 The BP ethanoic acid plant at Hull: processes based on naphtha oxidation and the methanol/carbon monoxide reaction are both in use on this site

2.20 Methanol, methanal and syngas

Methanol is a much sought after chemical feedstock. It is used to produce ethanoic acid, solvents, fuel components and, chiefly, **methanal** (formaldehyde).

$$\underset{\text{methanol}}{CH_3OH(1)} \xrightarrow{\text{oxidation}} \underset{\text{methanal}}{HCHO(g)}$$

The above conversion from primary alcohol to aldehyde can be brought about by catalytic oxidation using a copper or silver catalyst.

Methanal is a gas which irritates eyes, nose and throat. It is present in wood smoke and is partly responsible for the tears caused by smoke. Methanal dissolves in water to form a 40 per cent solution called **formalin**, which is used for preserving biological specimens.

2.20.1 Methanal polymers

Methanal forms important condensation polymers with the three monomers shown.

phenol (C_6H_5OH)

urea (carbamide) ($H_2N{-}C({=}O){-}NH_2$)

melamine (triazine ring with three NH_2 groups)

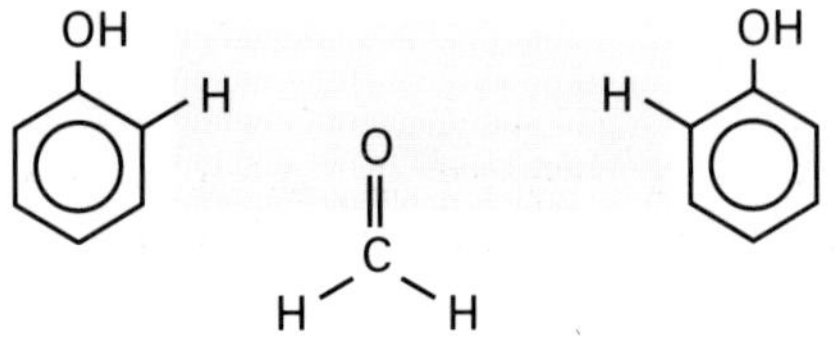

Figure 2.41 Condensation process

Phenol–methanal (phenol–formaldehyde) was the first fully synthetic plastic. It was patented in 1907, under the trademark 'Bakelite', by a Belgian chemist called Leo Baekeland working in New York. Two molecules of phenol join with one of methanol by eliminating water, as shown in Figure 2.41. This condensation process is repeated to form linear chains which are then cross-linked to another to form a three-dimensional network (Figure 2.42).

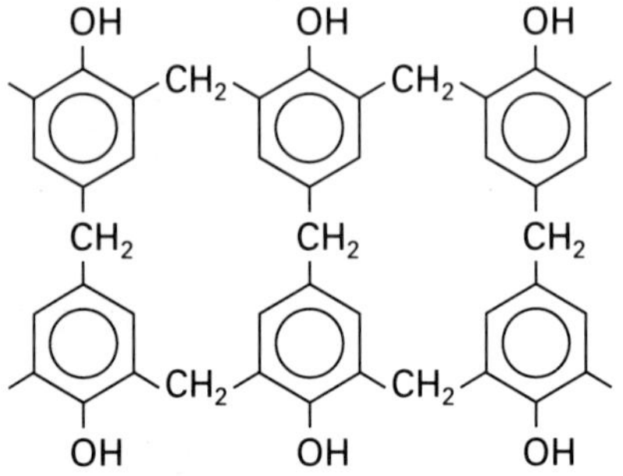

Figure 2.42 Cross-linking stage

The cross-linking stage is carried out in a heated mould under pressure so that the final reaction forms the finished product, which cannot then be softened by further heating.

Similar polymerisation occurs between methanal (formaldehyde) and the other monomers, **urea** and **melamine**. Thanks to their cross-linked structures, phenol–formaldehyde (PF), urea–formaldehyde (UF) and melamine–formaldehyde (MF) resins are all **thermosetting** plastics (not softened by heat).

The methanal thermosets are used in a variety of products such as kitchen worktops, chipboard adhesive, electrical plugs and sockets, unbreakable dinnerware, saucepan handles and printed circuit boards.

2.20.2 The manufacture of methanol

Methanol is in demand as a feedstock as it is a C_1 compound which is much more reactive than the C_1 hydrocarbon, methane CH_4.

The C_2, C_3 and C_4 building block molecules are the reactive alkenes ethene, propene and butenes (also butadiene). There is, of course, no C_1 alkene and so a different approach is required in the synthesis of reactive C_1 compounds.

When conducting research into ammonia synthesis, chemists from the German company BASF noticed the formation of methanol. This led to the development in 1923 of a process for making methanol from a mixture of carbon monoxide and hydrogen called **synthesis gas** or **syngas** for short.

$$CO + 2H_2 \longrightarrow CH_3OH$$

This BASF process has now been replaced by an ICI process, introduced in 1966, using lower pressure and a different catalyst, one based on copper. The conversion per pass (each passage through the reactor) is only about 2.5 per cent, but by constantly recycling the unreacted gases an overall yield of well over 90 per cent can be achieved.

2.20.3 The production of syngas

Synthesis gas (syngas) is now one of the most important feedstocks for making fuels and other chemicals. It may be made from coal, oil or gas.

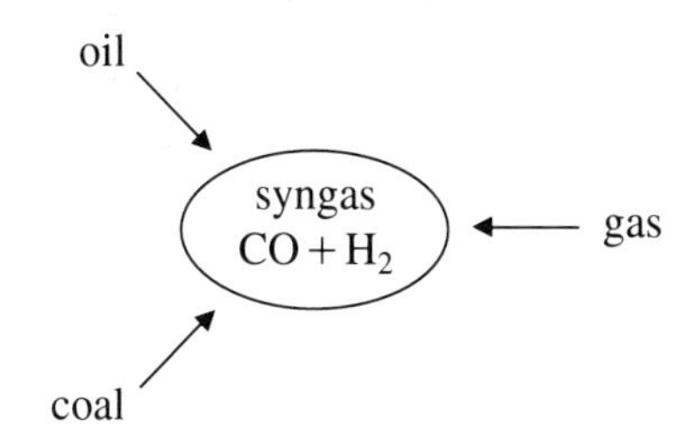

Syngas from coal

Coal gasification is a means of producing syngas. Coal and steam enter a reactor where they undergo a complex series of reactions at high temperatures. The effect of the steam and heat is to bring about a rearrangement of carbon-containing molecules. The process is therefore described as steam reforming of coal. The net change may be simply summed up by the equation below.

$$\underset{\text{coke}}{C} + \underset{\text{steam}}{H_2O} \longrightarrow \underset{\text{syngas}}{CO + H_2}$$

Coal gasification is of major importance in countries with large coal stocks such as South Africa, the USA and Germany. Syngas can, however, be generated from various other hydrocarbons, such as those present in naphtha or natural gas.

Syngas from methane

In Britain, methane from North Sea gas (rather than naphtha from North Sea oil) is used to make syngas, by steam reforming. The mixture of methane and steam, under pressure and at high temperatures, is passed over a catalyst which helps bring about the rearrangement.

$$CH_4 + H_2O \longrightarrow CO + 2H_2$$

The $CO{:}H_2$ ratio in the syngas mixture can be adjusted to suit whatever product is wanted. To make methanol, for example, a 1:2 ratio is best, but for other products made from syngas feedstock a different ratio may well be needed. Adjustment to the proportions is achieved by various chemical processes.

2.21 Fuels

At the surface of the earth there is a constant inflow and outflow of energy. More than 99 per cent of the energy input is **solar radiation**, i.e. radiation from the sun. It is supplemented by small amounts of heat from the interior of the earth (geothermal energy) and tidal energy from the gravitational system of the earth, moon and sun.

Most of the incoming solar radiation is reflected as short-wave radiation or absorbed by the earth and converted to heat. A small fraction produces winds and waves, and an even smaller fraction is captured by plants and stored chemically by photosynthesis.

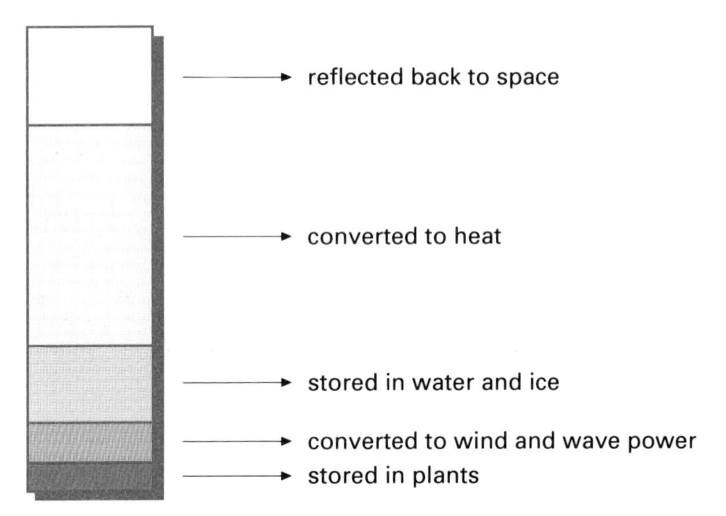

Figure 2.43 Distribution of solar radiation received

The solar radiation trapped by the chlorophyll of plant leaves becomes the energy supply of the photosynthetic process and ultimately the energy supply of the plant and animal world.

This biologically stored energy is released by oxidation (respiration) at roughly the same rate as it is stored. However, a tiny fraction is deposited in oxygen-deficient environments, under conditions that prevent complete decay to CO_2 and H_2O (and loss of energy).

During the past 600 million years, the plant and animal remains which are buried under sedimentary sands, muds and limestones have produced fossil fuels such as coal, natural gas and petroleum oil.

2.22 Natural gas and coal as fuels

Natural gas, which is mainly methane, and coal, which is mainly carbon, can be used directly as fuels.

North Sea gas, first piped ashore into Britain in 1965, is distributed throughout the country by British Gas for public and industrial consumption. The bulk of British coal is used in power stations to generate electricity: the heat from burning coal is used to produce the steam, which drives the turbines, which in turn spin the generators.

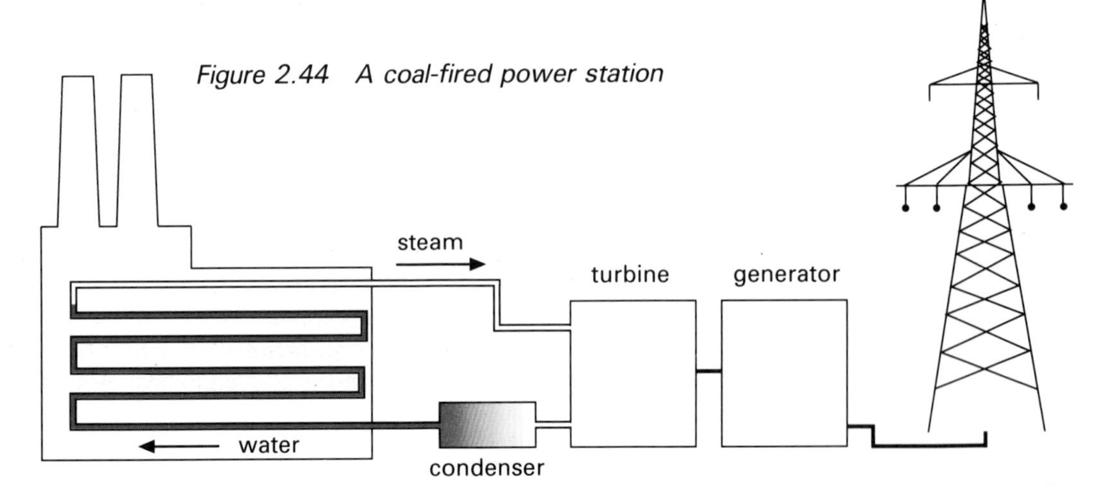

Figure 2.44 A coal-fired power station

In the long term it seems likely that, due to the much larger reserves of coal compared to gas, supplies of gas will have to be obtained from the processing of coal.

2.23 Fuels from petroleum oil

Less than 10 per cent of a barrel of oil is used in manufacturing chemicals for consumer products such as plastics, etc. The main use is in the production of fuels. Crude oil itself has no value as a direct fuel but from it can be made the three liquid fuels in demand for transporting people and freight around: **petrol**, **aviation fuel** and **diesel**.

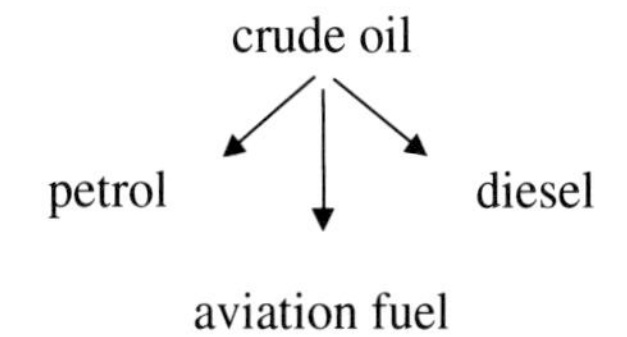

The first operation in any oil refining is separating the mixture into various fractions. The liquid fractions which correspond most closely in physical characteristics to petrol, aviation fuel and diesel are **gasoline**, **kerosine** and **gas oil** respectively. The problem for refineries is how to convert the raw oil fraction into a liquid fuel which behaves exactly as demanded by the automobile, aircraft, train or shipping industry.

gasoline $\xrightarrow{?}$ petrol

kerosine $\xrightarrow{?}$ aviation fuel

gas oil $\xrightarrow{?}$ diesel

The conversion of kerosine fraction to aviation fuel is relatively straightforward. Kerosine contains some acids and sulphur compounds which are undesirable because of their corrosiveness and nasty smell. These compounds must be removed and oxidised respectively in what is called a 'sweetening' process before selling as aviation fuel.

The production of petrol is more complicated. No 'straight' oil fraction will perform satisfactorily as a fuel in a modern petrol engine.

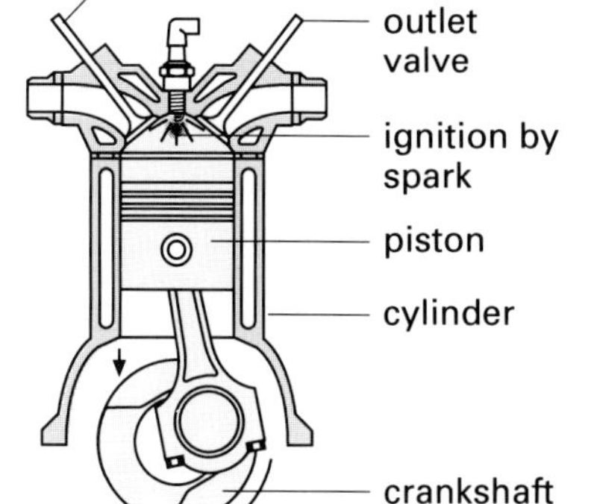

Figure 2.45 Internal combustion in a petrol engine cylinder

2.24 Petrol

In a normal car engine, petrol mixed with air enters a cylinder where it is ignited by means of a spark plug. The petrol burns, its flame enlarging progressively. The heat energy produced by the combustion expands the gases and pushes down a piston, thereby producing mechanical power to drive the engine.

2.24.1 Octane number

The higher the compression ratio of an engine, the better its fuel economy and power output. A compression ratio of say 10:1 means that the mixture of petrol and air is compressed (squeezed), by the rising piston, to one tenth of its original volume in the cylinder before it is ignited by the spark.

Compressing a gas raises its temperature. The more a gas is compressed the hotter it becomes. Increasing the compression ratio in an engine beyond a certain limit would bring about **premature** ignition or detonation of the gas – this is called 'knocking'. When the petrol–oil mixture is compressed in the cylinder and ignited by the spark, the resulting flame travels down from the spark plug. Detonation is caused by unburnt vapour spontaneously igniting and exploding before the expanding flame in the cylinder reaches it. Such premature detonation is harmful to the engine and its operation. A petrol's 'anti-knock' ability is expressed in terms of **octane number**. The higher the octane number of the petrol the less likely it is to knock.

Straight chain hydrocarbons are more prone to self-ignite (ignite without application of spark or flame) and cause knocking in a cylinder than **branched chain** hydrocarbons of similar size. The octane number of a petrol is measured by comparing its potential to self-ignite or detonate with that of a mixture of two liquids: a branched alkane, iso-octane, and a straight chain alkane, normal heptane.

$$\begin{array}{c} \quad CH_3 \quad\; CH_3 \\ \quad | \qquad\quad | \\ CH_3CH_2CH_2CH_2CH_3 \\ \quad | \\ \quad CH_3 \end{array}$$

iso-octane
(2,2,4-trimethylpentane)

assigned octane number 100

$$CH_3CH_2CH_2CH_2CH_2CH_2CH_3$$

heptane

assigned octane number 0

The **percentage of iso-octane** in the mixture with heptane that gives the same performance as the petrol fuel, in a standard test engine, is the **octane number** of that petrol.

A typical petrol has an octane number of around 95, i.e. it behaves like a mixture of 95 per cent of iso-octane, 5 per cent heptane.

2.24.2 Petrol blending

The octane numbers assigned to some individual hydrocarbons are given in Table 2.9.

Hydrocarbon	Carbon backbone	Octane number
octane	 O–O–O–O–O–O–O–O	–19
heptane	 O–O–O–O–O–O–O	**0**
2-methylheptane	O | O–O–O–O–O–O–O	21.7
2-methylhexane	O | O–O–O–O–O–O	42.4
2,4-dimethylpentane	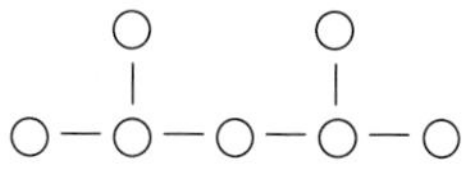 O O | | O–O–O–O–O	65.2
cyclohexane	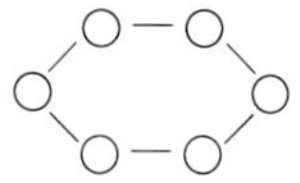 O–O O O O–O	83.0
pent-1-ene	 O=O–O–O–O	90.9
4-methylpent-1-ene	O | O=O–O–O–O	95.7
2,2,4-trimethylpentane	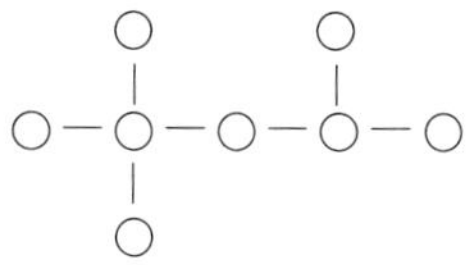 O O | | O–O–O–O–O | O	**100**
cyclopentane	O–O O | O–O	100.3
benzene	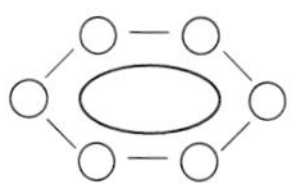 O–O O O O–O	105.8
methylbenzene	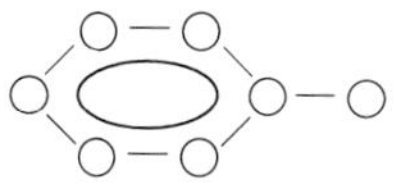 O–O O O–O O–O	117.8

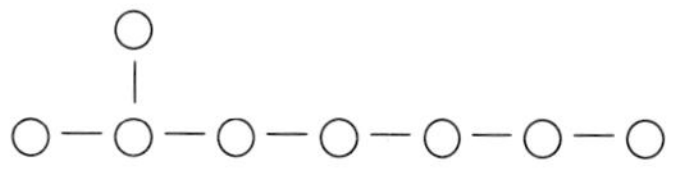

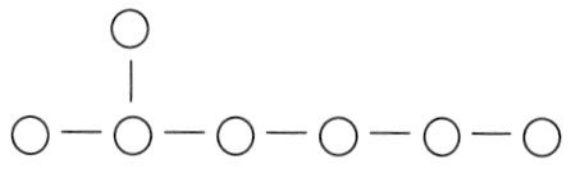

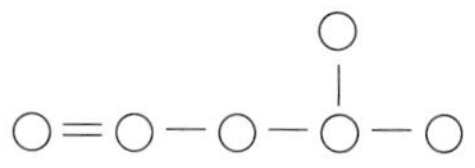

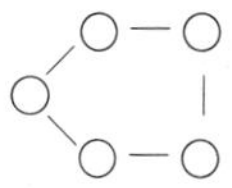

Table 2.9 Octane numbers of selected hydrocarbons

Up until 1910, the petrol produced commercially was simply a mixture of the hydrocarbons which were originally present in the crude oil. Petrol was the gasoline fraction. It was soon discovered that cracking heavier liquids, such as gas oil, produced smaller molecules suitable for petrol. As the demand for more and better quality petrol increased, the refining industry had to devise processes by which this high grade petrol could be produced. Today's petrol is a blend of outputs from various refinery processes, the most important of which is **catalytic reforming**.

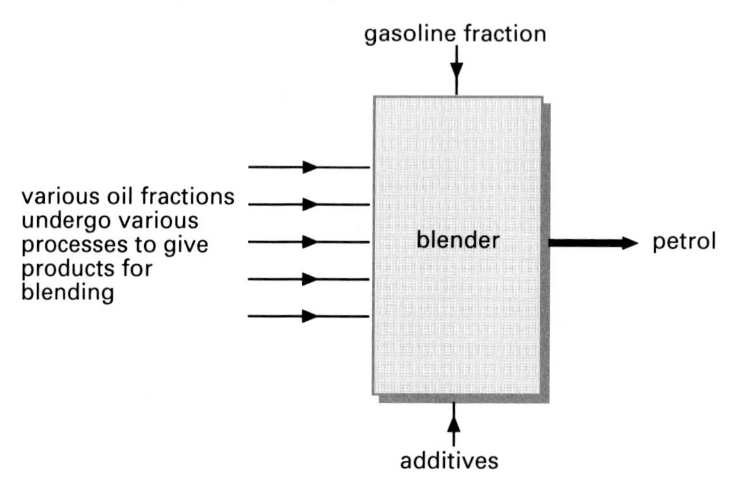

Figure 2.46 Blending of hydrocarbons and additives to give petrol

2.25 Reforming naphtha

Simple aromatics are C_6–C_8 hydrocarbons and have high octane numbers. They are therefore more useful components for petrol than C_6–C_8 straight chain alkanes and cycloalkanes which have lower octane numbers. The crude oil distillate, naphtha, is approximately a C_6–C_{10} hydrocarbon mixture, usually with low aromatic content, unsuitable for direct inclusion in petrol. It is therefore a suitable mixture for treatment. The treatment is **catalytic reforming**, a complex set of processes which effectively rearrange molecules without necessarily altering their size (as happens in cracking).

Reforming is a major process in oil refineries throughout Britain. In a catalytic reformer, such as the one at the BP refinery in Grangemouth, naphtha feedstock is passed over a platinum catalyst at high temperature and pressure. Typical rearrangements of the molecular structures are shown in Figure 2.47. Straight chain molecules are rearranged to form either branched molecules or aromatic molecules, both of which have higher octane numbers. Cyclic molecules are converted to aromatic molecules.

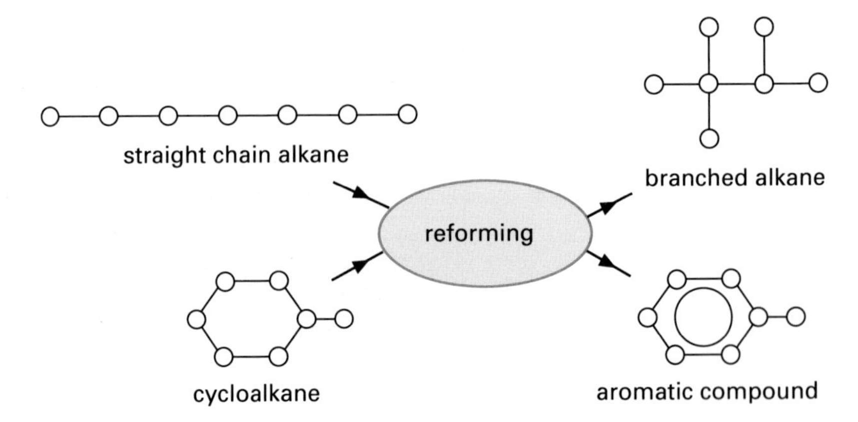

Figure 2.47 Reforming naphtha

Reformed naphtha is therefore richer in branched alkanes and aromatics and poorer in straight alkanes and cycloalkanes.

naphtha	reformed naphtha
straight chain alkanes cycloalkanes branched alkanes aromatics	fewer straight alkanes fewer cycloalkanes more branched alkanes more aromatics

2.25.1 Petrol additives

Before 1990, compounds such as tetramethyl-lead (TML) and tetraethyl-lead (TEL) were added to petrols to raise octane ratings. Legislation to control harmful emissions from automobile exhaust has forced a gradual switch from 'leaded' to 'unleaded' petrol. Lead would poison the catalysts present in catalytic converters. These are fitted to cars with the purpose of reducing pollutants in the exhaust gas.

The problem of increasing octane values is now tackled by increasing the proportion of branched or aromatic molecules present and/or adding other octane-raising compounds. Molecules containing carbon, hydrogen and generally one oxygen atom have high octane numbers, and such compounds can be added to the hydrocarbons in petrol to upgrade the quality.

2.26 Diesel fuel

Diesel engines for lorries, buses, trains, taxis and cars are the principal users of diesel fuel. A diesel engine operates quite differently from a petrol engine. In the latter type a mixture of petrol and air is drawn into a cylinder; there it is compressed by the rising piston before being ignited by a spark.

A diesel engine, on the other hand, draws in and compresses only air. The diesel engine has no spark plugs. Instead the piston, as it rises, squeezes the air in the cylinder and heats it up to about 870 K. The diesel is then sprayed in as a very fine mist and the hot air–diesel mixture ignites **spontaneously**. The combustion forces the piston back down the cylinder and powers the engine.

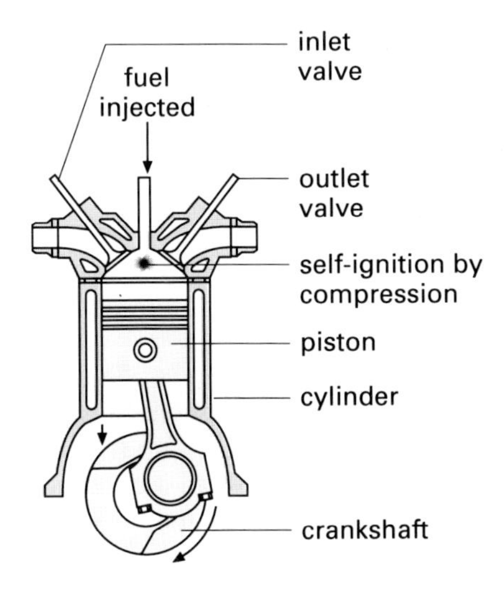

Figure 2.48 Internal combustion in a diesel engine cylinder

A diesel engine works therefore by compression-ignition while a petrol engine operates by spark-ignition. The key factor is controlled ignition and combustion of the diesel fuel. The period of time between the start of the fuel injection and its self-ignition is the most vital part of the combustion process. This interval is called the ignition delay and it is important not to have too much of a lag.

fuel injected | time --------- delay | fuel ignites

The efficiency of the ignition process depends on the hydrocarbon composition of the fuel. Straight chain alkanes have a greater tendency to self-ignition than branched alkanes or aromatics, so what was bad for knocking in petrol engines is good for rapid self-ignition in diesel engines. The type of hydrocarbon that has a low octane number has good diesel ignition qualities and vice versa.

2.26.1 Cetane number

The quality of diesel fuel is expressed as a **cetane number**. The cetane number of a diesel fuel is measured by comparing its ignition rate with that of a mixture of two liquids, both of molecular formula $C_{16}H_{34}$:

(a) a straight chain alkane called **cetane** which self-ignites quickly and is assigned a cetane number of 100; and

(b) a branched alkane, heptamethylnonane, which is much slower to ignite and is assigned a cetane number of 15.

$$CH_3CH_2CH_2CH_2CH_2CH_2CH_2CH_2CH_2CH_2CH_2CH_2CH_2CH_2CH_2CH_3$$

cetane

$$CH_3-\underset{CH_3}{\overset{|}{CH}}-\underset{CH_3}{\overset{|}{CH}}-\underset{CH_3}{\overset{|}{CH}}-\underset{CH_3}{\overset{|}{CH}}-\underset{CH_3}{\overset{|}{CH}}-\underset{CH_3}{\overset{|}{CH}}-\underset{CH_3}{\overset{|}{CH}}-CH_3$$

heptamethylnonane (HMN)

Just as octane number indicates the quality of a petrol fuel, so cetane number indicates the quality of a diesel fuel. The higher the cetane number of a diesel fuel, the more rapid its ignition.

2.26.2 Diesel blending

Fuel for diesel engines is **blended** from a range of oil fractions. A high-speed diesel engine for cars, buses and trains needs 'lighter' diesel fuel, with a higher cetane number, than the slower-speed diesel engines for ships.

The lighter diesel is taken mainly from gas oil fractions whereas denser, less volatile diesel may be obtained from heavier fractions such as fuel oil.

2.27 Fuels from natural gas

Natural gas can directly provide fuels such as liquefied natural gas (LNG) and liquefied petroleum gas (LPG), which are liquids, either at temperatures below normal or at pressures above normal. Petrol, which is usually derived from crude oil, can also be made, indirectly, from natural gas.

2.27.1 Liquefied natural gas (LNG)

Natural gas varies in composition from being 'dry', where it is virtually all methane, to 'wet' where a small proportion of heavier alkanes, C_5 up to even C_8, may be present. The southern North Sea gasfields off England are mainly 'dry', whereas gas from the northern sector off Scotland, Orkney and Shetland is 'wet'.

Name	Alkane size	bp/K
methane	C_1	112
ethane	C_2	184
propane	C_3	231
butane	C_4	265*
pentane	C_5	305*
hexane	C_6	342*
heptane	C_7	371*
octane	C_8	399*

* average of isomers

Table 2.10 The alkanes in natural gas

A large number of natural gas reserves are separated from potential customers by great expanses of land or sea, across which it is not feasible to lay pipelines. A more transportable form of the gas is required. Natural gas becomes a liquid, at atmospheric pressure, when its temperature is brought down below 112 K. A large refrigeration plant (liquefaction plant) is necessary to achieve this low temperature. The heavier alkanes (C_5 and above) are solids at this temperature and are removed completely.

Liquefied natural gas (LNG) can be transported in purpose-built insulated ships in which the LNG is kept below 112 K. On arrival at its destination, the LNG is off-loaded into refrigerated or pressurised storage tanks for later regasification and use in the natural gas grid of the country.

Britain is self-sufficient in piped gas from the North Sea. Not all countries are so fortunate. Japan, for example, has to import almost all its natural gas, as LNG, from countries such as Indonesia, Malaysia and Australia.

Figure 2.49 A natural gas liquefaction plant in Algeria

2.27.2 Liquefied petroleum gas (LPG)

The alkanes in natural gas which are heavier than methane are collectively known as **natural gas liquids (NGL)**. This is because they can, with the possible exception of ethane, be liquefied with reasonable ease, by lowering their temperature and/or raising their pressure.

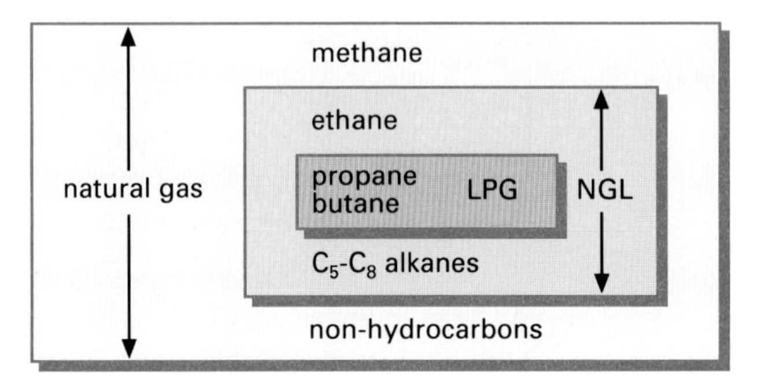

Figure 2.50 Natural gas liquids (NGL)

In Scotland, the methane is separated off at St Fergus for national gas supplies, and the NGL piped to Mossmorran where the other compounds can be separated.

The gases **propane** and **butane**, produced either during refining of oil or in the separation of natural gas components, can be readily stored in liquid form. **Liquefied petroleum gas (LPG)** is propane or butane or a mixture of propane and butane.

Bottled gas is LPG in a pressurised steel container. It is familiar as a mobile fuel pack for campers, sailors, caravanners, etc. Japan and many other countries have vehicles designed to run on LPG fuel.

2.27.3 Petrol from natural gas

New Zealand has natural gas reserves but no oil. To provide a liquid fuel for cars, etc., natural gas is converted into petrol. The conversion involves reforming methane into syngas which is then catalytically changed into methanol. The last process, methanol into petrol, is brought about with the help of a special catalyst called a zeolite.

$$\underset{\text{methane}}{CH_4} \longrightarrow \underset{\text{syngas}}{CO + H_2} \longrightarrow \underset{\text{methanol}}{CH_3OH} \longrightarrow \underset{C_6\text{–}C_8 \text{ hydrocarbons ('petrol')}}{-(CH_2)_n-}$$

The New Zealand 'gas-to-gasoline' plant went into production in 1985.

2.28 The products of combustion

Combustion in stationary installations such as boilers and furnaces leads to almost complete conversion of the fuel into CO_2 and H_2. Any sulphur compounds not previously removed from the fuel are converted into sulphur dioxide, SO_2. At the high temperatures reached during combustion some of the nitrogen and oxygen in the air supply will combine to form nitrogen monoxide gas.

$$N_2 + O_2 \longrightarrow 2NO$$

The conditions for combustion in an internal combustion engine are different from those in boilers or furnaces. The supply of air is more restricted and combustion is generally incomplete. A petrol-fuelled vehicle produces three major pollutants.

(a) Carbon monoxide (CO) – a poisonous gas which, if inhaled in a confined space, can cause death within minutes.

(b) Oxides of nitrogen (NO_x) – these include colourless NO which rapidly converts in air to brown NO_2, a poisonous gas which destroys lung tissue. NO_2 (along with SO_2) dissolves in water vapour to form acid rain which damages trees, marine life and limestone buildings.

(c) Hydrocarbons (HC) – these are present as a result of unburnt fuel. When mixed with sunlight and oxides of nitrogen they can form substances which irritate the mouth, throat, lungs, etc. Some hydrocarbons are carcinogenic; they may cause cancer.

In diesel-powered vehicles combustion is more efficient. Less CO and hydrocarbons are released than in a petrol engine. On the other hand diesel exhausts give out more sooty material.

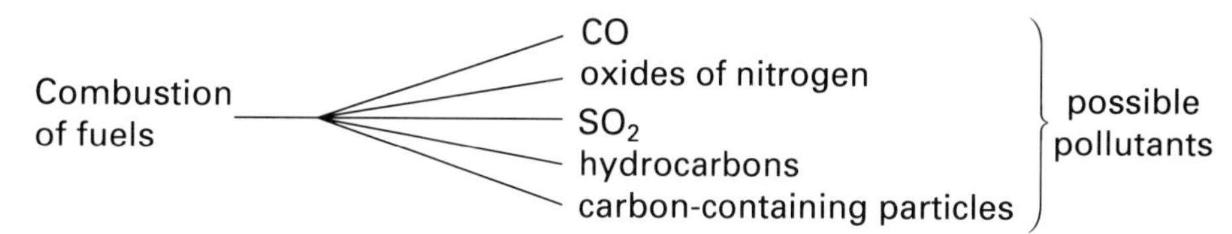

In most countries legislation exists to control emissions from both petrol and diesel engines. Catalytic converters, fitted to exhausts, complete the oxidation of CO and unburned hydrocarbons to CO_2 and H_2O. In petrol exhausts the oxides of nitrogen are simultaneously reduced to N_2. In a diesel engine combustion takes place with excess air and the exhaust gas contains a surplus of oxygen. This makes it impossible for three-way catalytic converters to reduce NO_x to N_2 in diesel exhausts. Dust traps can be added to the exhaust systems to catch sooty particles.

A fuel like ethanol, C_2H_5OH, has oxygen available in its chemical structure and generally burns more thoroughly to give only CO_2 and H_2O. Ethanol is therefore described as a 'clean' fuel compared to petrol or diesel. Methanol is also cleaner in combustion but is itself toxic.

The principal product in all fuel combustions, whether in vehicles or power stations, is **carbon dioxide**. The burning of fossil fuels over many years has not only drastically reduced natural fuel reserves but also released huge amounts of CO_2 into the atmosphere. This carbon dioxide is not directly harmful but it contributes to the greenhouse effect and global warming.

The only genuine 'environment friendly' fuel is the element **hydrogen**. Combustion of hydrogen gives only water as a product.

$$H_2 + \tfrac{1}{2}O_2 \longrightarrow H_2O$$

The idea of using hydrogen is not new. It was suggested by Jules Verne, as a universal energy source, as long ago as 1870 and its importance as a potential fuel has been emphasised ever since. So far, however, the only major use of H_2 as a fuel has been in the *Apollo* spacecraft. The main obstacles to its use are its high flammability and the expense of producing the gas. If hydrogen could be produced efficiently and economically from water, the so-called 'hydrogen economy' would be up and running.

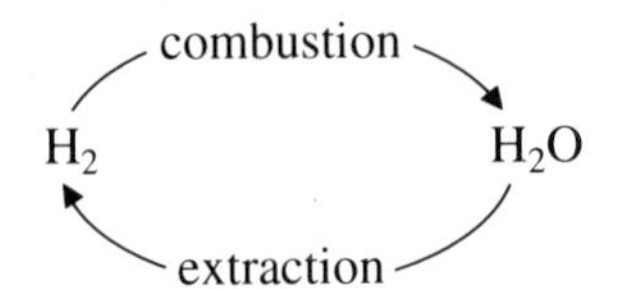

2.29 Fuels from biomass

Figure 2.51 A wood-burning stove made from clay, designed to save fuel

The material of plants is called biomass. It is a carbon-based material that, either in combustion or in natural metabolic processes, combines with oxygen to produce heat.

The biomass may be used directly as a fuel. Over half the world's population relies on wood or other biomass for cooking and heating, though deforestation has created enormous problems of fuel supply in developing countries. To make matters worse, when wood is scarce animal waste is used as a fuel rather than as a fertiliser and this reduces agricultural efficiency.

Biomass may be transformed by chemical or biological processes, to produce **biofuels** which may be solid (e.g. charcoal), liquid (e.g. ethanol) or gas (e.g. methane). The initial energy of the biomass–oxygen system is derived from photosynthesis. When this energy is released in combustion or respiration, the CO_2 also released can be recycled. There is therefore no net change in the balance of CO_2, a very important consideration if we hope to reduce the output of CO_2 into the atmosphere.

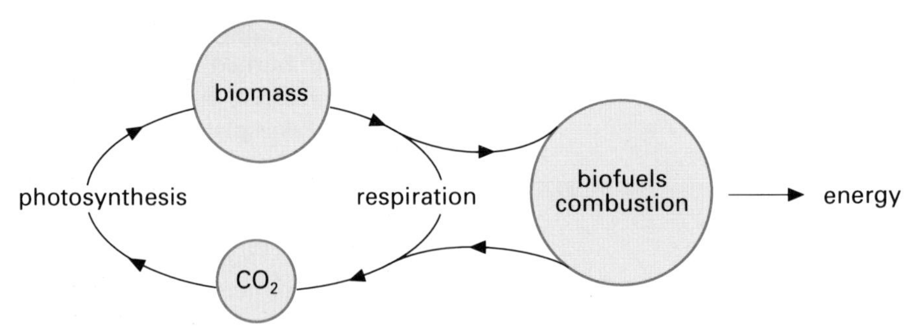

Figure 2.52 The biomass/biofuels CO_2 cycle

2.29.1 Bioethanol

Biomass covers a wide field of feedstocks, mainly carbohydrates, and ranging from wood to sugar. The major biofuel is ethanol which is offered as a substitute for petrol. The ethanol is obtained through fermentation of sugars as an aqueous solution (about 10 per cent ethanol), which must be distilled and dried to get rid of the water mixed with it. 100 per cent (anhydrous) ethanol can be used neat as a fuel but it is more commonly mixed with petrol. **Gasohol**, a mixture of 20 per cent ethanol and 80 per cent petrol, can be used in petrol engines and is a common motor fuel in countries such as Brazil, Kenya, Malawi and the United States.

The fact that petrol can be replaced as a motor fuel by bioethanol has encouraged chemists to develop liquid biofuels which may replace diesel. Certain plant oils, notably coconut, palm and sunflower oils, can now be chemically modified and then used in diesel-cycle engines.

2.29.2 Biogas

In the absence of free oxygen from the air, certain micro-organisms obtain their own energy supply by reacting with organic material, to produce some CO_2 and mostly methane, CH_4. The process is called **anaerobic** ('without air') **digestion** because, although it is a fermentation, it is similar to what goes on in the digestive system of animals such as cows.

The mixture of gases evolved is known as **biogas**. Nutrients such as soluble nitrogen compounds remain in the waste mixture, thus providing good humus and fertiliser.

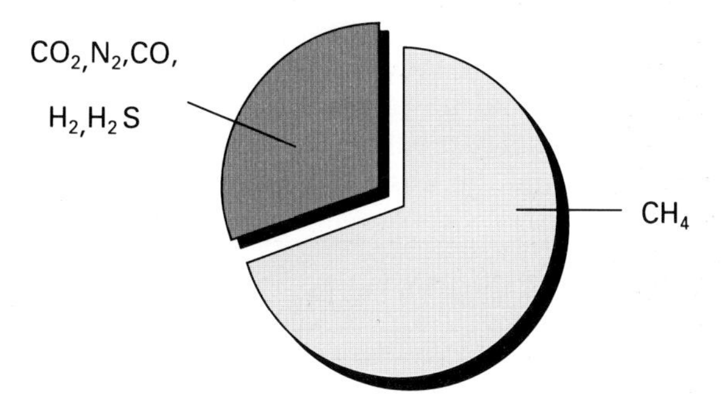

Figure 2.53 Typical composition of biogas

Biogas can be generated from the decomposition of a wide variety of waste material such as is produced in sewage plants, abattoirs, refuse dumps, food processing plants, farms and so on. In sewage works, for example, the sewage is decomposed in sludge digestion tanks and the methane produced is often used to provide power for the sewage plant.

Biogas generation is particularly well suited to farming where there is a wide variety of plant and animal waste. In Third World countries where incomes are low and energy either scarce or expensive, biogas production has great potential. In China this potential has already been developed. There are nearly nine million biogas pits in use, most of them generating methane for the cooking and lighting needs of individual households.

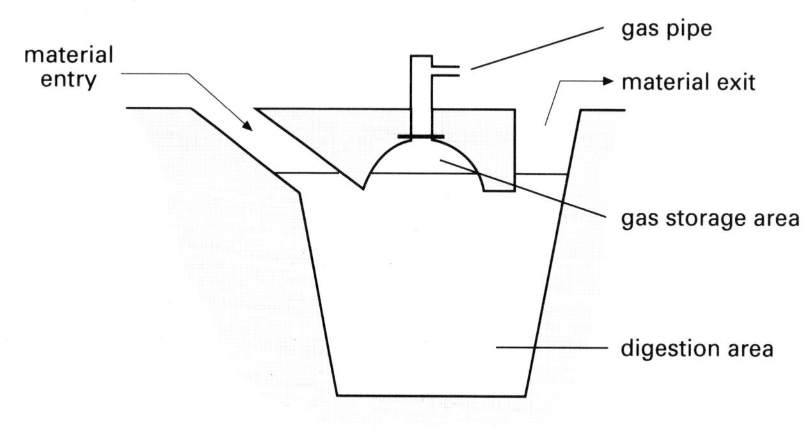

Figure 2.54 Biogas digester (Chinese design)

Biogas, when burned, releases more heat energy per unit mass than wood and other primitive fuels. In this respect it also compares favourably with coal but it has a lower energy value than other fossil fuels.

Fuel	Energy value/MJ kg^{-1}
seasoned wood	13
dry vegetation	15
rice husks, sugar cane residue	15
cow dung	15
peat	15
methanol	23
coal	27
biogas	28
ethanol	30
charcoal	32
diesel	46
petrol	47
methane	55

Table 2.11 Heat output from fuels

At present there is a dilemma. Do we grow biomass for energy production (energy farming) or for food? One possible way of resolving this problem would be to construct more biotechnology plants which handle food and energy-producing biomass together. The food component could be extracted and the waste biomass then converted to biofuel.

Summary of Unit 2

Having read and understood the information and ideas given in this unit, you should now be able to:

explain the competition between fuels and feedstocks for the limited supply of organic chemicals from coal, oil, gas and biomass

identify some simple functional groups and types of reaction

describe the production (from various sources) and uses of ethene, aromatic hydrocarbons like benzene, and ethanol

classify alcohols as primary, secondary or tertiary

categorise carbonyl compounds as aldehydes or ketones and compare their preparation from alcohols

contrast the molecular compositions of petrol and diesel, and relate these structures to combustion requirements within engines

outline the separation of natural gas into methane, ethane and LPG

identify the harmful effects of the products from combustion

draw and name compounds (up to C_8) which may be alkanes, alkenes, alkanols, alkanals, alkanones and alkanoic acids, and also primary amines (up to C_4)

identify the benzene ring/phenyl group

▲ describe the production of propene from oil and gas, and the use of molecules with two functional groups to form polymers

▲ outline the syntheses of ethanoic acid from biomass and naphtha

▲ describe the use of methanal for plastics and its production from coal or methane via synthesis gas and methanol

▲ explain the stability of the benzene ring in terms of resistance to addition reaction

▲ describe the manufacture of petrol and diesel from oil fractions with reference to reforming and blending processes

▲ outline the production of biogas.

PROBLEM SOLVING EXERCISES

1. The boxes in the grid below each contain a hydrocarbon and its octane number.

A	B	C
oct-l-ene 35	2-methylheptane 24	octane −15
D	**E**	**F**
2,4-dimethyl-hexane 70	pent-1-ene 77	3-methylheptane 35

The following statements are generalisations about molecular structures and octane numbers.

(i) Octane numbers rise as carbon chain length decreases.

(ii) With branched alkanes, octane number rises as the branch point moves nearer to the middle of the molecule.

(iii) Alkenes have higher octane numbers than corresponding alkanes.

When their information is combined, state which **two** boxes provide the most valid evidence for:

(a) relating octane number with chain length, as in generalisation (i)

(b) relating octane number with position of branching, as in generalisation (ii)

(c) comparing octane numbers of alkenes and alkanes, as in generalisation (iii).

(PS skill 7)

2. Polystyrene is made by polymerising the liquid, styrene. Styrene is manufactured by the following process.

Ethene and benzene are fed into a reactor where they are catalytically combined to form ethylbenzene. The ethylbenzene is then passed to a cracking furnace where it is decomposed into hydrogen and styrene.

The styrene is separated from hydrogen and unconverted ethylbenzene; the ethylbenzene is recycled.

Draw a flow diagram to outline the production of styrene.

(PS skill 2)

3. A flow diagram for the production of acrylonitrile (monomer for acrylic fibre polymer) is given below.

(a) Name the five feedstocks.

(b) Identify the product of the Monsanto process.

(c) Write the formulae of the **by-products** which are recycled.

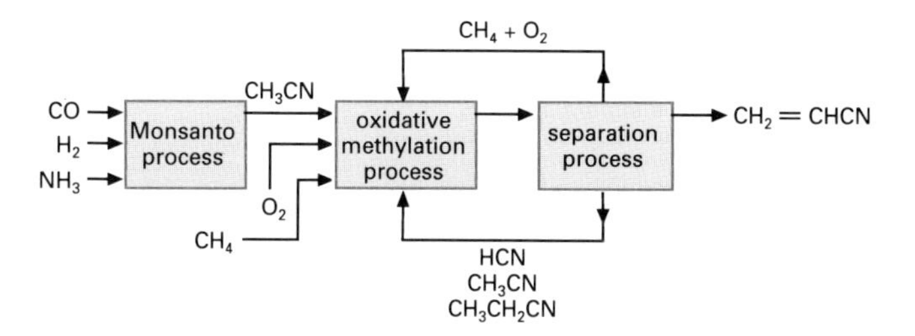

Figure 2.55

(PS skill 1)

4. Naphtha can be converted to ethanoic acid by air oxidation. The liquid flowing from the reactor is separated into two streams:

(i) unreacted hydrocarbons (gas) and

(ii) a mixture of methanoic, ethanoic and propanoic acids (liquid).

(a) What can be done with the reacted hydrocarbons?

(b) What process could be used to separate the acids from one another?

(PS skill 5)

5. The group of **alkene gases** includes the following six members:

(i) ethene (ii) propene (iii) but-1-ene

(iv) but-2-ene (v) methylpropene (vi) butadiene

Construct a branched diagram which subdivides the group, eventually to the individual compounds listed above.

(PS skill 3)

PROBLEM SOLVING EXERCISES

6. Tests were carried out on two cars to compare the amounts of carbon monoxide, hydrocarbons (HC) and oxides of nitrogen (NO_x) emitted from the exhaust. One car was fitted with a catalytic converter; the other car had no converter. Results are tabulated below.

Pollutant	Mass of pollutant/g per test	
	converter present	no converter
CO	15	62
HC	3	8
NO_x	1	6

Present the above information in the form of a bar graph, identifying each bar clearly.

(PS skill 2)

7. The shortened structural formulae of two useful solvents are given below.

$CHCl.CCl_2$ (A) $\quad$ $CH_3.CCl_3$ (B)

Suggest a practical method of distinguishing between these liquids.

(PS skill 5)

8. Compounds A, B, C and D are linked by reactions as shown below.

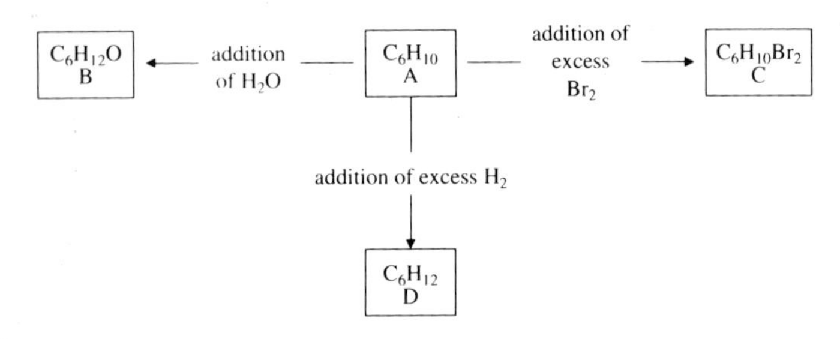

From the information given above, deduce the structures of A, B, C and D.

(PS skill 8)

9. Aldehydes are readily oxidised to carboxylic acids. An aldehyde can be prepared either by (a) oxidation or (b) catalytic dehydrogenation of the corresponding primary alcohol.

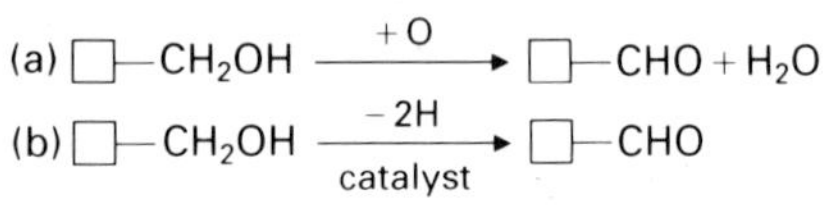

(a) $\square\!-\!CH_2OH \xrightarrow{+O} \square\!-\!CHO + H_2O$

(b) $\square\!-\!CH_2OH \xrightarrow[\text{catalyst}]{-2H} \square\!-\!CHO$

Explain why catalytic dehydrogenation (removal of hydrogen) is preferred to oxidation (addition of oxygen) as a means of preparing an aldehyde.

(PS skill 9)

10. Methyl *t*-butyl ether, MTBE, is a petrol additive and has the shortened structural formula shown.

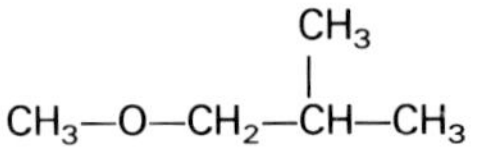

$CH_3\text{—}O\text{—}CH_2\text{—}CH(CH_3)\text{—}CH_3$

It is formed by the reaction between methanol and methylpropene. Ethyl *t*-butyl ether, ETBE, is a substitute for MTBE in petrol. It is made by reaction between ethanol and methylpropene.
Write a shortened formula for ETBE.

(PS skill 10)

11. Compounds with C=C bonds may react with oxygen of the air to form peroxides. These break down rapidly to form aldehydes with unpleasant smells.

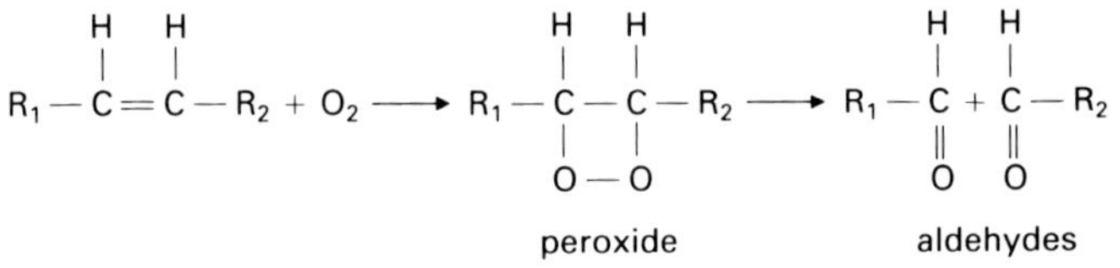

$R_1\text{—}CH{=}CH\text{—}R_2 + O_2 \longrightarrow R_1\text{—}CH(O\text{—})\text{—}CH(\text{—}O)\text{—}R_2 \longrightarrow R_1\text{—}CH{=}O + O{=}CH\text{—}R_2$

peroxide $\quad$ aldehydes

A compound X reacted with oxygen to produce propanal and 2-methylbutanal.
Write a shortened structural formula for X.

(PS skill 8)

12. The Oxo process is a method of making primary alcohols from alkenes. H and CHO groups are added across the C=C double bond.

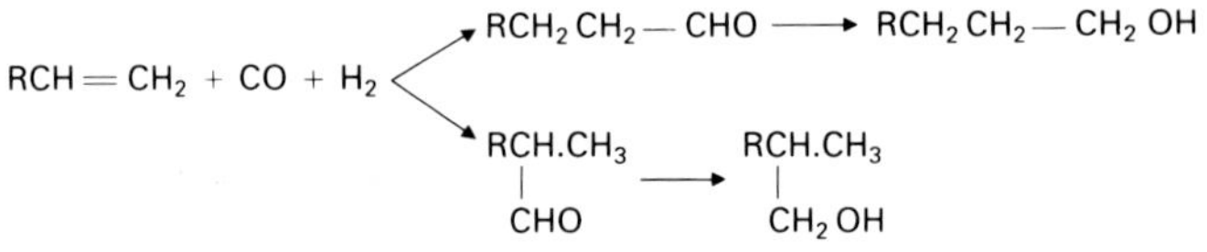

$RCH{=}CH_2 + CO + H_2 \rightarrow RCH_2CH_2\text{—}CHO \longrightarrow RCH_2CH_2\text{—}CH_2OH$

$RCH{=}CH_2 + CO + H_2 \rightarrow RCH(CH_3)CHO \longrightarrow RCH(CH_3)CH_2OH$

An important product made in this way is 2-ethylhexan-1-ol. It is manufactured in large amounts for use as a plasticiser in PVC.

Name the alkene from which it is made and indicate the position(s) of the double bond.

(PS skill 10)

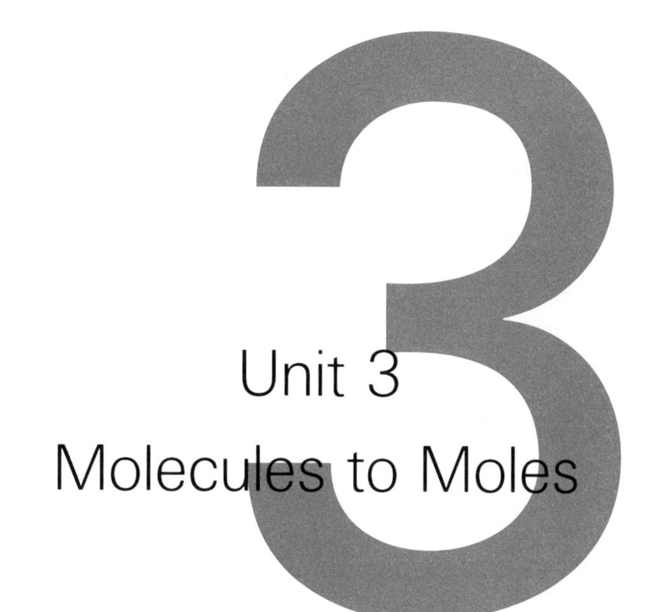

Unit 3
Molecules to Moles

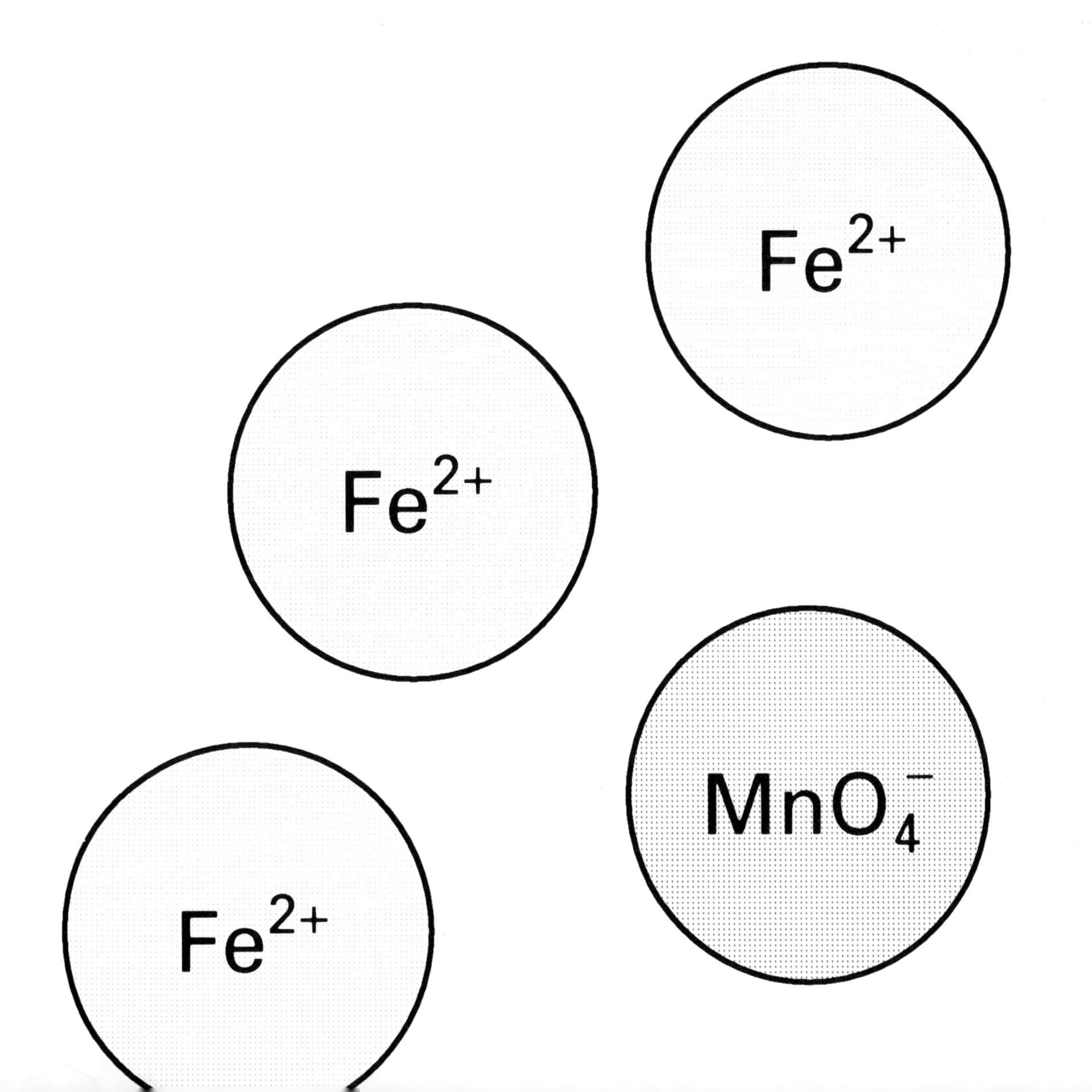

ASSUMED KNOWLEDGE AND UNDERSTANDING

Before starting on Unit 3 you should know and understand:

how to write simple balanced equations

how to convert mass to moles and moles to mass

how to work with moles, mass and volume when dealing with concentrations of solutions

the method of determining the concentration of an acid or alkali from the results of a volumetric titration

oxidation and reduction in terms of electron loss and gain

how to work out the empirical formula of a compound from its composition by mass

how to calculate percentage composition from a molecular formula

the process of electrolysis

the use of electrolysis as a way of extracting certain metals from their compounds

the use of electrolysis in purifying certain metals.

OUTLINE OF THE CORE AND EXTENSION MATERIAL

As you progress through Unit 3 you should, at least, try the fundamental content listed under **core** and, if possible, pick up the extra and/or more difficult content listed under **extension**.

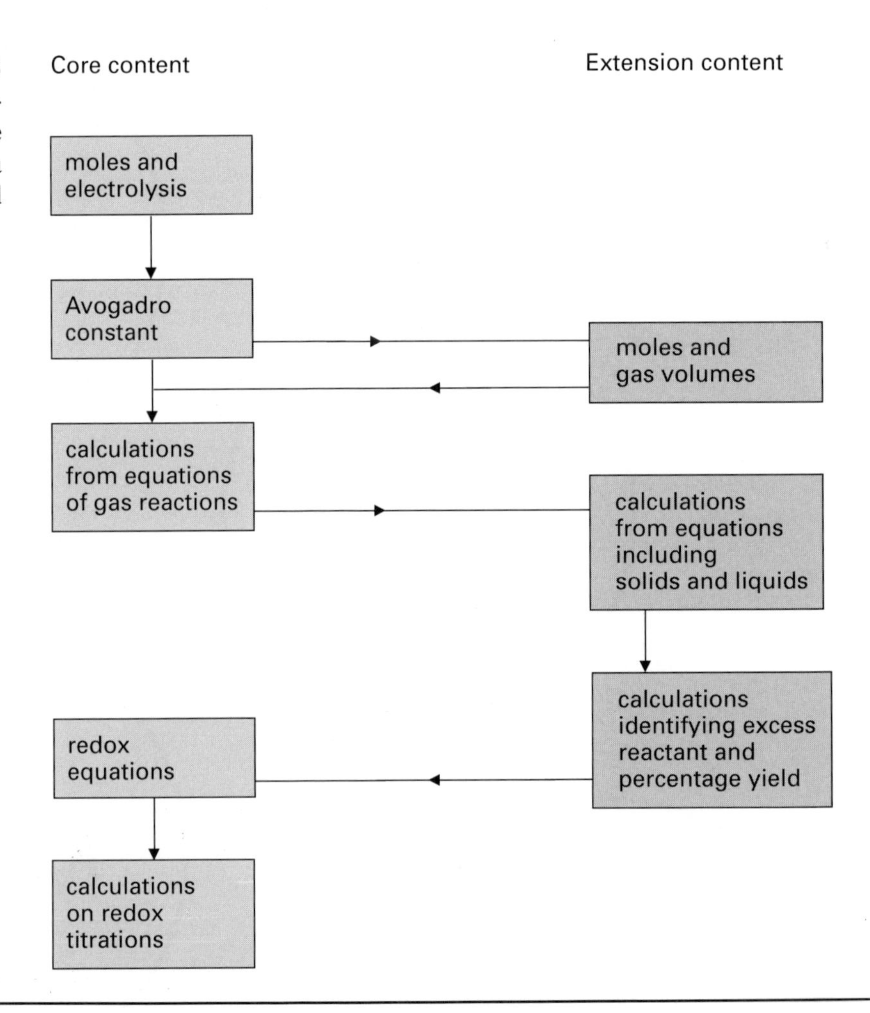

3 Molecules to Moles

3.1 Amount of substance – the mole

Scientists have to work with physical quantities like length, mass, time, and so on. Each physical quantity is measured in specified **units**, e.g. length is in metres, mass is in kilograms and time is in seconds. The choice of unit is a matter of convenience.

Chemists deal constantly with **amounts of substance** and so a unit had to be obtained for this physical quantity. Reactions involve rearrangements between particles. Consequently it is the relative **numbers** of particles and not their relative masses with which chemists are primarily concerned.

In comparing amounts of substance the chemist wants to compare **equal** numbers of particles. Equal 'amounts' of substance are therefore chosen to contain the same number of particles. The paper industry provides a good example of a convenient unit of amount. Paper is supplied in amounts called reams. You can buy reams of all kinds of paper but in every case you get the same number of sheets (500).

Figure 3.1 Reams of A4 and A3 paper

Instead of the ream, chemists use the **mole** (symbol **mol**) as a unit for amount of substance. The word 'mole' has nothing to do with a black furry animal which burrows underground, but is in fact derived from the Latin word meaning a collection, mass or pile. In 1971, the 'mole' was adopted as the seventh base unit of the Système Internationale (SI) system. The seven base units in SI are listed in Table 3.1.

Name of unit	Symbol	Physical quantity measured
the metre	m	length
the kilogram	kg	mass
the second	s	time
the ampere	A	electric current
the kelvin	K	temperature
the candela	cd	luminous intensity
the mole	**mol**	**amount of substance**

Table 3.1 SI units

3.1.1 Choosing a standard for the mole

The standard for the kilogram is the mass of a cylinder of platinium–iridium alloy kept in Paris. In the same way chemists had to have a standard mole against which all other moles could be compared.

The standard for comparing atomic masses of different elements is the common carbon isotope $^{12}_{6}C$, i.e. carbon-12. The atomic mass of this common carbon isotope was chosen as 12 atomic mass units (12 u). 12 grams (0.012 kg) of carbon was therefore considered a suitable standard for amount of substance.

One mole of carbon-12 is that amount of carbon-12 which weighs exactly 12.000 g.

3.1.2 Moles of elements

A ream of paper is made up of paper sheets. Similarly, a mole of carbon is in effect a mole of carbon **atoms**. Just as reams of paper sheets, whether the sheets be thick, thin, small or large, coloured or plain, must contain the same number of sheets, so moles of different elements must contain the **same number** of particles. For most elements this particle is the atom.

How can we measure out a mole of any element? What pile of magnesium powder, for example, contains the same number of magnesium atoms as there are carbon atoms in a 12 g pile of carbon powder? Counting the actual atoms is a non-starter, so chemists employ a technique used by bank tellers. Just as the cashier weighs different piles of coins to estimate relative numbers of coins and corresponding cash value, so the chemist **weighs** masses of elements to estimate **numbers** of atoms.

From the relative atomic mass scale we know that magnesium atoms weigh twice as much as carbon atoms (24 u compared to 12 u). It follows therefore that 24 g of magnesium will contain the **same number** of atoms as 12 g of carbon. To measure out one mole of atoms of any element, it is only necessary to measure out an amount in grams equal to the relative atomic mass of that element.

Table 3.2 provides data on some selected elements (all are solids at room temperature). Figure 3.2 indicates the relative volumes of one mole of these different elements.

Element	Relative atomic mass	Mass of 1 mol atoms of the element
carbon	12	12 g
magnesium	24	24 g
iron	56	56 g
copper	64	64 g
sulphur	32	32 g

Table 3.2 Masses of 1 mol of some elements

Figure 3.2 One mole of each of five different elements

The independent individual particle in some elements is not the atom but the **molecule** (usually two atoms joined together). Uncombined hydrogen is a gas made up of diatomic H_2 molecules. In **compounds** containing hydrogen, however, the element participates as H atoms. When we say 'one mole of hydrogen' we generally mean 'one mole of hydrogen gas'. We are, therefore, talking about one mole of H_2 molecules. In hydrogen-containing compounds such as hydrogen peroxide, H_2O_2, we are dealing with hydrogen atoms. One mole of H_2O_2 molecules, for example, contains two moles of hydrogen atoms, not one mole of hydrogen molecules.

Table 3.3 lists some elements where we may be dealing with atoms or molecules.

Table 3.3 Masses of 1 mol of atoms and molecules

Element	Relative atomic mass	Mass of 1 mol atoms	Molecular formula	Mass of 1 mol molecules
hydrogen	1	1 g	H_2	2 g
nitrogen	14	14 g	N_2	28 g
oxygen	16	16 g	O_2	32 g
chlorine	35.5	35.5 g	Cl_2	71 g
bromine	80	80 g	Br_2	160 g
iodine	127	127 g	I_2	254 g

3.1.3 Moles of compounds

Most chemical substances are of course not elements but compounds. The type of fundamental particle or unit varies according to the type of compound.

(a) Glucose, $C_6H_{12}O_6$, is a compound which is made up of separate molecules.
A mole of glucose is therefore a mole of $C_6H_{12}O_6$ **molecules**.

(b) Silicon dioxide, SiO_2, is a compound which exists as a giant network of silicon and oxygen atoms linked together by covalent bonds and

containing two oxygen atoms for every one silicon atom.
A mole of silicon dioxide is therefore a mole of SiO_2 **atomic building blocks**.

(c) Calcium chloride, $Ca^{2+}(Cl^-)_2$ is a compound which exists as a lattice of calcium and chloride ions linked by electrostatic (+/−) attraction and containing two chloride ions for every one calcium ion.
A mole of calcium chloride is therefore a mole of $Ca^{2+}(Cl^-)_2$ **ionic building blocks**.

These three examples are white solids. Their outward appearances may be similar but in terms of chemical structure they are composed of entirely different building blocks.

Figure 3.3 Samples of molecular, covalent network and ionic solids

3.1.4 Molar mass

Atoms, molecules, atomic building blocks and ionic building blocks tend to be the elementary structural units or **entities** in substances. The word 'entity' is a 'cover all' term, used to label the 'bits' that make up a substance. In a mole of raspberries the individual entity would be a single raspberry. In chemical substances the entity is usually described by the formula.

The mass of one mole of entities, whatever these entities might be, is called the **molar mass**. Molar mass has the units **grams per mole**, i.e. g mol^{-1} and is **numerically** equal to the formula mass.

Molar mass = mass of one mole of entities
= formula mass in grams

Table 3.4 lists the molar masses of some compounds.

Compound	Formula	Entity	Molar mass/g mol^{-1}
glucose	$C_6H_{12}O_6$	$C_6H_{12}O_6$ molecules	180
silicon dioxide	SiO_2	SiO_2 atomic blocks	60
calcium chloride	$CaCl_2$	$Ca^{2+}(Cl^-)_2$ ionic blocks	111

Table 3.4 Molar masses

Sample exercise

What is the molar mass of ammonium sulphate?

Method

Formula of ammonium sulphate = $(NH_4^+)_2SO_4^{2-}$
Ammonium sulphate is therefore made up of $(NH_4^+)_2SO_4^{2-}$ entities
Mass of 1 mol $(NH_4^+)_2SO_4^{2-}$ entities = $[2(14 + 4) + 32 + 4(16)]$ g
= 132 g
Therefore the molar mass of ammonium sulphate = 132 g mol^{-1}

3.1.5 Moles of ions

Nickel(II) chloride, $Ni^{2+}(Cl^-)_2$, has a molar mass of 130 g mol^{-1}. A pile of nickel chloride weighing 130 g would therefore represent a mole of $Ni^{2+}(Cl^-)_2$ ionic building blocks.

Nickel chloride dissolves in water to form nickel chloride solution. In doing so the ionic lattice in the solid disintegrates, allowing the nickel and chloride ions to exist as separate entities.

$$Ni^{2+}(Cl^-)_2(s) \longrightarrow Ni^{2+}(aq) + 2Cl^-(aq)$$

1 mol ionic blocks	1 mol free nickel ions	2 mol free chloride ions

The nickel and chloride ions in solution are mixed with one another and with water molecules. It is impossible to isolate and weigh a pile of nickel ions as one could isolate and weigh a pile of nickel atoms. Nevertheless, amounts of ions are still measured in moles.

Note Since electrons are of negligible mass, a mole of nickel ions is considered to weigh the same as a mole of nickel atoms, i.e. 59 grams.

Figure 3.4 One mole each of nickel(II) ions (in aqueous solution) and metallic nickel

Sample exercise

How many moles of iron and sulphate ions are present in 200 cm^3 of a solution of iron(III) sulphate, concentration 0.1 mol l^{-1}?

Method

Concentration of iron(III) sulphate = 0.1 mol in 1000 cm^3 solution
= 0.02 mol in 200 cm^3 solution

Therefore amount of iron(III) sulphate present = 0.02 mol

Ionic formula of iron(III) sulphate = $(Fe^{3+})_2(SO_4{}^{2-})_3$

When forming a solution,

$$(Fe^{3+})_2(SO_4{}^{2-})_3(s) \longrightarrow 2Fe^{3+}(aq) + 3SO_4{}^{2-}(aq)$$
$$1\ mol \longrightarrow 2\ mol + 3\ mol$$

Therefore 0.02 mol $\longrightarrow$ 0.04 mol + 0.06 mol

Consequently, 0.04 mol Fe^{3+} ions and 0.06 mol $SO_4{}^{2-}$ ions are present in the 200 cm^3 of this iron sulphate solution.

3.1.6 Definition of the mole

Any definition of a mole must cover the range of structural units or particles which make up chemical substances. The standard definition is given below.

> ***A mole is that amount of substance which contains as many elementary entities as there are carbon atoms in 0.012 kilogram (12 g) of carbon-12.***

When the mole is used it is essential to specify the elementary entities under consideration. These may be atoms, molecules, ions, electrons or other particles or groups of particles.

How **many** entities are there in a mole of entities? To find the answer, chemists turned their attention to electrolysis measurements and that tiny entity, the electron.

3.2 Amount of electrons

Michael Faraday, in his experiments on electrolysis in the 1830s, was the first person to relate, precisely, the amount of current passed to the amount of chemical change which took place at the electrodes.

Electrolysis enables us to measure the amount of electrical charge needed to deposit a certain mass of metal on an electrode. This relationship allows us to link electrons with atoms.

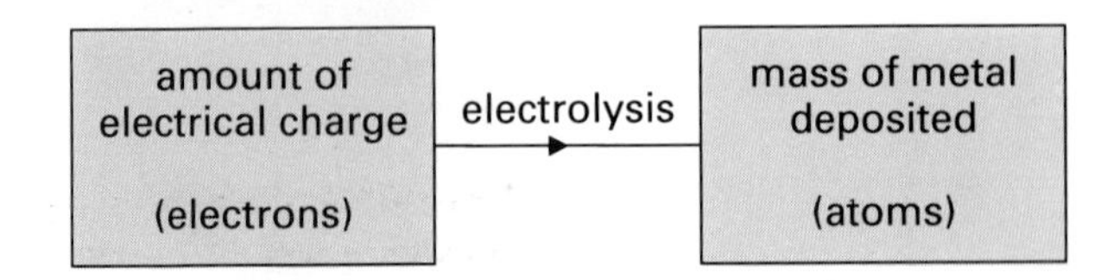

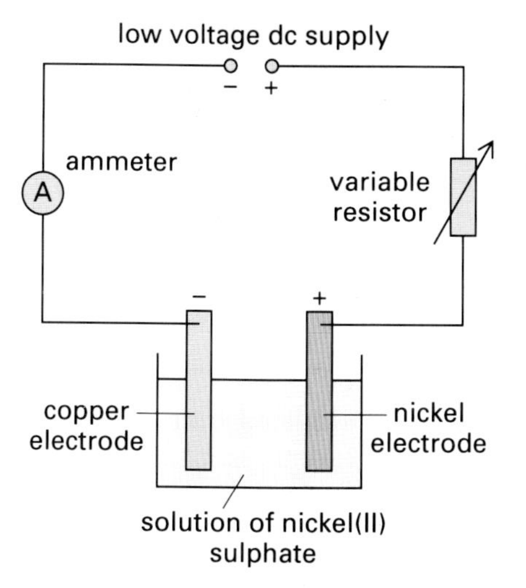

Figure 3.5 Apparatus for electrolysing nickel sulphate

Electrical charge is an amount of electrons. Charge (symbol Q) is measured by recording the current (I) and the time (t) for which current is passed.

Q	=	I	×	t
charge in **coulombs** (symbol C)	=	current in **amps** (symbol A)	×	time in **seconds** (symbol s)

3.2.1 Studying electrolysis

The relationship between the quantity of electricity passed during electrolysis and the mass of metal deposited may be investigated using the apparatus described in Figure 3.5. When the circuit is switched on, electric charge in the form of electrons is drawn from the negative terminal of the supply and builds up on the copper electrode. At the same time electrons are drained from the nickel electrode towards the positive terminal of the supply.

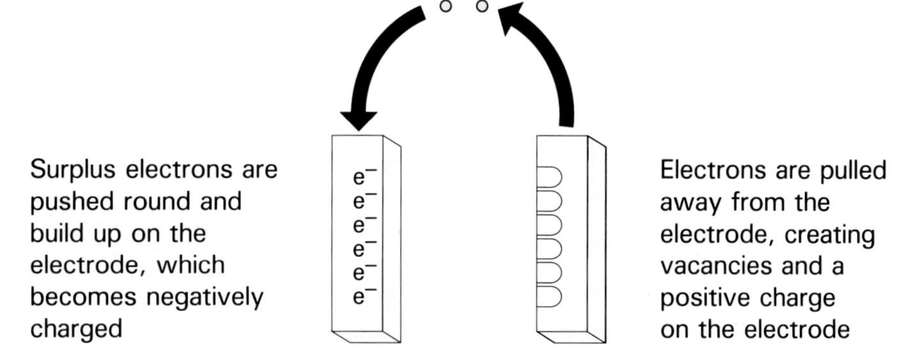

Figure 3.6 Electron movement

At the negatively charged copper electrode, nickel ions pick up electrons and are reduced to nickel atoms, forming a greyish metallic layer of nickel on the brown copper surface.

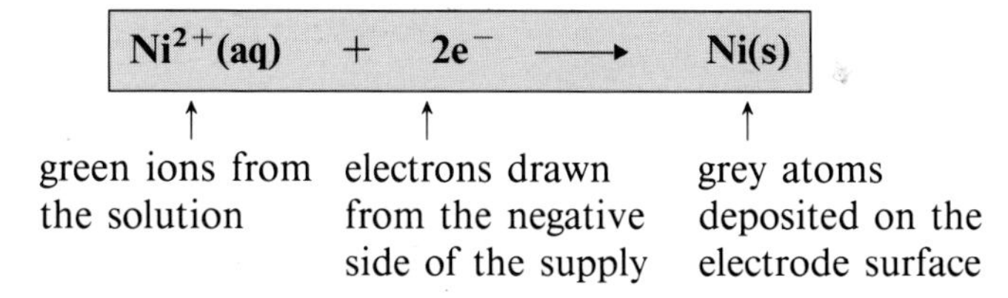

At the same time, nickel atoms, of the positively charged nickel electrode, go into the green solution as ions, leaving electrons behind. These electrons are then withdrawn from the electrode, by the positive terminal of the supply.

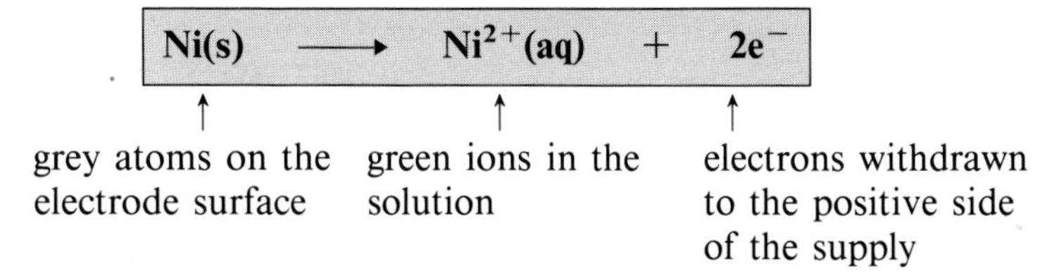

Thus nickel ions are being removed from the solution at the negative electrode and re-entering the solution at the positive electrode. As a result the concentration of nickel ions in the solution remains **constant**.

3.2.2 Coulombs and electrons

In the experiment described in Figure 3.5 the **mass** of **nickel deposited** can be measured by weighing the copper electrode before and after the electroplating with nickel, e.g.

$$\begin{aligned} \text{mass of copper electrode} &= 41.261 \text{ g} \\ \text{mass of copper electrode + nickel deposit} &= 41.536 \text{ g} \\ \text{therefore mass of nickel deposited} &= 0.275 \text{ g} \end{aligned}$$

The **quantity** of **electric charge passed** (Q) can be calculated from measurements of current and time. The current (ammeter reading) is maintained at a constant value throughout the electroplating, by adjustment of the variable resistor. The time (t) for which current (I) has passed is registered on a stopclock.

A current of 0.5 amperes (0.5 A), flowing for 30 minutes, will pass a quantity of electric charge, measured in coulombs (C), equal to the product of current, measured in amperes (A) and time, measured in seconds (s).

$$\begin{aligned} \text{Quantity of electric charge} &= \text{current passed} \times \text{time taken} \\ Q &= I \times t \\ \text{(coulombs)} &= \text{(amperes)} \times \text{(seconds)} \\ Q &= 0.5 \times 30 \times 60 \text{ coulombs} \\ &= 900 \text{ coulombs} \end{aligned}$$

The electrolysis results may be summarised as follows:

0.275 g of nickel was deposited by 900 coulombs of charge.

The molar mass of nickel is 59 g mol^{-1}.

59 g of nickel would therefore be deposited by $900 \times \frac{59}{0.275}$ coulombs of charge

$= \mathbf{193\ 000}$ **coulombs**

The results from the electrolysis of nickel sulphate can be placed alongside the results from electrolysing various ionic compounds, in molten or aqueous forms.

Elementary entities produced at electrode	Molar mass of element/g mol^{-1}	Number of coulombs required to produce molar mass of the element/C
Na atoms	23	96 500
H_2 molecules	2	193 000
Mg atoms	24	193 000
O_2 molecules	32	386 000
Cu atoms	64	193 000
Cl_2 molecules	71	193 000
Ag atoms	108	96 500
Al atoms	27	289 500

Table 3.5 Coulombs and molar mass

The results can also be presented in the form of a bar chart.

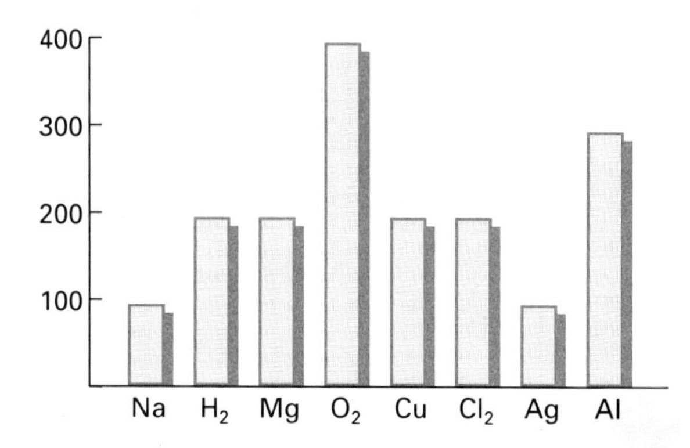

Figure 3.7 Number of kilocoulombs needed to deposit one mole of an element

What inference can we draw from these results? Apparently the number of coulombs required to deposit one mole of atoms or molecules of an element is either 96 500 or a multiple (2, 3 or 4) of 96 500.

The production of 1 mol atoms or molecules requires 96 500n coulombs where n = 1, 2, 3 or 4.

The significance of n becomes obvious when we examine the relevant electrode reactions (equations).

Electrode reaction	Value of n in number of coulombs required to produce 1 mol atoms
$Na^+ + e^- \longrightarrow Na$	1
$2H^+ + 2e^- \longrightarrow H_2$	2
$Mg^{2+} + 2e^- \longrightarrow Mg$	2
$2H_2O \longrightarrow O_2 + 4H^+ + 4e^-$	4
$Cu^{2+} + 2e^- \longrightarrow Cu$	2
$2Cl^- \longrightarrow Cl_2 + 2e^-$	2
$Ag^+ + e^- \longrightarrow Ag$	1
$Al^{3+} + 3e^- \longrightarrow Al$	3

Table 3.6 Electrode reactions and value of n

What conclusion can we draw from Table 3.6? The multiplying factor n can be equated to the number of electrons associated with the production of one atom or molecule of the element.

In the case of **metal** deposits, we are dealing with atoms and a general equation would be as follows:

$$M^{n+} + ne^- \longrightarrow M$$
$$n \text{ mol } e^- \longrightarrow 1 \text{ mol M atoms}$$
$$96\,500n \text{ coulombs} \longrightarrow 1 \text{ mol M atoms}$$

The quantity of electric charge, 96 500 coulombs, is now seen as being equivalent to the charge of one mole of electrons.

96 500 coulombs = 1 mole of electrons

The charge per mole of electrons is a constant and has the value of 96 500 coulombs per mole.

Sample exercise

Magnesium is produced industrially by electrolysing a molten mixture containing magnesium chloride, using a steel cathode (negative electrode).
Assuming 80 per cent efficiency, how long will it take a current of 20 000 amperes to produce 1000 kilograms of magnesium in each electrolytic cell?

Method

The reaction at the negative electrode is shown by the equation:

	electricity		mass deposited
	$Mg^{2+}(l) + 2e^-$	$\longrightarrow$	$Mg(l)$
Scaling up to moles	2 mol e^-	$\longrightarrow$	1 mol Mg atoms
Linking coulombs with grams	193 000 C	$\longrightarrow$	24 g Mg
Scaling up to kilograms	193×10^6 C	$\longrightarrow$	24 kg Mg
Therefore	$\left(193 \times 10^6 \times \frac{1000}{24}\right)$ C	$\longrightarrow$	1000 kg Mg
i.e.	8.04×10^9 C	$\longrightarrow$	1000 kg Mg

Quantity of charge supplied = 8.04×10^9 coulombs.

The current is quoted as 20 000 A but the efficiency is only 80 per cent. This means that 80 per cent of the current is used to deposit the magnesium and 20 per cent of the current is wasted, e.g. as heat.
Therefore **effective** current = $(0.8 \times 20\ 000)$ A
= 16 000 A

Applying the equation, time = $\dfrac{\text{charge supplied}}{\text{current}}$

$$t = \frac{Q}{I} \text{ seconds}$$

$$= \frac{Q}{I \times 60 \times 60} \text{ hours}$$

$$= \frac{8.04 \times 10^9}{16\,000 \times 60 \times 60} \text{ hours}$$

$$= 139.6 \text{ hours}$$

It would therefore take 139.6 hours to produce 1000 kg of magnesium under these conditions.

3.3 Avogadro constant

How many electrons are contained in one mole of electrons? This question was answered, using data collected by Robert Millikan of Chicago University, in 1909. Millikan determined the charge on a single electron by measuring the effect of an electric field on the rate at which charged oil droplets fall. He found that the charge on the oil drops was always an integral (whole number) multiple of 1.6×10^{-19} coulomb. He deduced that this was the charge of the electron.

Charge of one electron = 1.6×10^{-19} coulomb

Put the opposite way, this means that

$$\text{the charge of one coulomb} = \text{charge of } \frac{1}{1.6 \times 10^{-19}} \text{ electrons}$$

$$= \text{charge of } 6.24 \times 10^{19} \text{ electrons}$$

$$\text{The charge of 96 500 coulombs therefore} = \text{charge of } (6.24 \times 10^{19} \times 96\,500) \text{ electrons}$$

$$= \text{charge of } 6.02 \times 10^{23} \text{ electrons}$$

From electrolysis we know that 96 500 coulombs is equivalent to one mole of electrons (see 3.2.2). It follows, therefore, that there are **6.02×10^{23} electrons in one mole of electrons**.

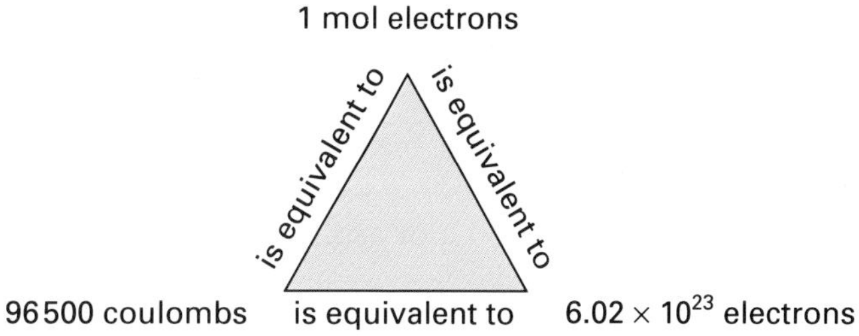

Figure 3.8 Relationship between moles, coulombs and electrons

We are now able to answer the question posed previously in 3.1.5. How many entities are in one mole of entities? The answer for electrons and all other particles, even raspberries, is 6.02×10^{23}.

There are, therefore, 6.02×10^{23} entities in 1 mol entities
6.02×10^{23} atoms in 1 mol atoms
6.02×10^{23} molecules in 1 mol molecules.

The number of elementary entities in one mole of any substance is the same and equal to 6.02×10^{23}.

6.02×10^{23} is a very large number indeed. Written out in numbers, 6.02×10^{23} is

602 000 000 000 000 000 000 000

It is equal to the **Avogadro constant** (symbol L or N_A), named after Amadeo Avogadro, an Italian, who made an immense contribution to this field of chemistry.

Avogadro constant, $L = 6.02 \times 10^{23}\ mol^{-1}$

The Avogadro constant is used to convert an amount expressed in moles to an actual number of entities, and vice versa.

Sample exercise

The population of the world is approximately six million million people. What would be the mass of the same number of water molecules?

Method

Molar mass of H_2O = 18 g mol^{-1}
i.e. 6.02×10^{23} H_2O molecules weigh 18 g

therefore 6×10^{12} H_2O molecules would weigh $18 \times \dfrac{6 \times 10^{12}}{6.02 \times 10^{23}}$ g

A world population of water molecules would in fact weigh 1.70×10^{-10}g (not a lot!).

Sample exercise

A tablet of aspirin, $C_9H_8O_4$, weighs, on average, 0.3 g. How many aspirin molecules are in each tablet (assuming the tablet is entirely aspirin compound)?

Method

Molar mass of aspirin = $(12 \times 9 + 8 \times 1 + 16 \times 4)$ g mol^{-1}
= 180 g mol^{-1}
180 g of aspirin contain 6.02×10^{23} molecules

0.3 g of aspirin contains $6.02 \times 10^{23} \times \dfrac{0.3}{180}$ molecules

= 1.00×10^{22} molecules
There are therefore 1×10^{22} aspirin molecules in each tablet.

3.4 Using equations to predict amounts and masses

A balanced chemical equation allows us to relate, first, amounts of substances (in moles) and, following that, masses. This can be demonstrated by means of a specific example.

Chromium metal is extracted commercially from chromium(III) oxide, by reaction with powdered aluminium. A familiar use of chromium is in the production of stainless steel. Incorporating 12 to 25 per cent chromium in the alloy gives the steel good corrosion resistance.

The balanced equation for the extraction process is:

$$Cr_2O_3 + 2Al \longrightarrow 2Cr + Al_2O_3$$

(a) What mass of chromium can be **theoretically** obtained from 10 tonnes of chromium oxide?
(b) How much aluminium will be required to bring about complete conversion?

The relative amounts (in moles) of reactants and products can be deduced from the balanced equation:

$$\begin{array}{llll} Cr_2O_3(s) + & 2Al(s) & \longrightarrow & 2Cr(s) + Al_2O_3(s) \\ 1\ \text{mol} & 2\ \text{mol} & & 2\ \text{mol} \end{array}$$

In this case we are not interested in the amount of aluminium oxide formed so, after balancing, it can be ignored.

To convert amounts in moles to amounts in grams, we have to work out the molar masses of the substances. (The relative atomic masses are: Cr = 52, Al = 27, O = 16.)

$$\begin{array}{lll} & Cr_2O_3 + 2\ Al & \longrightarrow 2\ Cr + Al_2O_3 \\ & 1\ \text{mol} + 2\ \text{mol} & \longrightarrow 2\ \text{mol} \\ \text{Converting to grams} & 152\ \text{g} + 54\ \text{g} & \longrightarrow 104\ \text{g} \end{array}$$

The mass of chromium oxide is given in tonnes rather than grams, so it is necessary to show the relative proportions of substances in tonnes.

$$152\ \text{tonnes} + 54\ \text{tonnes} \longrightarrow 104\ \text{tonnes}$$

The mass of chromium oxide actually supplied was not 152 tonnes but 10 tonnes. The other two relative masses therefore need to be scaled down accordingly.

$$\begin{array}{lll} Cr_2O_3 \quad + & 2Al & \longrightarrow 2Cr \\ 10\ \text{tonnes} + & (54 \times \frac{10}{152})\ \text{tonnes} & \longrightarrow 104 \times \frac{10}{152}\ \text{tonnes} \\ \text{i.e. } 10\ \text{tonnes} + & 3.55\ \text{tonnes} & \longrightarrow 6.84\ \text{tonnes} \end{array}$$

Assuming 100 per cent conversion, the mass of chromium obtained from 10 tonnes of chromium oxide would be 6.84 tonnes. In addition, 3.55 tonnes of aluminium would be required to convert all the oxide.

3.5 Moles of gases

Molar mass is the mass of one mole of entities. **Molar volume**, similarly, is the volume of one mole of entities. Figure 3.2 indicated the molar volumes of some solids. Figure 3.9 below shows the molar volumes of some liquids: water, ethanol and trichloroethane.

Clearly different solids and different liquids have different molar volumes. The question then arises: what about the volume of one mole of entities when the entities are in the gas phase?

The entities may be atoms or molecules. The noble gases helium, neon, argon, krypton, xenon and radon are composed of separate, unpaired, unreactive **atoms**. All other substances which are gases at room temperature are made up of molecules.

The volume of a gas depends on its temperature and pressure. As the temperature rises the volume of gas increases, and as the pressure is increased so the volume of the gas is diminished. We therefore have to state the temperature and pressure at which the volume or density of a gas is

Figure 3.9 One mole of each of three different liquids

measured. Conditions such as 298 K (25 °C) and one atmosphere pressure (1.01×10 N m^{-2}) are described as **normal temperature and pressure** (NTP for short).

$$\text{Volume can be calculated from:} \quad \text{volume} = \frac{\text{mass}}{\text{density}}$$

$$\text{therefore the volume of 1 mol gas particles} = \frac{\text{mass of 1 mol gas particles}}{\text{density of gas}}$$

$$\text{i.e. gas molar volume} = \frac{\text{molar mass}}{\text{density}}$$

Results from a range of gases are listed in Table 3.7.

Table 3.7 Gas molar volumes

Name of gas	Formula	Molar mass/ g mol^{-1}	Density at NTP/ g l^{-1}	Molar volume/ l mol^{-1}
ammonia	NH_3	17	0.71	23.9
argon	Ar	40	1.66	24.4
carbon monoxide	CO	28	1.15	24.4
carbon dioxide	CO_2	44	1.81	24.3
chlorine	Cl_2	71	2.99	23.8
ethane	C_2H_6	30	1.24	24.2
fluorine	F_2	38	1.58	24.0
helium	He	4	0.17	23.5
hydrogen	H_2	2	0.08	25.0
krypton	Kr	84	3.46	24.3
methane	CH_4	16	0.72	24.2
neon	Ne	20	0.84	24.0
oxygen	O_2	32	1.33	24.1
xenon	Xe	131	5.5	23.8

From the figures for molar volume given in Table 3.7, it is reasonable to conclude as follows:

> ***The molar volume of all gases at NTP is approximately the same (24 l mol^{-1}).***

It is also reasonable to generalise further, to include conditions other than normal.

> ***The molar volumes of all gases at the same temperature and pressure will be the same.***

Now we know that the molar volume of any gas contains 6.02×10^{23} molecules (or atoms in the case of noble gases). From this it is fair to infer, for gases other than the noble gases, that:

> ***Equal volumes of different gases, under the same conditions of temperature and pressure, contain equal numbers of molecules.***

For noble gases, equal volumes contain equal numbers of atoms.

The above proposition was made by Avogadro in 1811 and is known as **Avogadro's hypothesis**.

If equal volumes of gas (under similar conditions) contain equal numbers of molecules, then it is fair to claim that twice the volume will contain twice the number of molecules, and so on.

Avogadro's hypothesis

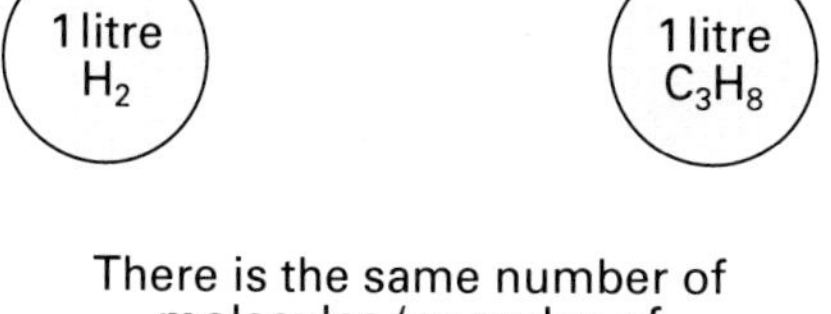

There is the same number of molecules (or moles of molecules) in each litre of gas (assuming same T and P)

Figure 3.10 Avogadro's hypothesis

v litres gas contain N molecules or M moles of molecules

$2v$ litres gas contain $2N$ molecules or $2M$ moles of molecules

$3v$ litres gas contain $3N$ molecules or $3M$ moles of molecules

Figure 3.11 Proportionality between volume and amount of molecules

Sample exercise

A sample of chlorine gas at normal temperature and pressure (NTP) has a volume of 400 cm^3.

Calculate (a) the amount of chlorine molecules (in moles)
(b) the mass of the sample
(c) the number of chlorine molecules in the sample.
(Assume molar volume at NTP is 24 l mol^{-1}.)

Method

(a) 24 litres of Cl_2 at NTP contain 1 mol Cl_2 molecules
Therefore 400 cm^3 of Cl_2 at NTP contain $\frac{400}{24\,000}$ mol Cl_2 molecules
i.e. amount of chlorine molecules = 0.017 mol
(b) 1 mol Cl_2 molecules weighs 71 g
Therefore 0.017 mol Cl_2 molecules weighs (71 × 0.017) g
i.e. mass of sample = 1.21 g
(c) 1 mol Cl_2 contains 6.02×10^{23} Cl_2 molecules
Therefore 0.017 mol Cl_2 contains ($6.02 \times 10^{23} \times 0.017$) Cl_2 molecules
i.e. number of Cl_2 molecules in the sample = 1.02×10^{22}

3.6 Using equations to predict amounts and volumes

Figure 3.12 A pair of propane cylinders

A balanced chemical equation involving gases allows us to relate amounts of substance (in moles) and, thereafter, volumes of gases. This can be demonstrated by means of a specific example.

A portable gas cylinder contains propane for burning. What volume of oxygen is needed to ensure the complete combustion of one litre of propane gas and what volume of carbon dioxide will be produced by this combustion? (Assume normal pressure.)

The balanced equation for the combustion reaction is:

$$C_3H_8(g) + 5O_2(g) \longrightarrow 3CO_2(g) + 4H_2O(g)$$

The equation shows that 1 mole of propane molecules reacts with 5 moles of oxygen molecules to form 3 moles of carbon dioxide molecules. It follows that the 1:5:3 ratio will also apply to volumes of the gases, i.e. 1 litre of propane gas will react with 5 litres of oxygen gas to give 3 litres of carbon dioxide gas.

The reasoning can be set out simply as follows:

$$C_3H_8(g) + 5O_2(g) \longrightarrow 3CO_2(g) + 4H_2O$$
$$1 \text{ mol} + 5 \text{ mol} \longrightarrow 3 \text{ mol}$$
$$1 \text{ l} + 5 \text{ l} \longrightarrow 3 \text{ l}$$

It is important to note that the moles/volumes relationship applies only to **gases** (where the particles can be thought of as moving independently of one another) and not to liquids or solids.

If the water vapour produced in the combustion of propane was allowed to cool below 373 K (the boiling point of water) it would condense to form drops of water liquid. The volume occupied by the condensed water is negligible and so the volume of H_2O is reduced effectively to zero.

at temperatures > 373 K $C_3H_8(g) \longrightarrow 4H_2O(g)$
1 l ⟶ 4 l water vapour
at temperatures < 373 K $C_3H_8(g) \longrightarrow 4H_2O(l)$
1 l ⟶ '0 l' condensed water vapour, i.e. zero volume.

3.7 Limiting and excess reactants

We use the relative numbers of moles of substances, as shown in balanced equations, to calculate the amounts of reactants needed or the amounts of products formed.

In the lab, in industry or in nature we cannot expect that the reactants will happen to be available in exactly the amounts required for the reaction. Almost always there will be less of one reactant than is needed to allow all of the reactants to combine. The **limiting reactant** is the substance that is fully used up and thereby limits the possible extent of the reaction. The other reactant (or reactants) is said to be **in excess** because some amount will be left over without reacting. Calculations based on the balanced equation must begin with the amount of the limiting reactant.

When aluminium is warmed with iron(III) oxide a vigorous reaction occurs. The equation and the theoretical reacting amounts are shown below.

$$2Al + Fe_2O_3 \longrightarrow Al_2O_3 + 2Fe$$

2 mol   1 mol
54 g   160 g

If we start with the reactants in the exact proportions, by mass, of 54:160, then reaction will be completed with both reactants fully used up. If we do not start with the two reactants in these proportions, then one reactant (the limiting reactant) will be fully used up and the reaction stopped while some of the other reactant remains unchanged.

To identify the limiting reactant take the following steps.

1. Look at the amount of one reactant (A).
2. See if you have enough of the other reactant (B) present to use up all of A.
3. If you do, then A is the limiting reactant.
4. If you do not, then B is the limiting reactant.

In the following four situations the **limiting** reactant is printed in bold type.

(a) $\dfrac{\text{54 g Al}}{\textbf{less than 160 g Fe}_2\textbf{O}_3}$ (b) $\dfrac{\textbf{54 g Al}}{\text{more than 160 g Fe}_2\text{O}_3}$

(c) $\dfrac{\textbf{less than 54 g Al}}{\text{160 g Fe}_2\text{O}_3}$ (d) $\dfrac{\text{more than 54 g Al}}{\textbf{160 g Fe}_2\textbf{O}_3}$

Sample exercise

Magnesium ribbon, weighing 1.5 g, is placed in 100 cm^3 of hydrochloric acid, concentration 1.0 mol l^{-1}.
Calculate the volume of hydrogen gas evolved from the reaction at NTP (assume molar volume of a gas at NTP = 24 l mol^{-1}.)

Method

The amounts of both reactants (in moles) can be calculated.
1 mol Mg atoms weighs 24 g.

Therefore amount of Mg atoms $= \frac{1.5}{24}$ mol $= 0.06$ mol

One litre (1000 cm^3) of 1 mol l^{-1} HCl contains 1 mol HCl entities.

Therefore amount of HCl entities $= \frac{100}{1000}$ mol $= 0.1$ mol

The balanced equation for the reaction is:

$$\underset{1\text{ mol}}{Mg} + \underset{2\text{ mol}}{2HCl} \longrightarrow MgCl_2 + \underset{1\text{ mol}}{H_2}$$

In this case the actual amount of acid (0.1 mol) is **less than twice** the amount of magnesium (0.06 mol) as required by the equation. The acid is therefore the limiting reagent in this reaction; the magnesium ribbon is in excess.
Calculation of the amount of product formed is therefore based on the amount of acid available.

2 mol HCl entities $\longrightarrow$ 1 mol H_2 molecules

0.1 mol HCl entities $\longrightarrow \left(1 \times \frac{0.1}{2}\right)$ mol H_2 molecules

$= 0.05$ mol H_2 molecules

1 mol H_2 molecules occupies 24 l at NTP

Therefore 0.05 mol H_2 molecules occupies $\left(24 \times \frac{0.05}{1}\right)$ l at NTP

$= 1.2$ l

The volume of hydrogen gas evolved is 1.2 litres.

3.8 Percentage yield

Frequently, the amount of a product actually obtained from a reaction is less than the amount calculated. There may be several reasons for this.
(a) Part of the reactant(s) may not react.
(b) Part of the reactant(s) may react in a different way, e.g. participating in side-reactions and creating side-products.
(c) Not all of the product may be recovered from the separation and purifying process.

The **percentage yield** relates the amount of product that is obtained in practice (the **actual** yield) to the amount that is predicted from a balanced equation (the **theoretical** yield).

$$\textit{Percentage yield} = \frac{\textit{actual yield}}{\textit{theoretical yield}} \times 100$$

Note Percentage yield is sometimes expressed as percentage 'conversion'.

Sample exercise

Titanium dioxide, TiO_2, is used on a large scale as the pigment in white paint. It is manufactured from the mineral ilmenite, $FeTiO_3$, in three stages.
If 45.1 kg of TiO_2 is obtained on average from each 100 kg of $FeTiO_3$, what is the percentage yield of the conversion? (RAM of titanium = 48)

Method

The conversion is	$FeTiO_3$	$\longrightarrow$	TiO_2
If yield was 100 per cent,	1 mol	$\longrightarrow$	1 mol
Converting to mass	152 g	$\longrightarrow$	80 g
Adjusting to kilograms	0.152 kg	$\longrightarrow$	0.08 kg
By proportion	100 kg	$\longrightarrow$	$\left(0.08 \times \frac{100}{0.152}\right)$ kg

$= 52.63$ kg

Therefore theoretical yield $= 52.63$ kg TiO_2

$$\text{Percentage yield} = \frac{\text{actual yield}}{\text{theoretical yield}} \times 100$$

$$= \frac{45.1}{52.63} \times 100 \text{ per cent}$$

$$= 85.7 \text{ per cent}$$

3.9 Redox reactions

Redox reactions include reactions which involve loss and gain of electrons: one reactant donates (gives) its electrons to another reactant. The reactant giving away the electrons is called the **reducing agent** (reducer): it is itself oxidised. The reactant taking the electrons is called the **oxidising agent** (oxidiser): it is itself reduced.

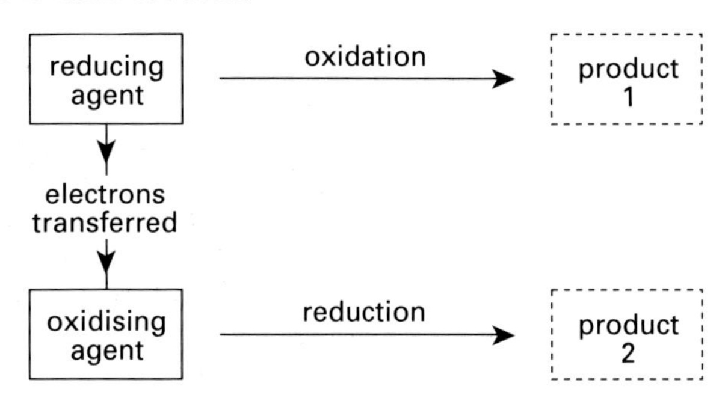

3.9.1 Half-reactions

Although oxidation and reduction take place simultaneously, it is convenient to consider them as separate steps. The extraction of silver illustrates this point. Silver metal is isolated as an element by converting insoluble silver ores into soluble silver cyanide. Zinc dust is then added to the solution of silver cyanide and silver is precipitated out of solution. (More reactive metal displacing less reactive metal from solution.) The word equation is:

zinc(s) + silver cyanide(aq) ⟶ zinc cyanide(aq) + silver(s)
The reaction consists of two processes.

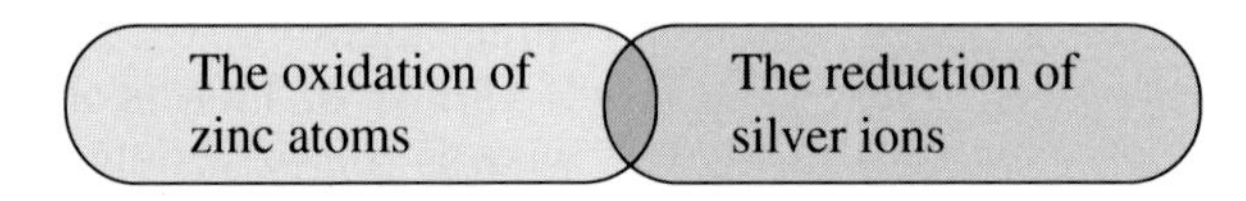

(a) Oxidation: $Zn(s) \longrightarrow Zn^{2+}(aq) + 2e^-$
(b) Reduction: $e^- + Ag^+(aq) \longrightarrow Ag(s)$

The equations above are described as **ion–electron** or **half-reaction** equations.

The number of electrons lost in the oxidation half-reaction must equal the number of electrons gained in the corresponding reduction half-reaction. When this condition is met and the half-reactions are balanced accordingly, the two half-reactions are added and electrons cancelled to give the full redox equation.

$$Zn(s) \longrightarrow Zn^{2+}(aq) + 2e^-$$

$$2e^- + 2Ag^+(aq) \longrightarrow 2Ag(s)$$

$$\boxed{\mathbf{Zn(s) + 2Ag^+(aq) \longrightarrow Zn^{2+}(aq) + 2Ag(s)}}$$

Redox reactions are not concerned with spectator ions and these do not appear in the equation. In the above reaction cyanide ions are not **directly** involved in the transfer of electrons and are therefore omitted.

3.9.2 Balancing redox equations

The zinc–silver equation above can easily be balanced by inspection. However, in many redox reactions, an inspection or trial and error method is not a method of balancing the equation. The reaction between aqueous iron(II) sulphate, $FeSO_4$, and an acidified solution of potassium permanganate, $KMnO_4$, is a case in point. When the pale green solution of iron(II) sulphate is added to the purple solution of acidified potassium permanganate, the purple colour disappears and the resulting solution takes on a pale brown colour.

The first step is to identify where oxidation and reduction have occurred. The pale brown colour formed is due to $Fe^{3+}(aq)$: the result of oxidation of $Fe^{2+}(aq)$. The disappearance of the purple colour shows that $MnO_4^-(aq)$ has been chemically altered – reduced to $Mn^{2+}(aq)$, which is such a pale pink as to be virtually colourless. Here the reducer is the iron(II) ion and the oxidiser is the permanganate ion. The sulphate and potassium ions are merely spectator ions and do not take part in the oxidation–reduction.

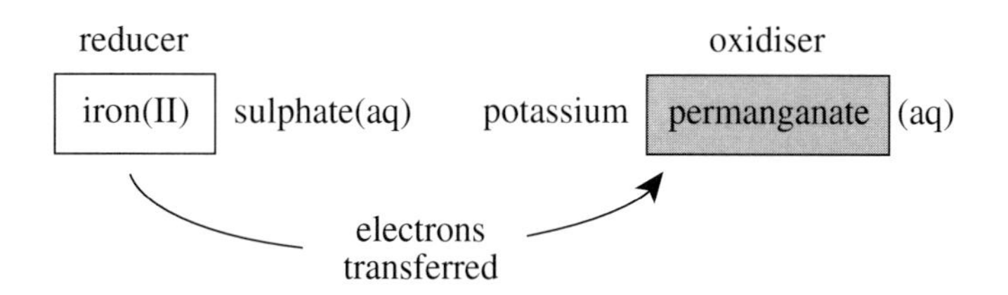

The iron(II) ions oxidise to iron(III) ions and donate electrons to the permanganate ions, which accept them and are thereby reduced to manganese(II) ions. The rough unbalanced equation is:

$$\underset{\text{pale green}}{Fe^{2+}(aq)} + \underset{\text{purple}}{MnO_4^-(aq)} \longrightarrow \underset{\text{pale/brown}}{Fe^{3+}(aq)} + \underset{\text{colourless}}{Mn^{2+}(aq)}$$

Any aqueous solution contains water molecules, hydrogen and hydroxide ions. In acidic solutions the predominant particles are H_2O and H^+. In alkaline solution they are H_2O and OH^-.

When writing half-equations for aqueous redox reactions we can use these ions or molecules to achieve balance.

The **oxidation** half-reaction in this case is straightforward.

$$Fe^{2+}(aq) \longrightarrow Fe^{3+}(aq) + e^-$$

The **reduction** half-reaction is more complicated. It involves the change:

$$MnO_4^-(aq) \longrightarrow Mn^{2+}(aq)$$

Mn is already in balance. O atoms are balanced by adding four H_2O molecules to the right-hand side.

$$MnO_4^-(aq) \longrightarrow Mn^{2+}(aq) + 4H_2O(l)$$

This introduces an imbalance of eight H atoms. They are balanced by adding eight H^+ ions to the left-hand side.

$$8H^+(aq) + MnO_4^-(aq) \longrightarrow Mn^{2+}(aq) + 4H_2O(l)$$

At this point all the atoms match up but the charge needs to be balanced. The total charge on the reactants is 7+ while that of the products is 2+. To balance the half-reaction electrically we must add five electrons to the reactants side.

$$5e^- + 8H^+(aq) + MnO_4^-(aq) \longrightarrow Mn^{2+}(aq) + 4H_2O(l)$$

The number of electrons released by the oxidation step must be the same as the number of electrons picked up in the reduction step. The final stage in compiling the overall redox equation is, therefore, to multiply each half-equation by an appropriate factor; so that the number of electrons lost by Fe^{2+} ions is equal to the number of electrons gained by MnO_4^- ions.

In this case the Fe^{2+}/Fe^{3+} half-reaction must be multiplied by five. The half reaction can then be added together and electrons cancelled to give the final equation.

$$5Fe^{2+}(aq) \longrightarrow 5Fe^{3+}(aq) + 5e^-$$

$$5e^- + 8H^+(aq) + MnO_4^-(aq) \longrightarrow Mn^{2+}(aq) + 4H_2O(l)$$

$$\mathbf{5Fe^{2+}(aq) + 8H^+ + MnO_4^-(aq) \longrightarrow 5Fe^{3+}(aq) + Mn^{2+}(aq) + 4H_2O(l)}$$

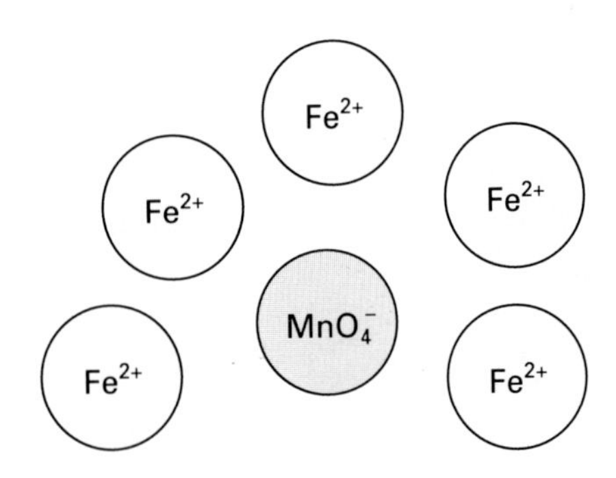

Figure 3.13 Relative amounts of reactants

The introduction of H^+ and H_2O into the equation is not an artificial device for balancing the atoms. These particles participate fully in the reactions as indicated. In the absence of acid, $H^+(aq)$, the MnO_4^- ion will not change to Mn^{2+} but will instead react to form a precipitate of manganese dioxide, $MnO_2(s)$.

It is necessary to have a balanced redox equation in order to establish the amount of oxidiser which reacts completely with an amount of reducer. In this case we are now in a position to say that five moles of Fe^{2+} ions react fully with one mole of MnO_4^- ions.

Sample exercise

Sodium hypochlorite, $Na^+ClO^-(aq)$, is used in household bleaches. It oxidises coloured compounds to colourless ones and is itself reduced to $Na^+Cl^-(aq)$.
Write an equation for the reduction step showing the uptake of electrons.

Method

$$Na^+ClO^-(aq) \longrightarrow Na^+Cl^-(aq)$$

We are told:
the Na^+ ions are merely spectating and can be omitted from equations; the oxidation of the dye brings about the reduction of the hypochlorite ion.

Balancing the O $\quad ClO^-(aq) \longrightarrow Cl^-(aq)$
with an H_2O $\quad ClO^-(aq) \longrightarrow Cl^-(aq) + H_2O(l)$

Balancing the Hs
with H^+ ions $\quad 2H^+(aq) + ClO^-(aq) \longrightarrow Cl^-(aq) + H_2O(l)$

Finally balancing the charge

$$2e^- + 2H^+(aq) + ClO^-(aq) \longrightarrow Cl^-(aq) + H_2O(l)$$

3.9.3 Redox titrations

Titration is a technique for measuring the concentration of a solution. You are already familiar with acid–base titrations which involve acidic and basic solutions. At the equivalence-point of an acid–base titration, precisely enough H^+ ions have been added to turn all the OH^- ions initially present into water molecules.

A second type of titration is the **redox titration**, which involves solutions of reducing and oxidising agents. At the equivalence-point of a redox titration, precisely enough electrons have been removed to oxidise all of the reducing agent (or precisely enough electrons have been donated to reduce all of the oxidising agent).

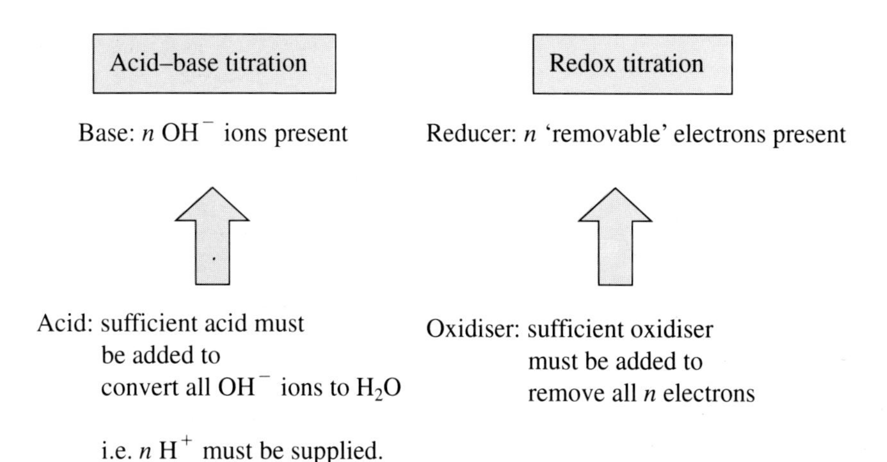

Titration is a procedure which allows chemists, knowing the concentration of one solution, to calculate the concentration of another solution reacting with it. For example, the concentration of iron(II) ions, Fe^{2+}(aq), in a solution can be determined by titrating the iron(II) solution with a standard solution (one of known concentration) of acidified permanganate ions, MnO_4^-(aq).

The solution of MnO_4^-(aq) (e.g. potassium permanganate) is dispensed from a burette into a fixed volume of the solution of Fe^{2+}(aq) (e.g. iron(II) sulphate) until the reaction is completed.

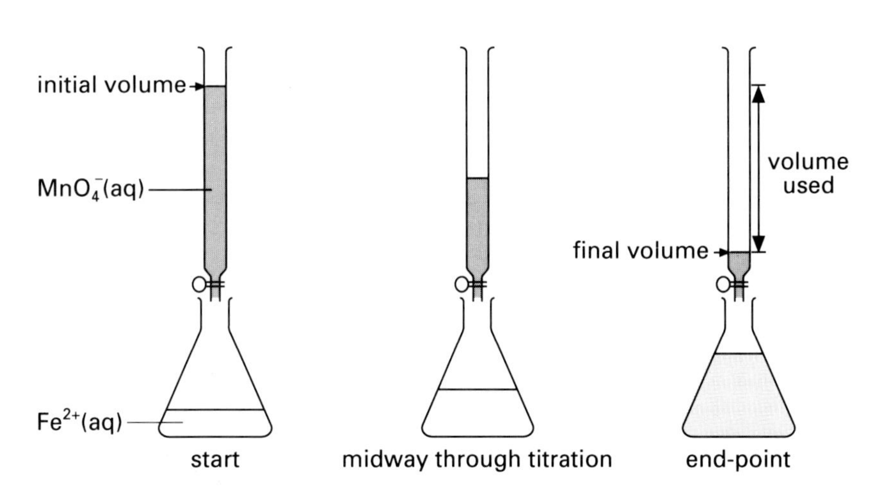

Figure 3.14 Titration set-up

The colour of MnO_4^- ions in aqueous solution is purple. When the titration is started, incoming MnO_4^- ions are reduced and decolorised by Fe^{2+} ions present in the flask. The purple colour disappears on shaking the flask.

However, when **all** the Fe^{2+} ions in the flask are oxidised to Fe^{3+} ions, **one additional drop** of the MnO_4^- solution will colour the whole solution in the flask permanently purple. This is because there are no Fe^{2+} ions left to reduce the additional MnO_4^- ions added from the burette.

The first **permanent** appearance of the purple colour in the flask indicates the equivalence-point (end-point) of the titration.

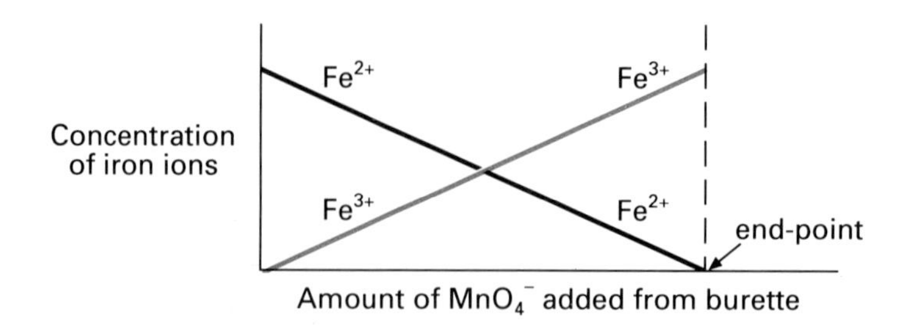

Figure 3.15 Complete oxidation of iron(II) ions by permanganate ions

3.9.4 Quantitative analysis

In qualitative analysis we try to show the presence or absence of a particular ion. In **quantitative** analysis we try to find **how much** of a given ion is present. This can be calculated from the results of redox titration. Here is an example.

It takes 20 cm^3 of 0.1 mol l^{-1} permanganate solution to react completely with 25 cm^3 of an iron(II) solution. What is the concentration of iron(II) ions?

The first step is to calculate the amount (in moles) of permanganate used in the reaction.

Concentration of permanganate = 0.1 mol MnO_4^- in 1 litre solution
= 0.1 mol MnO_4^- in 1000 cm^3 solution
= 0.002 mol MnO_4^- in 20 cm^3 solution
Therefore amount of permanganate used = 0.002 mol $MnO_4^-(aq)$
The full redox equation is:

$$\boxed{5Fe^{2+}} + 8H^+ + \boxed{MnO_4^-} \longrightarrow 5Fe^{3+} + Mn^{2+} + 4H_2O$$

This shows that every 1 mole of MnO_4^- ions oxidises 5 moles of Fe^{2+}ions.
Therefore amount of iron(II) ions oxidised = (5 × 0.002) mol $Fe^{2+}(aq)$
i.e. amount of iron(II) ions used = 0.01 mol $Fe^{2+}(aq)$
This 0.01 mol Fe^{2+} was contained in a volume of 25 cm^3 solution.
Therefore concentration of $Fe^{2+}(aq)$ = 0.01 mol Fe^{2+} in 25 cm^3 solution
= 0.4 mol Fe^{2+} in 1000 cm^3 solution
= 0.4 mol Fe^{2+} in 1 litre solution

Concentration of iron(II) solution = 0.4 mol l^{-1}

Sample exercise

A 0.5 g sample of iron ore was dissolved in dilute sulphuric acid and titrated with 42 cm^3 of 0.02 mol l^{-1} potassium permanganate solution.
What is the percentage mass of iron in the ore?

Method

As the iron ore dissolves in sulphuric acid, a solution of iron(II) sulphate is formed. We assume that the amount of Fe^{2+} ions formed will equal the amount of Fe atoms in the ore.

Concentration of MnO_4^- used

$= 0.02$ mol MnO_4^- in 1 litre solution

$= 0.02$ mol MnO_4^- in 1000 cm^3 solution

$= \left(0.02 \times \frac{42}{1000}\right)$ mol MnO_4^- in 42 cm^3 solution

Therefore amount of MnO_4^- used $= 0.0094$ mol MnO_4^-

The redox equation is:

$$5Fe^{2+} + MnO_4^- + 8H^+ \longrightarrow 5Fe^{3+} + Mn^{2+} + 4H_2O$$

Therefore amount of Fe^{2+} present $= (5 \times 0.000\,84)$ mol Fe^{2+} ions

$= 0.0042$ mol Fe^{2+} ions

Amount of Fe atoms in sample $= 0.0042$ mol Fe atoms

Since molar mass of Fe is 56 g mol^{-1},

mass of Fe atoms in sample $= (0.0042 \times 56)$ g Fe

$= 0.235$ g Fe

Therefore percentage iron in the sample of ore $= \frac{0.235}{0.5} \times 100$

$= 47$ per cent

Summary of Unit 3

Having read and understood the information and ideas given in this unit, you should now be able to:

define and explain the term 'mole'

explain the significance of the quantity of charge, 96 500 coulombs

carry out calculations relating quantity of electricity passed to the mass of a substance deposited

use the Avogadro constant to connect the amount of a substance with the number of elementary entities present

state that the molar volume for all gases under the same conditions is the same

carry out calculations involving volumes of gases reacting or produced

identify, by calculation, reactants that are in excess

calculate percentage yield

write a balanced equation for a redox reaction by first compiling half-reactions and then balancing the electrons transferred

calculate the concentration of an oxidising or reducing agent in solution from the results of volumetric titration

▲ carry out calculations involving molar volume and mass or amount of substance (in moles)

▲ carry out calculations involving balanced equations and volumes of gaseous reactants and products, recognising that solid or liquid species have negligible volume compared to gases

▲ given the molar volume of a gas, use balanced equations to predict volumes of gaseous reactants or products, and masses

▲ carry out calculations which involve identification of reactants that are in excess

▲ carry out calculations to determine percentage yield.

PROBLEM SOLVING EXERCISES

1. Two electrolysis cells were set up in series in an electrical circuit as shown.

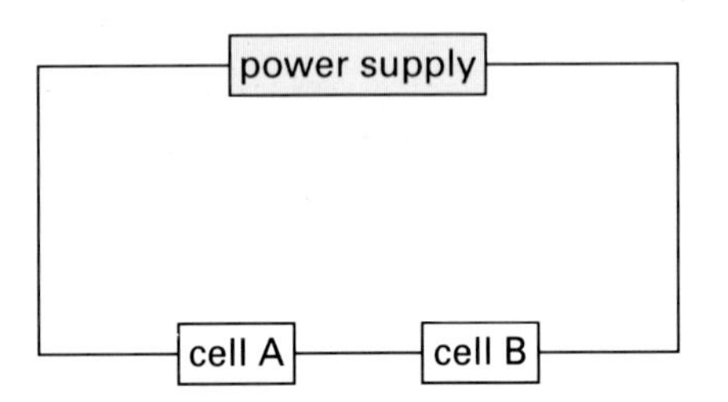

Figure 3.16

Cell A contained copper electrodes and 1 mol per litre of $Cu^{2+}(aq)$.
Cell B contained silver electrodes and 1 mol per litre of $Ag^{+}(aq)$.
Which of the quantities in the grid below would be the same in each cell after a period of electrolysis?

A	B
number of atoms deposited at negative electrode	mass of metal deposited at negative electrode
C	**D**
number of electrons transferred between electrodes	number of positive ions discharged at negative electrode

(PS skill 6)

2. Syngas is a mixture of CO and H_2 and is a very important feedstock. The proportion of CO to H_2 may have to be adjusted to suit the product being manufactured.
What ratio of CO to H_2 is required when syngas is converted to ethanediol, CH_2OHCH_2OH?

(PS skill 8)

3. An aluminium smelter has a line of 180 electrolysis cells and produces 110 kg of aluminium per cell per day (24 hours).
If the yield of aluminium is 85 per cent, calculate the average current being continuously supplied to each cell.

(PS skill 4)

4. Many fuels are burned in limited supplies of air. It is therefore important to know what mass of air is required for the complete combustion of a given mass of the fuel.
What mass of air would be required to completely burn 1 g of heptane?
Assume the fraction of oxygen in air, by mass, is 0.23.

(PS skill 4)

5. The passage of 441 coulombs of charge causes 0.3 g of gold to be deposited in an electrolysis cell containing a gold compound.
Calculate the charge on the gold ions.

(PS skill 4)

6. The grid below contains a selection of factors relating to gases.
Which of these variables will affect the volume of 1 mol of a gas?

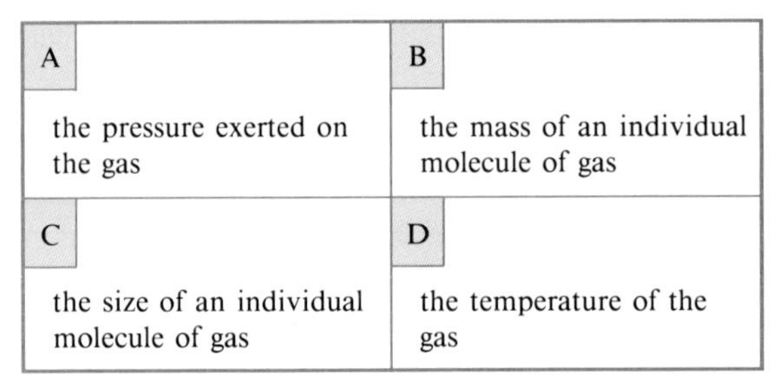

A	B
the pressure exerted on the gas	the mass of an individual molecule of gas
C	**D**
the size of an individual molecule of gas	the temperature of the gas

(PS skill 7)

7. A hydrocarbon gas has a density of 1.75 g l^{-1} at normal temperature and pressure.
If we assume molar volume at NTP is 24 l mol^{-1}, what is the molecular formula of the hydrocarbon?

(PS skill 8)

8. The *Apollo* spaceship used a fuel cell for electric power.
Hydrogen and oxygen gases were supplied to porous carbon electrodes separated by an electrolyte, KOH. The electrode reactions are as shown.

$$\text{Oxidation: } H_2 + 2OH^- \longrightarrow 2H_2O + 2e^-$$
$$\text{Reduction: } O_2 + H_2O + 2e^- \longrightarrow \tfrac{1}{2}O_2 + 2OH^-$$
$$\text{Overall: } H_2 + \tfrac{1}{2}O_2 \longrightarrow H_2O$$

What volume of hydrogen, at NTP, would be used up during the transfer of 1000 coulombs of electric charge?
Assume molar volume is 24 l mol^{-1} at NTP.

(PS skill 4)

PROBLEM SOLVING EXERCISES

9. Trichloromethane, $CHCl_3$, is used to make tetrafluoroethene, the monomer for PTFE polymerisation. Trichloromethane is formed by the reaction below.

$$CH_4(g) + Cl_2(g) \longrightarrow CHCl_3(g) + HCl(g)$$

What should be the ratio of methane to chlorine, by volume, in the reacting mixture, to ensure trichloromethane is formed?

(PS skill 8)

10. Sketch a diagram of suitable apparatus for electrolysing water and collecting the hydrogen and oxygen gases given off at opposite electrodes.

(PS skill 5)

11. The reaction between sodium carbonate and sulphuric acid is described by the equation:

$$Na_2CO_3 + H_2SO_4 \longrightarrow Na_2SO_4 + H_2O + CO_2$$

In an investigation, 0.5 mol, 1.0 mol, 1.5 mol and 2.0 mol Na_2CO_3 were separately added to 1.0 mol H_2SO_4 and the total amount (in moles) of CO_2 measured. Copy and complete the bar chart below.

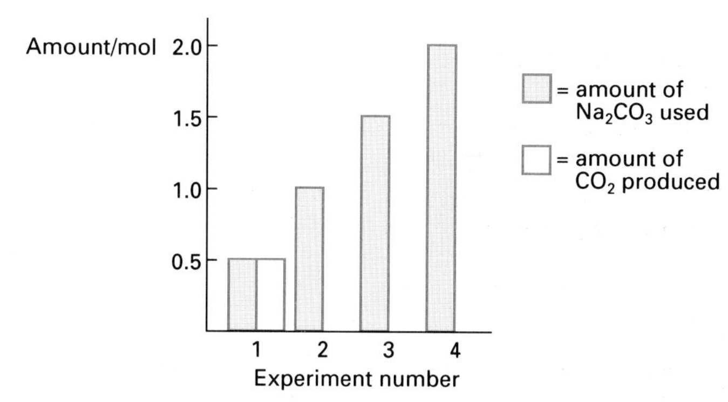

Figure 3.17

(PS skill 6)

Unit 4
Biomolecules

ASSUMED KNOWLEDGE AND UNDERSTANDING

Before starting on Unit 4 you should know and understand:

the nature of unsaturated hydrocarbons

the addition of hydrogen to alkenes

condensation polymerisation

the nitrogen cycle

the importance of carbohydrates as a source of energy for the body

the function of enzymes

the general structures of alcohols

the general structures of carboxylic acids

OUTLINE OF THE CORE AND EXTENSION MATERIAL

As you progress through Unit 4 you should, at least, try the fundamental content listed under **core** and, if possible, pick up the extra and/or more difficult content listed under **extension**.

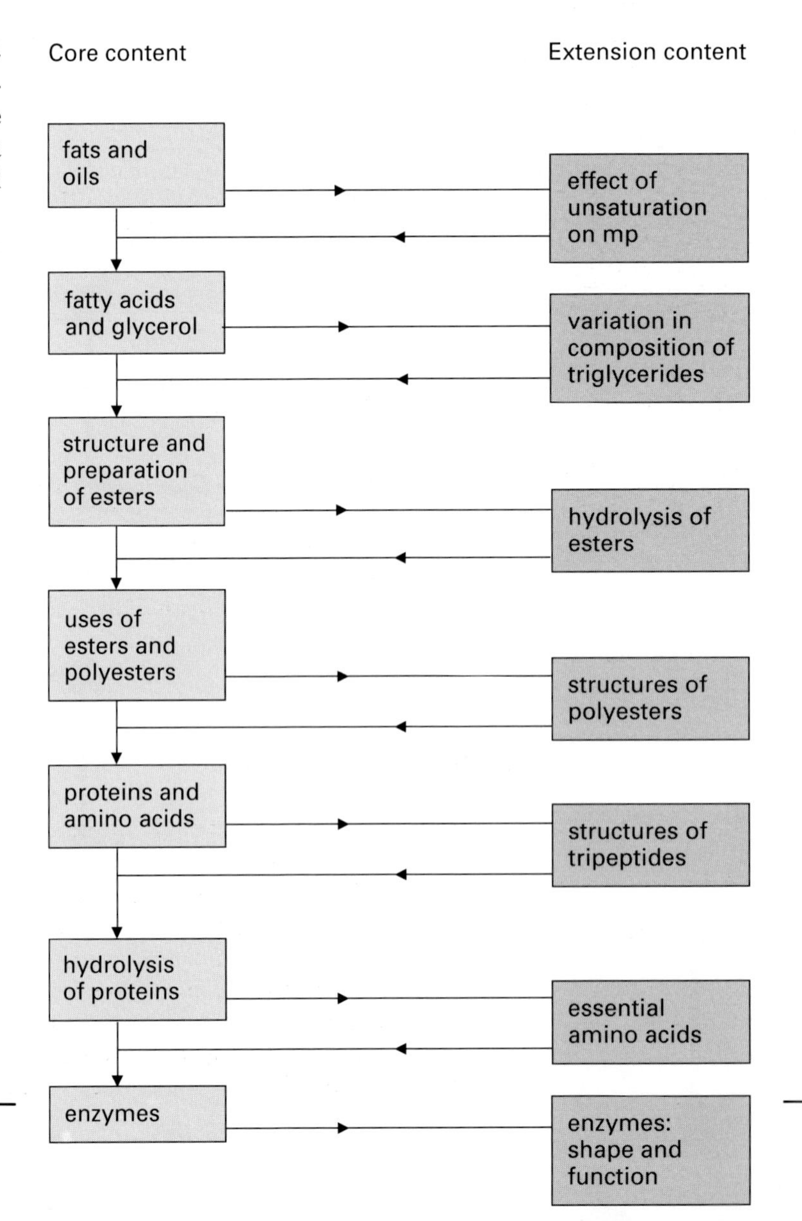

4 BIOMOLECULES

4.1 Fats and oils

A simple division of foods, by chemical type of their constituents, is into carbohydrates, proteins and fats. Another way of classifying foods, however, is in terms of **source**, e.g. land animal (meat), vegetable (plant) and marine animal (fish). Fats (and oils) may be readily classified by source.

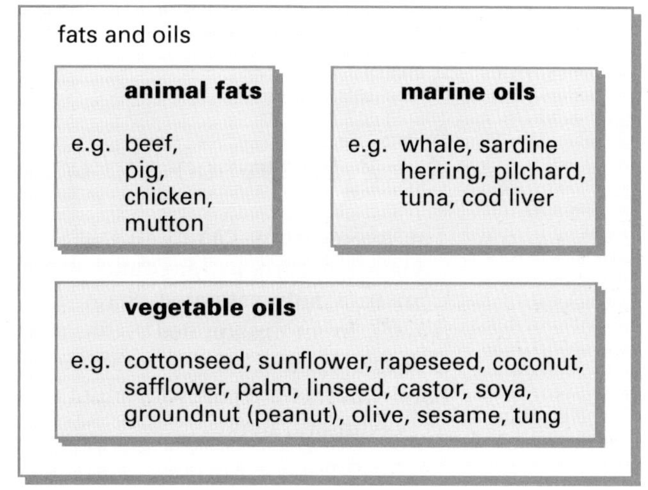

Figure 4.1 Classification of fats and oils

Figure 4.2 Almost 90 per cent (by mass) of the fruit of the Malaysian oil palm consists of oil

Animal fats are one of the many by-products from the slaughtering of animals for human consumption. Marine oils are obtained by the extraction of oil from fish. Vegetable oils are derived from the seeds of plants which grow in many parts of the world, but chiefly in tropical regions.

Fats and oils belong to a larger group of naturally occurring substances which tend to be insoluble in water and soluble in organic solvents like trichloromethane (chloroform). Other members of this large family include waxes, such as beeswax, and steroids, such as cholesterol.

4.1.1 Triglycerides

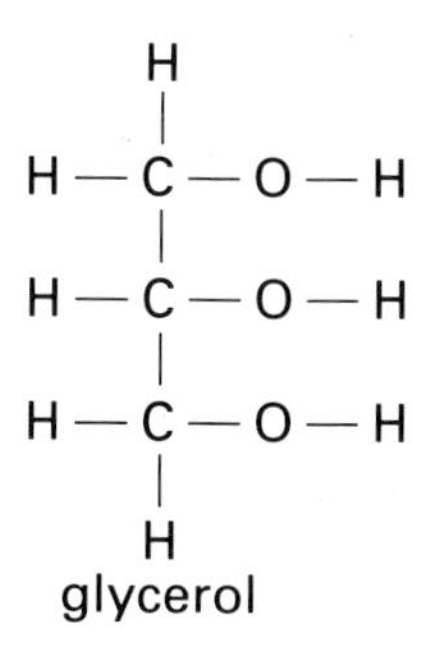

glycerol

Fats and oils are a range of substances all based on the compound **glycerol** (propane-1,2,3-triol).

Glycerol is a colourless viscous liquid which mixes well with water and is present in many sweet wines. If the H atom on each OH group in glycerol (also known as glycerine) is replaced by a nitro group, $—NO_2$, we obtain a molecule of nitroglycerine. Nitroglycerine is very unstable and when absorbed in a porous material it produces **dynamite** – invented by Alfred Nobel.

Each OH group of glycerol can combine chemically with one carboxylic acid molecule. The resulting molecules are of fats and oils, and are described as triacylglycerols or **triglycerides**. Triglycerides have, typically, the molecular structure shown in Figure 4.3.

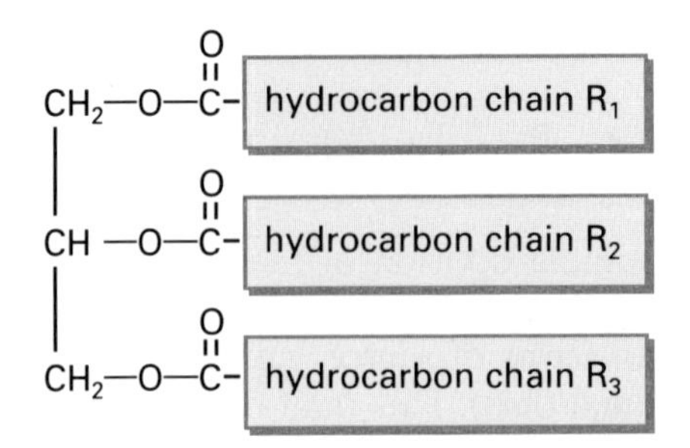

Figure 4.3 Molecular structure of triglycerides

In a triglyceride molecule, the hydrocarbon chains R_1, R_2 and R_3 may be either the same or different. A wide variety of triglyceride compounds is therefore possible.

4.1.2 Saturation and unsaturation

The hydrocarbon chains in triglycerides may vary in length (from four to twenty-four carbons long) and also in the degree of saturation or unsaturation. In a **fully saturated** chain, all carbon–carbon bonds are **single bonds** (C–C). In an **unsaturated** chain, at least one of the carbon–carbon bonds is a **double bond** (C=C). A mono-unsaturated chain contains only one C=C bond whereas a chain with more than one double bond would be described as polyunsaturated.

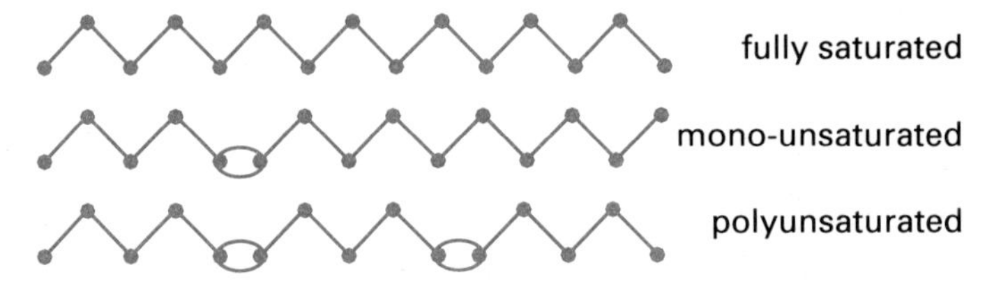

Figure 4.4 Saturated and unsaturated hydrocarbon chains

The introduction of C=C bonds, however, causes **kinks** (of about 42°) in otherwise straight hydrocarbon chains. This kinking affects the shape of the whole triglyceride molecule as Figure 4.5 illustrates.

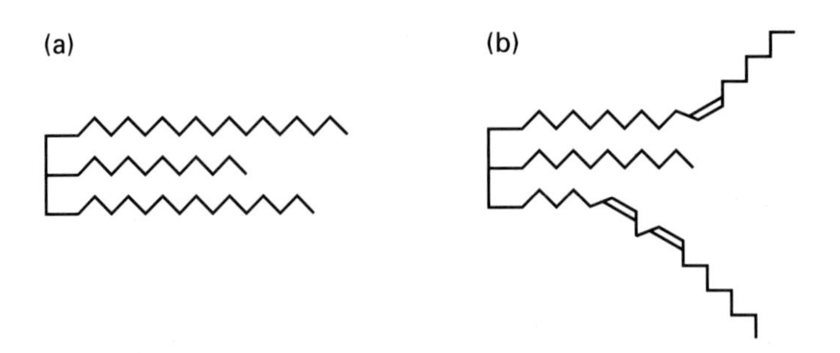

Figure 4.5 (a) A saturated triglyceride molecule
(b) An unsaturated triglyceride molecule

The effect of this difference in shape is shown in melting point data. Unsaturated triglyceride molecules do not pack together so closely and compactly and can therefore be separated more easily. Consequently, unsaturated triglycerides have lower melting points than triglycerides that are

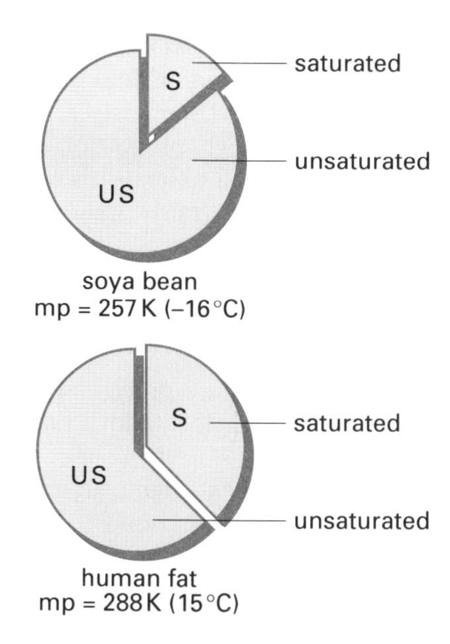

Figure 4.6 Relative proportions of saturated and unsaturated triglycerides

more highly saturated. Figure 4.6 shows the relative proportions of saturated and unsaturated triglycerides in soya bean oil and human fat, with their respective melting points.

Natural fats and oils are **mixtures** of triglyceride compounds, each compound having its own melting point. The fat or oil, therefore, does not have a sharp precise melting point but rather a **melting range**.

If we took normal temperature as 298 K (25 °C), then any mixture of triglycerides melting below this temperature will be **liquid** and any mixture with a melting range above 298 K will remain in **solid** form.

There is no deep fundamental difference between a fat and an oil. The distinction between a fat and an oil is simply that, at normal temperatures, any triglyceride mixture which is solid is classed as fat and any triglyceride mixture which is liquid is classed as oil.

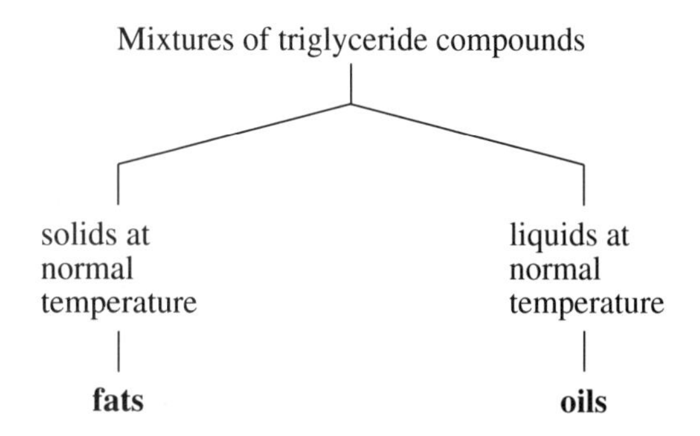

This distinction is rather vague since it depends on what you regard as 'normal' temperature. A change in climate may change an oil into a fat (or vice versa). For example, coconut oil is liquid when the coconut is hanging from a tropical coconut palm but solid when used on a coconut shy in a British fairground.

The link between degree of unsaturation and melting point suggests that oils are likely to be more unsaturated than fats. One measure of unsaturation is **iodine number**. Iodine reacts with carbon double bonds and so it is possible to estimate the degree of unsaturation in a fat or oil by measuring the number of grams of iodine that is absorbed by 100 grams of fat/oil (its iodine number). The higher the iodine number, the greater the degree of unsaturation. The iodine numbers of some fats and oils are shown in Figure 4.8.

Figure 4.7 Harvesting coconuts in Tamil Nadu, southern India

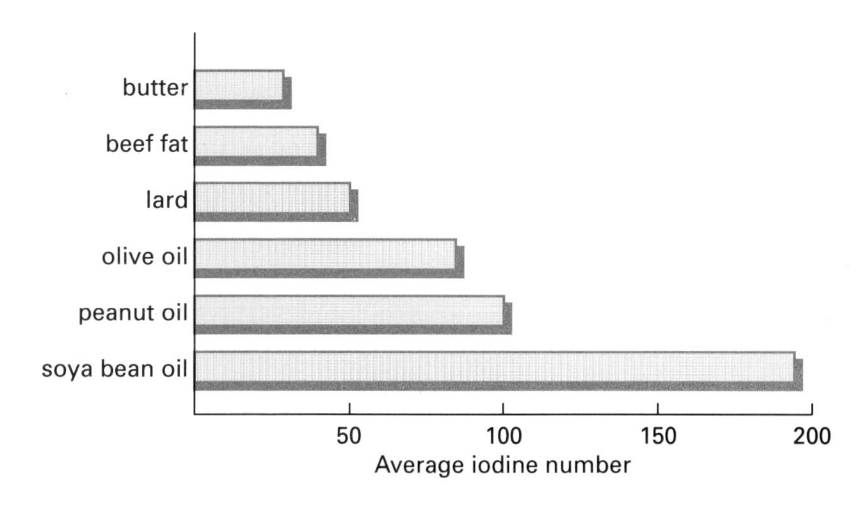

Figure 4.8 Iodine numbers of some fats and oils

Triglycerides have to be transported around the plant or animal which contains them. They must therefore be in a reasonably liquid or fluid form.

Plants and fish may have to cope with low temperatures. Their triglyceride supply is therefore in the form of oils containing a high proportion of unsaturated triglycerides. Humans and other warm-blooded animals operate at a higher body temperature. In these circumstances even fairly saturated fats are reasonably fluid and present no transport problems.

4.1.3 Conversion of oils to fats

While some fats and oils may be eaten directly, most require processing to remove undesirable colours, flavours and smells, and to 'tailor' their properties to a wide range of uses.

Highly unsaturated fats and oils, when subjected to heat, light, air and moisture, are prone to undergo oxidation at their double bonds. This oxidation leads to splitting of molecules at the double bond position and formation of various aldehydes, ketones and carboxylic acids, many of which have unpleasant smells. A fat containing these smelly compounds is said to be **rancid**.

To stop their flavour 'going off', cheap and abundant unsaturated oils are partially **hydrogenated**. The hydrogenation is carried out by bubbling hydrogen gas through the heated oil containing a catalyst, finely divided nickel.

$$-\overset{\displaystyle H}{\overset{|}{C}}=\overset{\displaystyle H}{\overset{|}{C}}- + H_2 \xrightarrow{\text{catalyst}} -\underset{\displaystyle H}{\underset{|}{\overset{\displaystyle H}{\overset{|}{C}}}}-\underset{\displaystyle H}{\underset{|}{\overset{\displaystyle H}{\overset{|}{C}}}}-$$

before hydrogenation after hydrogenation

The effect of hydrogenation is to reduce the number of double bonds present. This raises the melting point and produces a stiffer mixture. If all the C=C bonds in a liquid oil were hydrogenated, the resulting solid fat would be hard and brittle. In practice the degree of hydrogenation or **hardening**, as it is called, is carefully controlled to produce a product with the desired texture.

In 1869, Mege Mouries, a French chemist, patented an emulsion of animal fats as a cheap but crude substitute for butter. This was effectively the first-ever **margarine**. (An emulsion is one liquid dispersed in another liquid.)

The big step in the development of margarine was the introduction of hydrogenation as a technique. This eventually led to vegetable and fish oils replacing animal fats. Margarine is made from a water-in-oil emulsion, the aqueous (watery) phase being fat-free milk and the oil phase being a blend of oils. The two phases are brought together to form an emulsion.

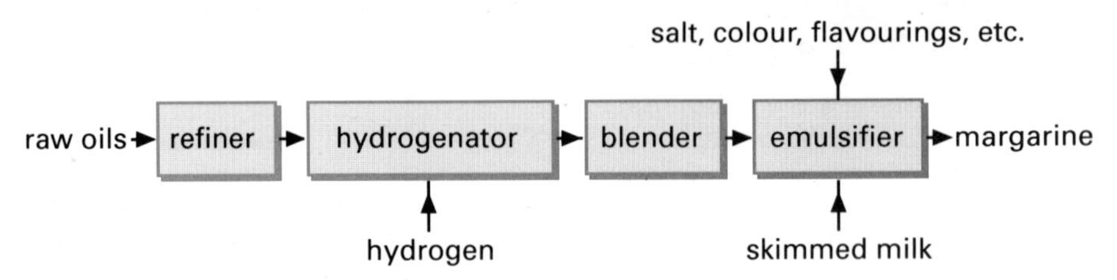

Figure 4.9 Manufacture of margarine

Cooking fats and cooking oils differ from margarine in that they are pure fat or pure oil mixtures rather than an oil/fat–water emulsion.

4.1.4 Fat in the diet

Although a great deal of energy is stored as carbohydrate by both animals and plants, an alternative storage material is fat or oil.

Triglyceride molecules are insoluble in water and groups of such molecules tend to come together and form a large droplet, separated from the aqueous phase. This is how fat is stored in tissue called adipose tissue. Fat is a good heat insulator. Adipose tissue can also act as a form of cushioning under the skin and around vital organs such as the liver and heart.

Complete oxidation of a fat produces carbon dioxide and water. It also releases about 40 kilojoules of energy per gram of fat. This is more than twice the amount of energy released when a gram of carbohydrate is fully oxidised and is due to the fact that fats contain a lower proportion of oxygen than carbohydrates. This gives fats more scope of forming C=O bonds by combustion (CO_2) and giving out energy.

The proportion of fat in some foods is shown in Figure 4.10.

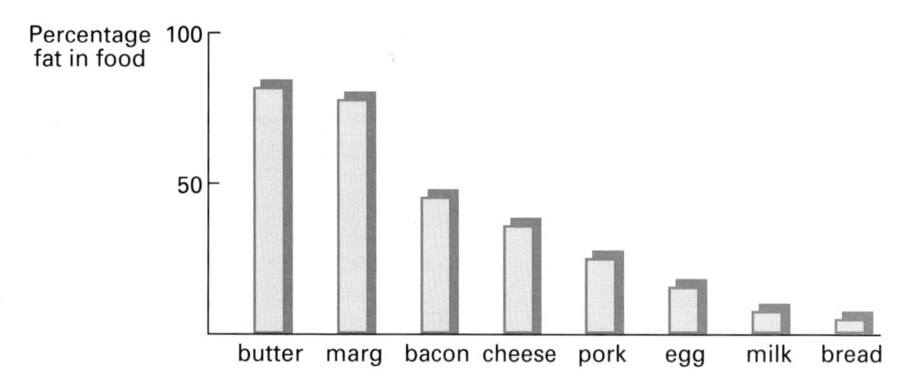

Figure 4.10 Average percentage fat in some foods

While fat is used by the body as the most compact and effective source of energy, there is much debate about the best **type** of fat to include in one's diet. It is believed that excessive intake of saturated fat pushes up the level of **cholesterol** in the bloodstream. This in turn causes fatty deposits (atheroma) on the inner walls of the arteries, restricting the flow of blood and putting the heart at risk.

Foods which contain **polyunsaturated** fats are considered to be less potentially harmful to the heart. Consequently, medical opinion generally advises us to reduce our total fat consumption and, where possible, switch from eating foods with saturated fats to foods with polyunsaturated fats. The argument, however, is far from resolved and opinion remains divided on several key points.

Figure 4.11 The Inuit people ('Eskimos') have healthy hearts in spite of their high 'fat' diet; the 'fat' is largely a polyunsaturated fish oil

Saturated fats tend to be found in animal, coconut, palm oils and hydrogenated vegetable oils.

Polyunsaturated fats tend to be found in most vegetable oils, fish oils and nut oils.

4.1.5 Fatty acids

For humans and animals to be able to digest fats, the insoluble triglyceride molecules must be broken down into smaller soluble molecules. The first step in this decomposition process is the splitting of triglycerides into glycerol and fatty acids; this splitting is catalysed by enzymes called lipases.

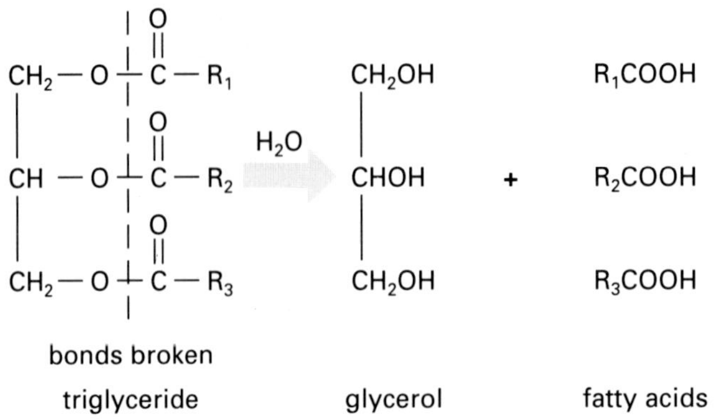

$$CH_2-O-\overset{O}{\overset{\|}{C}}-R_1 \qquad CH_2OH \qquad R_1COOH$$
$$CH-O-\overset{O}{\overset{\|}{C}}-R_2 \xrightarrow{H_2O} CHOH + R_2COOH$$
$$CH_2-O-\overset{O}{\overset{\|}{C}}-R_3 \qquad CH_2OH \qquad R_3COOH$$

bonds broken

triglyceride glycerol fatty acids

Figure 4.12 Splitting a triglyceride into glycerol and fatty acids

The reaction, described as **hydrolysis**, involves the simultaneous breaking of bonds and addition of the components of water.

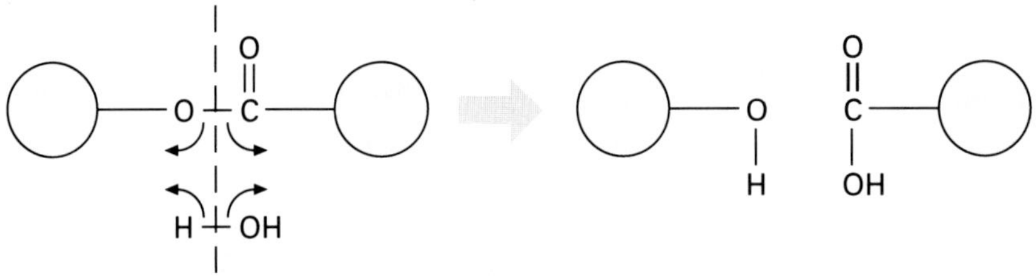

O—O—C(=O)—O + H—OH → O—O—H + HO—C(=O)—O

Figure 4.13 Hydrolysis

The breaking of two existing bonds and the forming of two new bonds leads to **splitting** of the triglyceride.

This hydrolysis reaction, which takes place in aqueous solution, is hampered by the fact that fats and oils are insoluble in water. As a result not much hydrolysis takes place in the stomach. The gall bladder releases compounds, called bile salts, into the small intestine to help the hydrolysis. These bile salts break up the larger globules of fats into an **emulsion** (a suspension of very small droplets) so that hydrolysis can be carried out more rapidly.

Complete hydrolysis of a triglyceride yields one molecule of glycerol and three molecules of fatty acid. **Fatty** acids are so called because they are obtained from the hydrolysis of fats. They are unbranched carboxylic acids, almost all ranging in size from C_4 to C_{24} but mainly C_{16} and C_{18} long.

Nearly all fatty acids have an even number of carbon atoms. This is due to the fact that these acids are synthesised in living organisms by extending the carbon chain **two** at a time.

Traditional name	Shortened structural formula	Systematic name
Saturated fatty acids		
butyric acid	$CH_3(CH_2)_2COOH$	butanoic acid
lauric acid	$CH_3(CH_2)_{10}COOH$	dodecanoic acid
myristic acid	$CH_3(CH_2)_{12}COOH$	tetradecanoic acid
palmitic acid	$CH_3(CH_2)_{14}COOH$	hexadecanoic acid
stearic acid	$CH_3(CH_2)_{16}COOH$	octadecanoic acid
Unsaturated fatty acids		
palmitoleic acid	$CH_3(CH_2)_5CH{=}CH(CH_2)_7COOH$	hexadec-9-enoic acid
oleic acid	$CH_3(CH_2)_7CH{=}CH(CH_2)_7COOH$	octadec-9-enoic acid
linoleic acid	$CH_3(CH_2)_4(CH{=}CHCH_2)_2(CH_2)_7COOH$	octadec-9,12-dienoic acid
linolenic acid	$CH_3CH_2(CH{=}CHCH_2)_3\ (CH_2)_6COOH$	octadec-9,12,15-trienoic acid

Table 4.1 Some common fatty acids

4.16 Saturated and unsaturated fatty acids

Fatty acids may be liquids or solids at normal temperatures, depending on their molecular size and degree of saturation/unsaturation.

Stearic acid and **oleic** acid are similar in size and mass. The presence of a C=C bond in the oleic molecule and its absence from the stearic molecule produce a clear difference in melting point.

Acid	Extended molecular formula	Formula mass	Melting point
Stearic acid	$C_{17}H_{35}COOH$	284	343 K (70 °C)
Oleic acid	$C_{17}H_{33}COOH$	282	287 K (14 °C)

Table 4.2 Comparing melting points

The saturated molecular chains of stearic acid are flexible enough to roll into compact balls. Stearic acid is therefore a **solid** at room temperature.

On the other hand, the unsaturated molecular chain of oleic acid cannot be twisted around the C=C double bond. The chain is much less flexible, cannot roll up into a ball and cannot pack together with neighbouring chains so well. Oleic acid is therefore **liquid** at room temperature.

4.1.7 Composition of triglyceride molecules

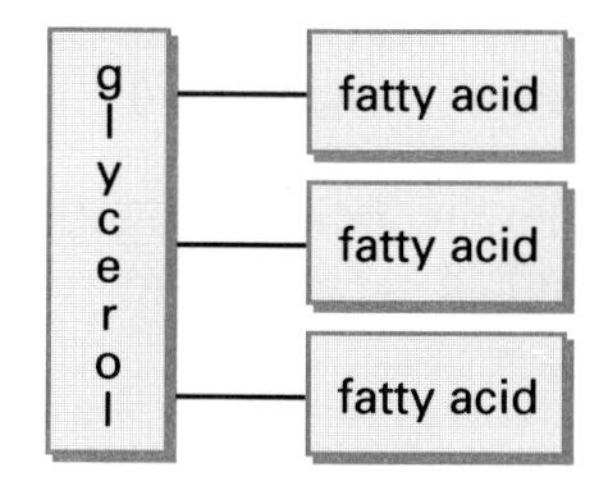

Figure 4.14 A triglyceride

Glycerol is a trihydric alcohol with three functional groups. A **triglyceride** molecule is formed through the condensation of **one glycerol molecule** with **three fatty acid molecules**. This gives a molecule whose general composition is shown in Figure 4.14. The properties of a fat or oil depend on the fatty acids which are combined with glycerol in each triglyceride.

Almost all triglycerides in nature are **mixed** in type, i.e. involving more than one fatty acid. Congo palm oil, for example, is a mixture of over 30 different triglycerides. These triglycerides contain various combinations (permutations) of five fatty acids. The five acids are myristic (M), palmitic (P), oleic (O) stearic (S) and linoleic (L). The most abundant triglyceride in Congo palm oil has the composition shown in Figure 4.15(a). The proportions of other triglycerides are shown in Figure 4.15(b).

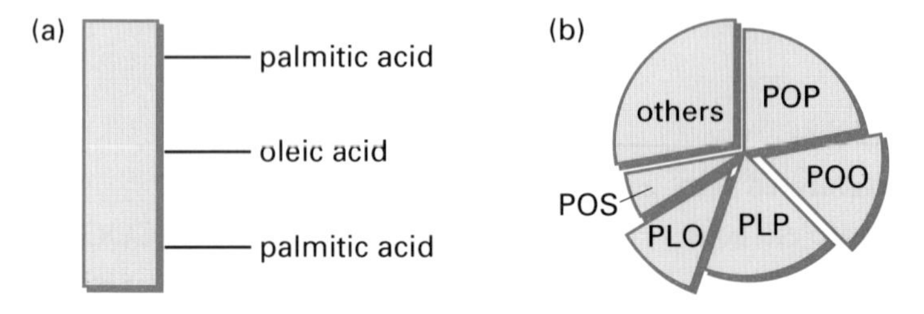

Figure 4.15
(a) A 'POP' triglyceride
(b) Triglycerides in Congo palm oil

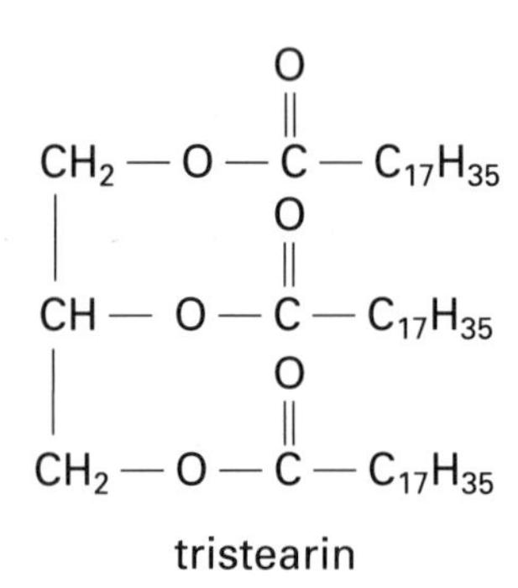

One of the few triglycerides which is made up of **similar** fatty acids is **tristearin**. Here glycerol is anchored to three stearic acid molecules. Tristearin is one of the chief components of beef fat. The role of fat in animals and plants is mainly as a reserve fuel supply (ultimately oxidisable to CO_2 and H_2O).

The camel's hump of fat (Figure 4.16) is a rather neat way of storing both energy and water – both released by oxidation.

$$\text{fat} + O_2 \longrightarrow CO_2 + H_2O + \text{energy}$$

Figure 4.16 The camel's hump is its portable larder

Plant stems and leaves can store their fuel reserves as carbohydrate. However, their seeds need to support the developing embryo until it is self-sufficient. For this a compact fuel reserve is needed, and consequently oils are found in seeds and nuts rather than in stems or leaves.

4.2 Esters

Esters are the products of condensation reactions between **carboxylic acids** and **alcohols**. Triglycerides, the compounds in fats and oils, are formed when long chain carboxylic acids (fatty acids) combine with the triple alcohol (triol) we know as glycerol. Triglycerides are therefore **esters**.

The **condensation** of carboxylic acid with a simple monohydric alcohol (one OH group per molecule) can be shown by the equation below.

(R = rest of molecule)

$$R_1COOH + R_2OH \longrightarrow R_1COOR_2 + H_2O$$

carboxylic acid + alcohol ⟶ ester + water

The condensation reaction is brought about by heating a mixture of carboxylic acid and alcohol, along with a little mineral acid (usually sulphuric) to act as catalyst.

The ester which is formed usually has a lower boiling point than either the carboxylic acid or alcohol. It can therefore be **distilled out** out of the reaction mixture and collected.

The mechanism of the condensation process involves removing the H and OH parts of a water molecule from the carboxyl (COOH) and hydroxyl (OH) groups respectively.

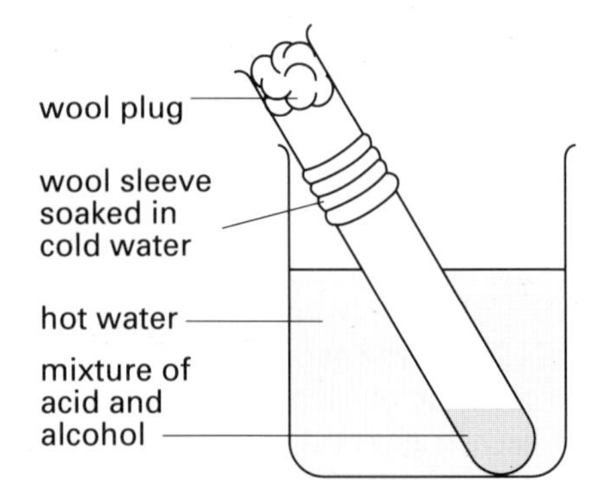

Figure 4.7 Forming an ester

A new functional group, the ester group $-\overset{\overset{\displaystyle O}{\|}}{C}-O-R$, is created.

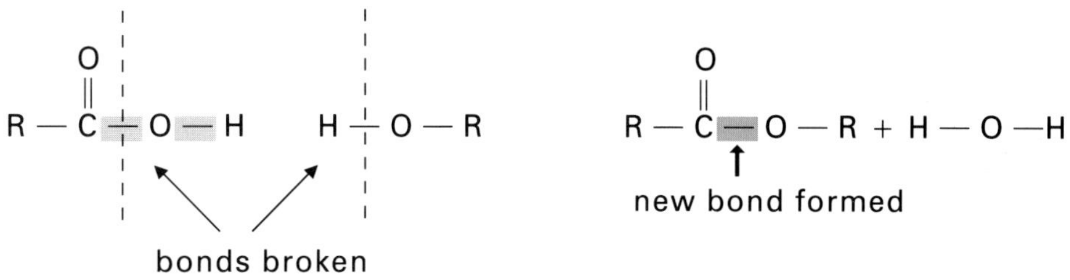

$$R-\overset{\overset{\displaystyle O}{\|}}{C}\,|\,O-H \quad H\,|\,O-R \qquad R-\overset{\overset{\displaystyle O}{\|}}{C}-O-R + H-O-H$$

bonds broken

new bond formed

Figure 4.18 Mechanism of the condensation process

Ethyl ethanoate is an ester and an important industrial solvent. It is used in the preparation of nail varnish and also as a solvent in the removal of varnish from fingernails. It is made by reaction between ethanoic acid and ethanol.

$$\underset{\text{ethanoic acid}}{CH_3COOH} + \underset{\text{ethanol}}{CH_3CH_2OH} \longrightarrow \underset{\text{ethyl ethanoate}}{CH_3COOCH_2CH_3} + H_2O$$

The full structural formula for ethyl ethanoate is given below.

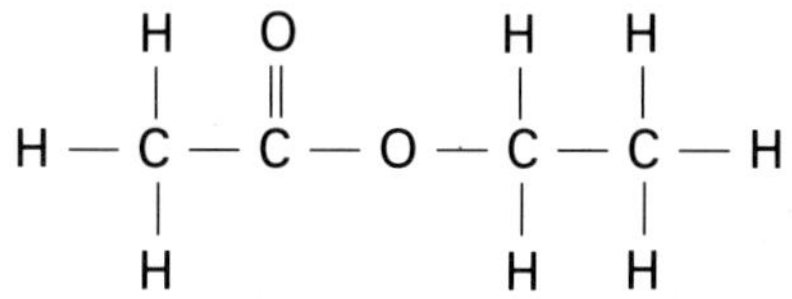

$$H-\underset{\underset{\displaystyle H}{|}}{\overset{\overset{\displaystyle H}{|}}{C}}-\overset{\overset{\displaystyle O}{\|}}{C}-O-\underset{\underset{\displaystyle H}{|}}{\overset{\overset{\displaystyle H}{|}}{C}}-\underset{\underset{\displaystyle H}{|}}{\overset{\overset{\displaystyle H}{|}}{C}}-H$$

Perhaps the most distinctive characteristic of esters is their strongly fruity smells or flavours. The particular taste and smell of most fruits comes from a complicated mix of different esters along with alcohols, acids, aldehydes, ketones, etc. Esters are therefore synthesised industrially and used as components in artificial flavourings. Some 'fruity' esters with names and structures are given in Table 4.3.

Fruit	Shortened structural formula	Systematic name
Oranges	$CH_3COOCH_2(CH_2)_6CH_3$	octyl ethanoate
Bananas	$CH_3COOCH_2CH_2\underset{\underset{\displaystyle CH_3}{|}}{C}HCH_3$	3-methylbutyl ethanoate
Pears	$CH_3COOCH_2CH_2CH_3$	propyl ethanoate
Peaches	$CH_3COOCH_2C_6H_5$	benzyl ethanoate
Raspberries	$CH_3COOCH_2\underset{\underset{\displaystyle CH_3}{|}}{C}HCH_3$	2-methylpropyl ethanoate
Pineapples	$CH_3CH_2CH_2COOCH_2CH_3$	ethyl butanoate
Apples	$CH_3CH_2CH_2COOCH_3$	methyl butanoate
Apricots	$CH_3CH_2CH_2COOCH_2CH_2CH_2CH_2CH_3$	pentyl butanoate
Plums	$CH_3CH_2CH_2COOCH_2CH_2\underset{\underset{\displaystyle CH_3}{|}}{C}HCH_3$	3-methylbutyl butanoate

Table 4.2 Some of the esters in selected fruits

As you can see from Table 4.3 an ester is named, systematically, as a derivative of the carboxylic acid from which it is made. The group (usually alkyl) attached to oxygen is named first. The ester is then named by dropping the suffix **-ic acid** and adding the suffix **-oate**.

Aromatic acids or alcohols can also form esters. For example, an ester which contributes to the flavour of grapes is methyl 2-aminobenzoate.

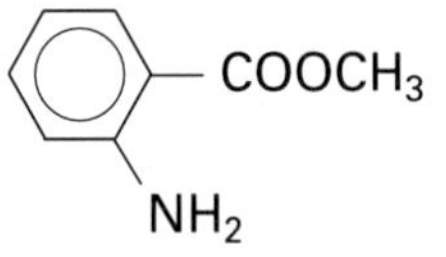

$C_6H_4(NH_2)COOCH_3$

methyl 2-aminobenzoate

Figure 4.19 The vines are loaded with grapes; esters contribute to their flavour

4.2.1 Hydrolysis of esters

The triglyceride esters in edible fats and oils are hydrolysed by **enzymes** in the body during the digestion process. Simple esters can be similarly hydrolysed, in the lab, back to their component alkanoic acids and alkanols.

$$\text{alkanoic acid} + \text{alkanol} \underset{\text{hydrolysis}}{\overset{\text{condensation}}{\rightleftharpoons}} \text{ester} + \text{water}$$

Ester formation is therefore a **reversible** reaction. The product (ester) can be decomposed into reactants (acid and alkanol).

The mechanism for hydrolysis is the reverse of the mechanism for condensation. The bonds broken and formed in the creation of the ester are the same bonds now formed and broken in the ester's decomposition.

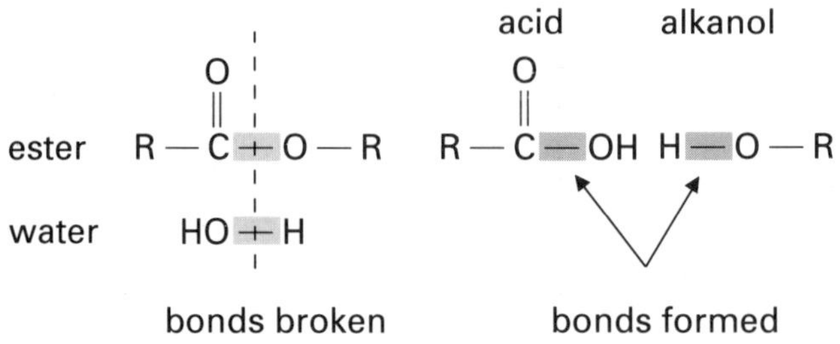

ester $R-C(=O)-O-R$ → acid $R-C(=O)-OH$ + alkanol $H-O-R$

water $HO-H$

bonds broken — bonds formed

Figure 4.20 Mechanism for hydrolysis

Ester hydrolysis is carried out in the lab (see Figure 4.22) by heating the ester, under reflux, with either dilute acid or alkali.

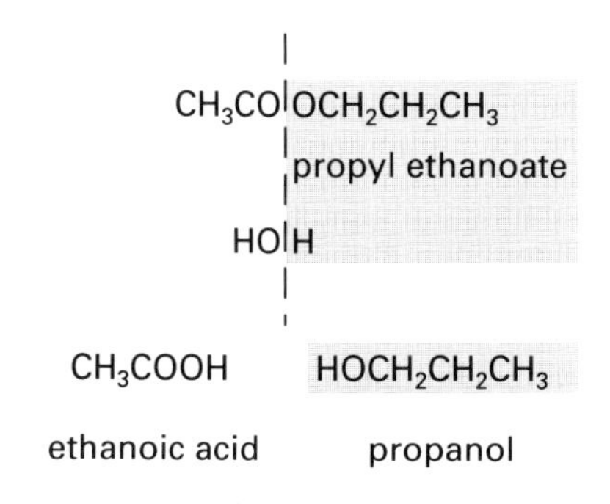

Figure 4.21 Acid hydrolysis of propyl ethanoate

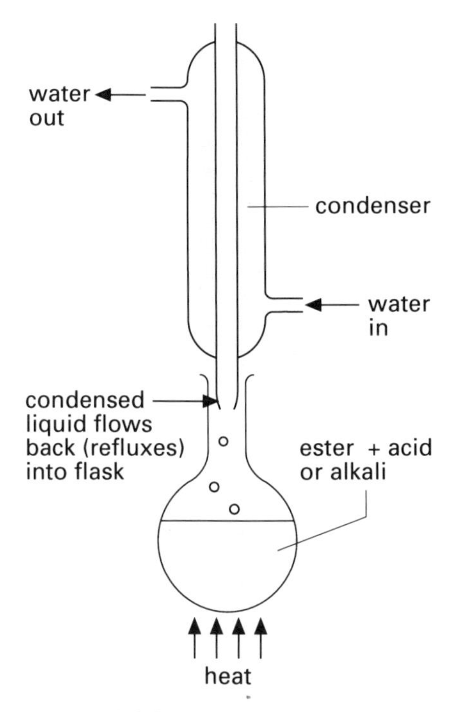

Figure 4.22 Hydrolysis of ester using reflux apparatus

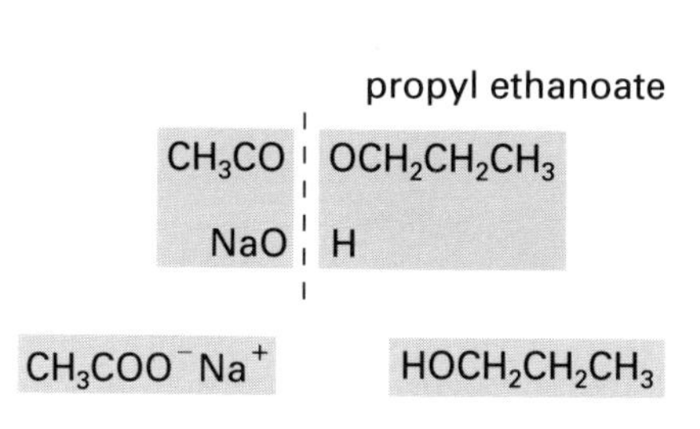

Figure 4.23 Alkali hydrolysis of propyl ethanoate

Acid hydrolysis provides the free acid and alkanol, as shown in Figure 4.21.

Dilute **alkali** is a more effective hydrolysing agent than acid, but produces the **metal salt** of the alkanoic acid rather than the free acid, as shown in Figure 4.23.

For all practical purposes, hydrolysis of an ester using aqueous alkali is **irreversible**. This is because the carboxylate ion, COO^-, unlike the carboxyl group COOH, has no tendency to react with an alcohol.

Hydrolysis of a fat or oil with aqueous sodium hydroxide gives glycerol and the **sodium salts** of the fatty acids involved. These salts are **soaps**.

The main components of ordinary soap are:

sodium stearate	$C_{17}H_{35}COO^- \; Na^+$
sodium oleate	$C_{17}H_{33}COO^- \; Na^+$
sodium palmitate	$C_{15}H_{31}COO^- \; Na^+$

Soaps are able to emulsify grease in water and clean material. This is due to the solubility of their long hydrocarbon chains in grease and the solubility of their ionic carboxylate groups in water.

$C_{17}H_{35}$	$COO^- \; Na^+$
soluble in grease	soluble in water

4.2.2 Polyesters

▢—COOH + ◯—OH ⟶ ▢—COO—◯

mono-acid · mono-ol · ester

If each reactant contains **two** functional groups, then a polymer containing many units of each component can be formed. This macromolecular compound is called a **polyester**.

HOOC—▢—COOH + HO—◯—OH ⟶ [—OC—▢—COO—◯—O—]$_n$

diacid · diol · polyester

Early polyester fibres, made from non-aromatic diacids and diols, were found to be unsuitable for textile use because their melting points were too low. Two English chemists, Whinfield and Dickson, reckoned that a greater **resistance to rotation** in the polymer backbone would stiffen the polymer, raise its melting point and thereby make it a more acceptable polyester fibre. To create stiffness in the polymer chain, a **benzene ring** was incorporated into the diacid molecule.

The polyester **fibre**, which is now manufactured under trademarks such as Terylene, Crimplene, Trevira and Dacron, is poly(ethene benzene-1,4-dicarboxylate) (polyethylene terephthalate, PET). It is made by polymerising a mixture of the aromatic diacid and diol below.

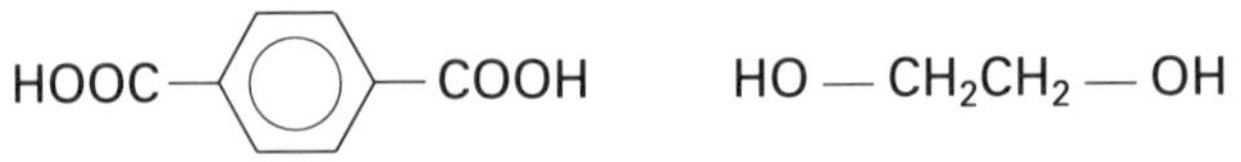

$HOOC-C_6H_4-COOH$ benzene-1,4-dicarboxylic acid (terephthalic acid)

$HO-CH_2CH_2-OH$ ethane-1,2-diol (ethylene glycol)

The molten polymer is extruded and the resulting fibre stretched. This uncoils the PET molecular chains and they lie together in an orderly arrangement.

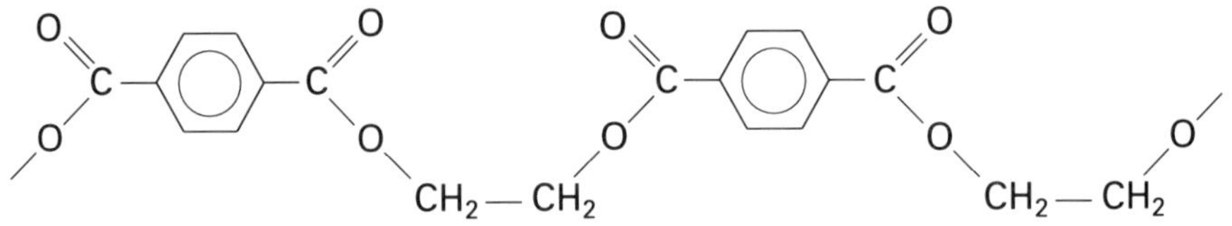

$-O-CO-C_6H_4-CO-O-CH_2-CH_2-O-CO-C_6H_4-CO-O-CH_2-CH_2-O-$

Figure 4.24 Part of the polyester molecule

The stiffness of the polyester molecule backbone is responsible for the crease resistance of 'no need to iron' polyester fabrics. Plastic film made from the polymer is used for cassette tape.

Figure 4.25 The manufacture of polyester film at the ICI plant at Dumfries

Polyester fibre or film is a thermoplastic linear polymer. However, if the dicarboxylic acid monomer is **unsaturated**, cross-linking between parallel molecular chains can be brought about. A thermosetting, cross-linked (cured) polyester resin is then formed.

Glass reinforced polyester (GRP) is manufactured by cross-linking the linear unsaturated polyester chains using phenylethene (styrene) molecules as bridges. The resulting resin is then mixed with glass fibre.

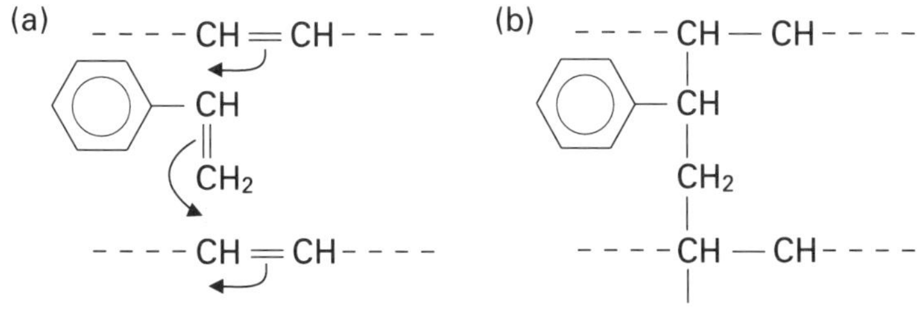

(a) $----CH{=}CH----$ with curved arrows: $C_6H_5-CH{=}CH_2$ bridging to $----CH{=}CH----$

(b) $----CH-CH----$ / C_6H_5-CH / CH_2 / $----CH-CH----$

Figure 4.26

(a) Mechanism for bridging parallel polyester chains *(b) Resulting cross-linked polyester*

A second method of producing a cross-linked polyester is by using a **trihydric** alcohol such as propane-1,2,3-triol (glycerol) as a monomer rather than the dihydric alcohol, ethane-1,2-diol.

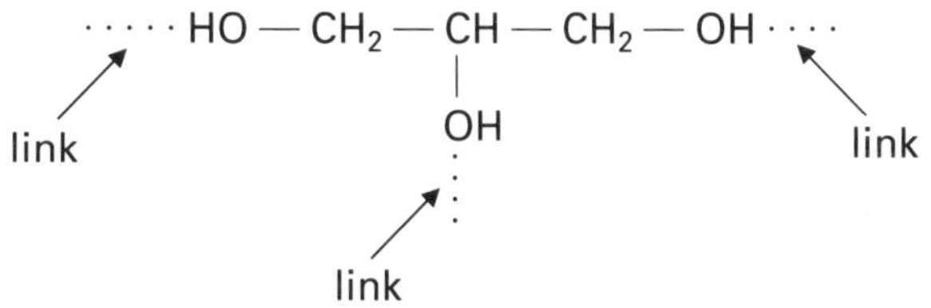

$$\cdots HO-CH_2-CH(OH)-CH_2-OH\cdots$$

link link link

Figure 4.27 Forming a 3-D polyester structure

Linking with the diacid can take place at three points rather than two, and a rigid three-dimensional polyester structure can be obtained. Polyesters of this type are used as binders in paints.

4.3 Proteins

The proteins are a leading group of macromolecules found in all living organisms. There are hundreds of thousands of protein compounds. Even a single cell of a living plant or animal may contain several thousand different proteins.

Spiders and silkworms secrete a thick solution of the protein **fibroin** which quickly solidifies into a very strong thread used to form webs or cocoons.

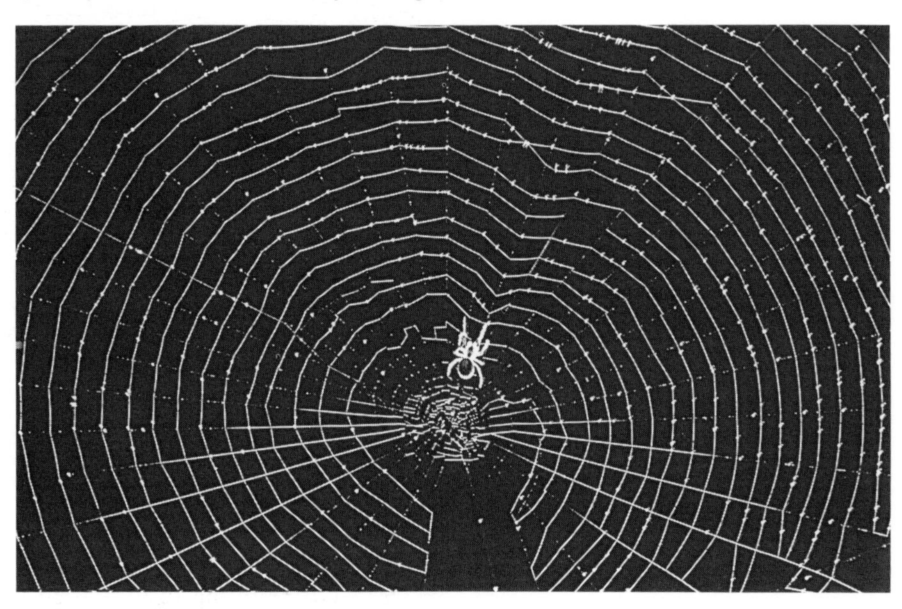

Figure 4.28 The protein filaments that make up a spider's web are stronger, weight for weight, than steel

There is no form of life without proteins. They are needed to carry out a wide range of biological tasks.

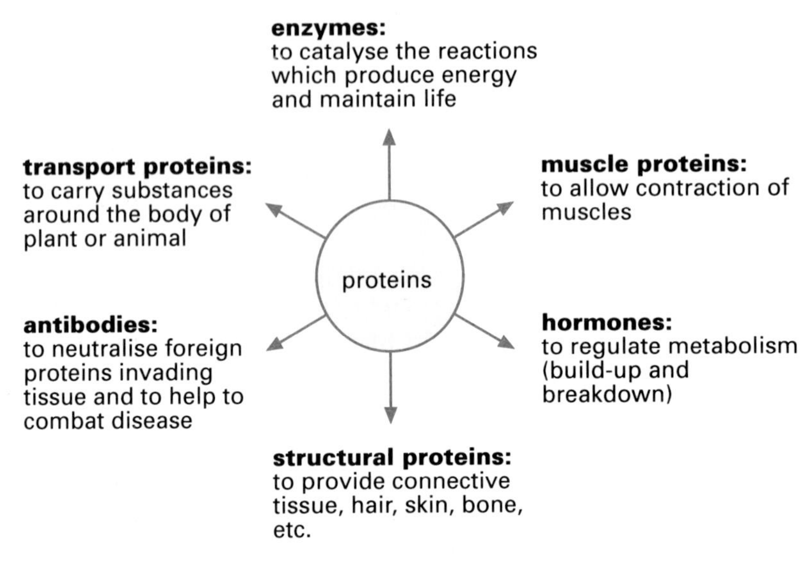

Figure 4.29 Proteins and their tasks

The chemical composition of proteins differs from that of carbohydrates and fats, in that proteins contain elements other than carbon, hydrogen and oxygen.

biochemical group	**elements present**
carbohydrates	C H O
fats and oils	C H O
proteins	C H O N S *

*occasionally other elements such as P, Cu, Mn, Fe, Mg or Zn may be present

Figure 4.30 Chemical composition of proteins, carbohydrates and fats

The most notable addition to the list of elements present in protein is that of **nitrogen**. The presence of nitrogen is an important feature as it gives proteins many of their characteristic properties.

Nitrogen is found in three main places:

(a) as uncombined nitrogen in the air (nitrogen gas)
(b) as combined nitrogen in plants (mainly protein)
(c) as combined nitrogen in soil (mainly nitrates).

Plants usually extract the nitrogen they need for building protein from soil. In some cases, e.g. legume plants such as peas, beans and clover, this nitrogen is removed directly from air. Animals and humans obtain their necessary supply of nitrogen by feeding on plants or, in a second-hand way, by feeding on other animals.

4.3.1 Amino acids

Like the carbohydrates starch, cellulose and glycogen, proteins are long chain polymers. Protein molecules vary greatly in size. The relative formula masses of some proteins are given in Table 4.4.

Protein	Relative formula mass	Biological task
insulin	6 000	hormone
pepsin	24 000	enzyme
haemoglobin	68 000	transport
myosin	540 000	muscle filament
immunoglobulin	960 000	antibody

Table 4.4 Relative formula masses of proteins

Proteins are built up by linking together monomers called **amino acids**. Unlike the carbohydrate polymers whose monomers are all glucose molecules, proteins are built up through combinations of up to twenty different kinds of amino acids. Just as linking together the 26 letters of the alphabet produces an enormous range of words, sentences, paragraphs, etc., so linking together amino acids in various ways creates a tremendous variety of protein molecules.

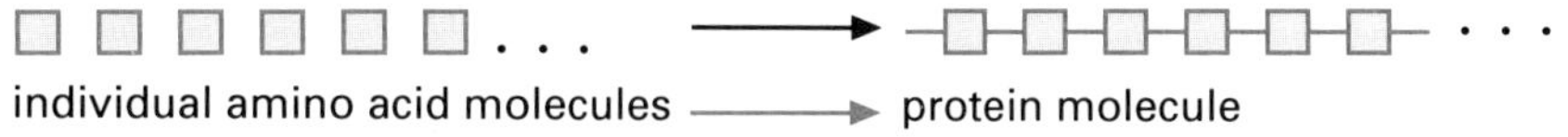

□ □ □ □ □ □ . . . ⟶ –□–□–□–□–□–□– . . .

individual amino acid molecules ⟶ protein molecule

Figure 4.31 Linking together amino acids

Amino acids are compounds where molecules contain the functional groups **amino** (—NH_2) and **carboxyl** (—COOH). All proteins are constructed from amino acids in which the amino group and carboxyl group are attached to the same carbon atom. The general formula is shown below.

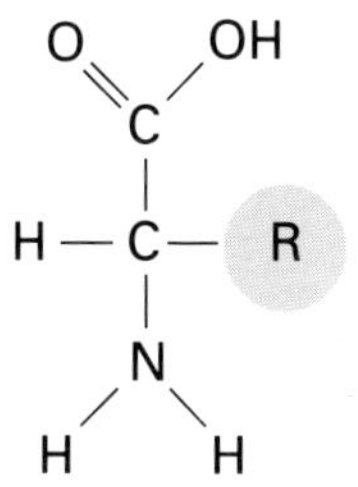

```
O     OH
 \\  /
   C
   |
H—C— R
   |
   N
  / \
 H   H
```

general formula of an amino acid

The structure of the side chain, R, can vary widely between different amino acids. R can be a hydrogen atom, alkyl group, a chain or a ring. Amino acids are different because they have different R groups.

Only twenty or so different amino acids are typically found in proteins. The first one to be isolated was **glycine** in 1820, and the most recent was **threonine** in 1935.

The structures of some common amino acids are given below, along with their abbreviated (three-letter) names.

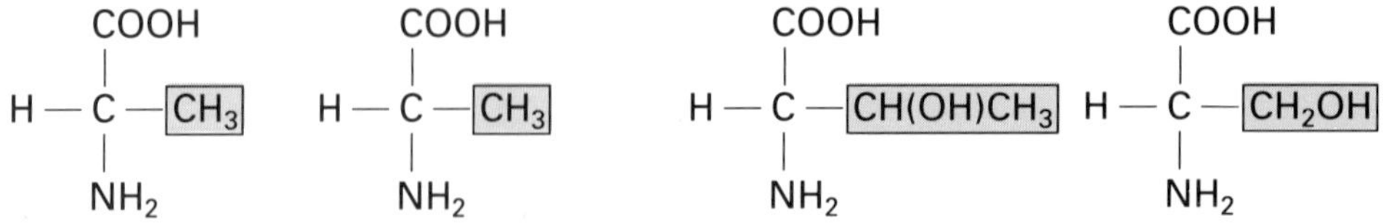

$$\begin{array}{c} COOH \\ | \\ H-C-\boxed{CH_3} \\ | \\ NH_2 \end{array} \quad \begin{array}{c} COOH \\ | \\ H-C-\boxed{CH_3} \\ | \\ NH_2 \end{array} \quad \begin{array}{c} COOH \\ | \\ H-C-\boxed{CH(OH)CH_3} \\ | \\ NH_2 \end{array} \quad \begin{array}{c} COOH \\ | \\ H-C-\boxed{CH_2OH} \\ | \\ NH_2 \end{array}$$

glycine (gly) | alanine (ala) | threonine (thr) | serine (ser)

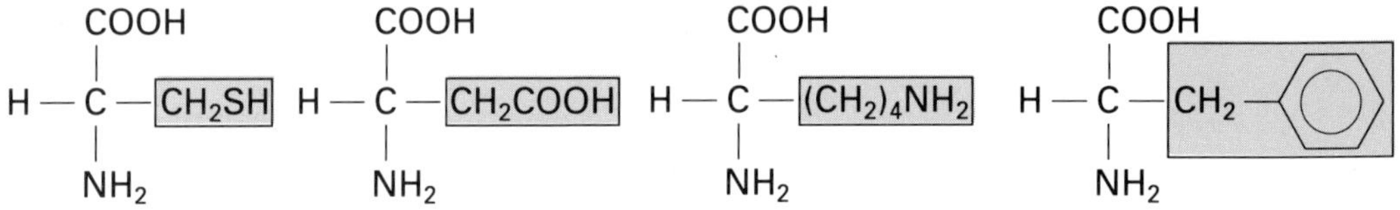

$$\begin{array}{c} COOH \\ | \\ H-C-\boxed{CH_2SH} \\ | \\ NH_2 \end{array} \quad \begin{array}{c} COOH \\ | \\ H-C-\boxed{CH_2COOH} \\ | \\ NH_2 \end{array} \quad \begin{array}{c} COOH \\ | \\ H-C-\boxed{(CH_2)_4NH_2} \\ | \\ NH_2 \end{array} \quad \begin{array}{c} COOH \\ | \\ H-C-\boxed{CH_2-C_6H_5} \\ | \\ NH_2 \end{array}$$

cysteine (cys) | aspartic acid (asp) | lysine (lys) | phenylalanine (phe)

4.3.2 Peptides

Intemediate in molecular complexity between the amino acids and proteins are the **peptides**, compounds formed by linking amino acids through the formation of peptide bonds. Peptide bonds are created by a process known as condensation. This involves the removal of OH and H groups from COOH (carboxyl) and NH_2 (amino) groups on adjacent molecules. The formation of a dipeptide, glycine-alanine, is shown below.

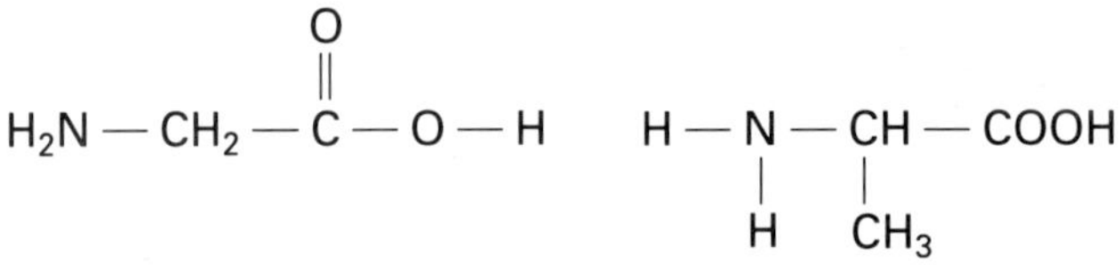

$$\begin{array}{c} \quad\quad\quad\quad O \\ \quad\quad\quad\quad \| \\ H_2N-CH_2-C-O-H \\ \\ \text{glycine} \end{array} \qquad \begin{array}{c} \\ \\ H-N-CH-COOH \\ \quad | \quad\;\; | \\ \quad H \quad CH_3 \\ \text{alanine} \end{array}$$

$$\begin{array}{c} \quad\quad\quad\quad O \\ \quad\quad\quad\quad \| \\ H_2N-CH_2-C-N-CH-COOH + HOH\ (H_2O) \\ \quad | \quad\;\; | \\ \quad H \quad CH_3 \end{array}$$

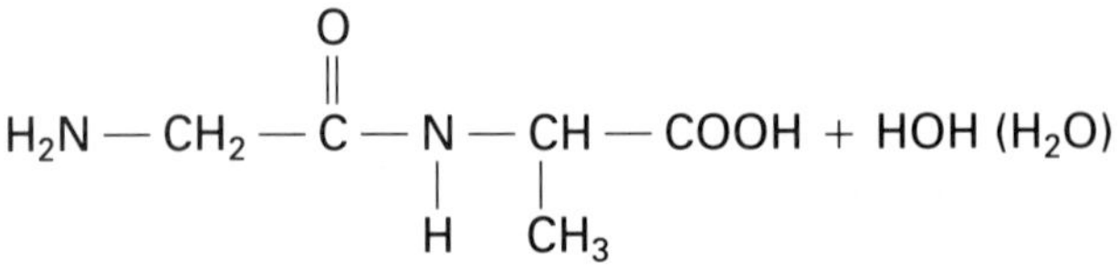

glycine-alanine dipeptide (gly-ala)

The arrangement $\begin{array}{c} O \\ \| \\ -C-N- \\ \quad | \\ \quad H \end{array}$, or (—CONH—), is called a peptide or amide group. The bond formed between the carbon and nitrogen atoms is called a peptide **bond**.

Tripeptides are built up from three amino acid molecules, while molecules containing four or more amino acid units are generally referred to as **polypeptides**.

A peptide chain has direction since it has different groups at either end. By convention, the terminal amino group is taken to be the beginning of the chain and the terminal carboxyl group to be the end of the chain. For example, a tripeptide composed of serine, aspartic acid and cysteine would be designated **ser-asp-cys** and have the following structure:

$$\boxed{H_2N}-\underset{H}{\overset{CH_2OH}{C}}-CONH-\underset{H}{\overset{CH_2COOH}{C}}-CONH-\underset{H}{\overset{CH_2SH}{C}}-\boxed{COOH}$$

side chains (R groups)

amino (start)

carboxyl (finish)

tripeptide: ser-asp-cys

4.3.3 Structure of proteins

Figure 4.32 A nurse teaching a diabetic child to inject herself with insulin

Some proteins are composed of a single polypeptide chain, but many consist of two or more polypeptide chains. The first successful unravelling of the structure of a protein was achieved by Frederick Sanger, at Cambridge in 1956. He tackled the relatively small protein called **insulin**. Insulin controls the concentration of glucose in the blood. People who are deficient in insulin suffer the condition known as 'diabetes', which shows as very high blood glucose levels. Treatment is by regulating diet or, in more severe cases, by daily injections of insulin.

Insulin is made up of two short polypeptide chains with 21 and 30 amino acid units respectively. Sanger determined the correct **sequence** of the amino acids in each chain and the position at which cross-linking occurred. For this achievement he was awarded the Nobel Prize in 1959. Proteins do not have branched chains but it is possible for a chain to bend back on itself or for two chains to become cross-linked to one another.

Insulin's two polypeptide chains (A and B) are bound together by cross-links as shown in Figure 4.33.

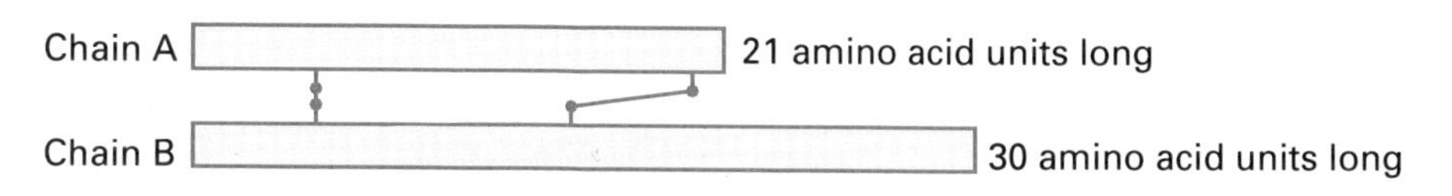

Figure 4.33 Linking of polypeptide chains in insulin

Proteins are not simply polypeptide chains which flap about in a haphazard way. On the contrary, proteins have a very definite three-dimensional structure, held together by interactions between the amino acid side chains. Each molecule of the same protein has the same **shape**. Proteins can be classified according to their shape into fibrous proteins and globular proteins.

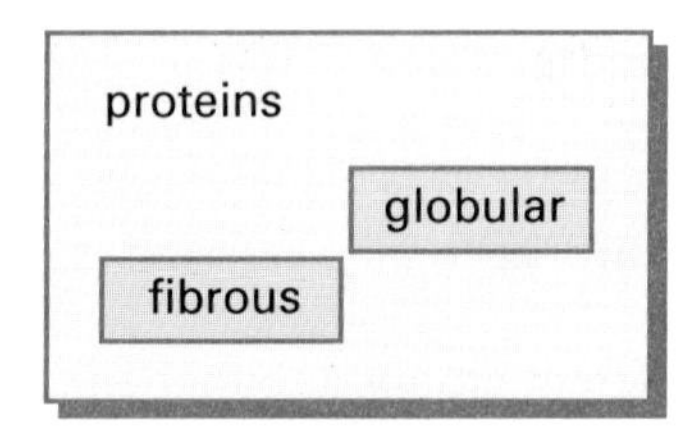

Figure 4.34 Classification of proteins by shape

Fibrous proteins have their molecular chains interwoven to form multi-strand cables. They are insoluble in water, fairly stable to changes in pH and temperature, tough and unreactive. With such properties they are equipped to act as molecules which hold cell walls together. Fibrous proteins, for example, provide the **structural** element of skin, bone, tendons, cartilage, hair, horn, nail, claw, feathers, scales, wool, blood and teeth.

The proteins which operate **inside** cells have to be soluble in water, sensitive to changes in pH and temperature, and generally highly reactive. These are the **globular** proteins such as ribonuclease (enzyme) and haemoglobin (transport function). In globular proteins the polypeptide chains are coiled together into irregular spherical or ball shapes.

Figure 4.35

A fibrous protein chain arrangement

A globular protein chain arrangement

Sometimes additional bits are attached to protein molecules to help the protein perform special functions. For example, the protein molecule globin has bound to it a single iron atom. It is this extra iron or 'haem' part which chemically grasps oxygen and enables **haemoglobin** to act as a carrier of oxygen in the bloodstream.

4.3.4 Digestion of proteins

Digestion is the name given to the chemical demolition process on food. The digestion of proteins in the food involves decomposition (hydrolysis) of the very large protein molecules into smaller fragments and eventually to the individual amino acids which formed its building bricks.

It is strange perhaps that the enzyme molecules which help to dismantle the incoming proteins are proteins themselves. These enzymes are classified as **peptidases**. A large number of enzymes are engaged in demolishing proteins, some attacking a range of peptide links and others being much more selective.

The digestion of protein begins in the stomach with the action of the enzyme **pepsin**. Hydrolysis produces a mixture of polypeptides which pass into the intestine where a team of enzymes completes the dismantling of the protein into separate amino acids.

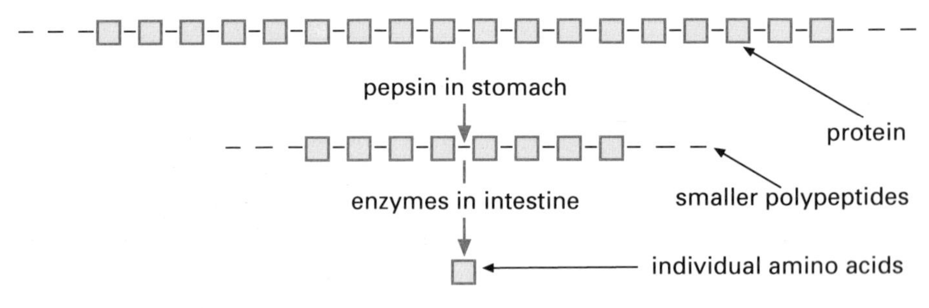

Figure 4.36 Breakdown of protein molecule

Enzymes enable water molecules to be 'added' across each peptide bond, C—N. As the carbon atom becomes bonded to an OH group and the nitrogen atom bonds to an H atom, the bond between the C and N atoms breaks. This process is described as **hydrolysis**.

$$-C(=O)-N(H)- \;+\; H-O\,|\,H \;\Rightarrow\; -C(=O)-OH \;\;\; H-N(H)-$$

Figure 4.37 Hydrolysis of the peptide bond

A tripeptide would be hydrolysed at its two peptide bonds, as shown in Figure 4.38.

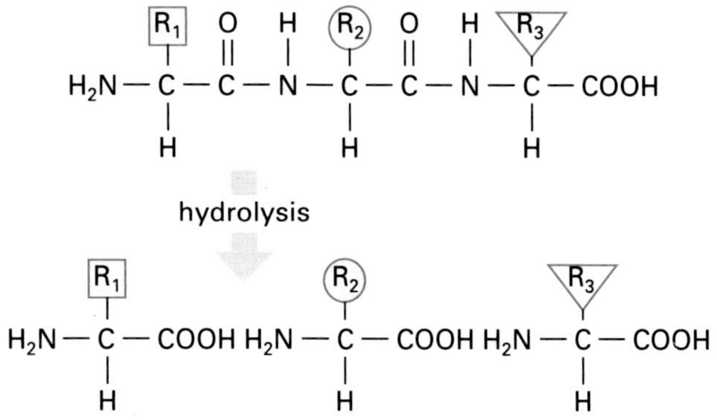

$$H_2N-\underset{H}{\overset{R_1}{C}}-\overset{O}{\overset{\|}{C}}-\overset{H}{N}-\underset{H}{\overset{R_2}{C}}-\overset{O}{\overset{\|}{C}}-\overset{H}{N}-\underset{H}{\overset{R_3}{C}}-COOH$$

hydrolysis

$$H_2N-\underset{H}{\overset{R_1}{C}}-COOH \quad H_2N-\underset{H}{\overset{R_2}{C}}-COOH \quad H_2N-\underset{H}{\overset{R_3}{C}}-COOH$$

Figure 4.38 Hydrolysis of tripeptide into three amino acids

Proteins are thus completely hydrolysed into amino acids which pass from the small intestine into the blood, from where they are taken up by all the cells of the body.

4.3.5 Synthesis of proteins

Most plants are capable of synthesising for themselves all of the 20 amino acids that normally make up proteins. These amino acids can then condense in various sequences to produce different polypeptides and finally different proteins.

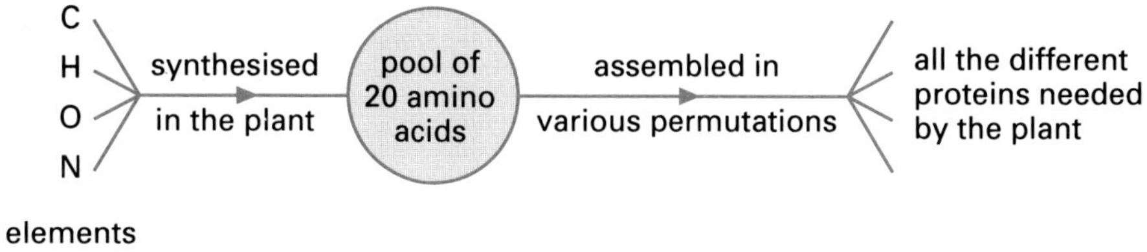

C
H
O
N
elements
synthesised in the plant
pool of 20 amino acids
assembled in various permutations
all the different proteins needed by the plant

Figure 4.39 Formation of proteins in plants

Humans and other higher animals are able to synthesise only eleven of the set of 20 amino acid building blocks. The other nine amino acids are, of course, equally necessary for building proteins. They have to be obtained through eating plants and other animals, i.e. from digesting food. The nine amino acids which have to be obtained in this second-hand way from diet (printed in green in the list below) are known as **essential** amino acids, because it is essential that they be included in the protein which forms part of our diet.

alanine	**threonine**	**phenylalanine**	**leucine**	glycine
cysteine	arginine*	tyrosine	proline	**lysine**
histidine	glutamic acid	asparagine	**tryptophan**	serine
methionine	**isoleucine**	**glutamine**	aspartic acid	valine

* may be essential for some individuals.

For **humans** the pool of amino acids is built up partly through internal synthesis in the body, and partly by hydrolysing the proteins taken in as food.

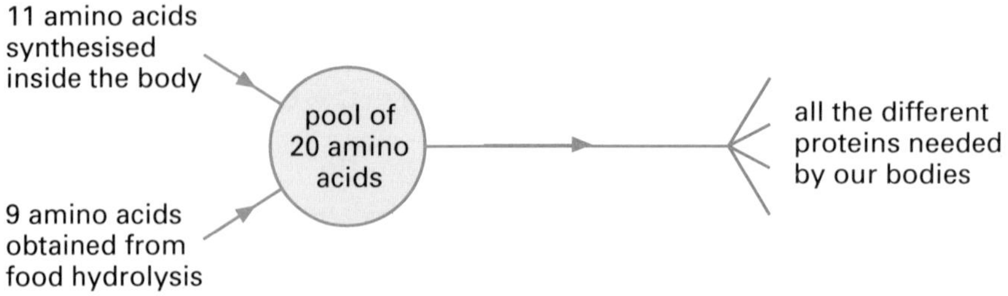

11 amino acids synthesised inside the body → pool of 20 amino acids

9 amino acids obtained from food hydrolysis → pool of 20 amino acids

pool of 20 amino acids → all the different proteins needed by our bodies

Figure 4.40 Formation of proteins in humans

The value of a protein-containing food is judged, therefore, not only by the amount of protein it contains but also the number and amounts of essential amino acids which it provides. Almost all protein molecules contain at least one unit of each of the 20 amino acids. The absence of even one amino acid therefore prevents formation of any protein.

Most proteins from animal sources, e.g. eggs, milk and meat, contain adequate amounts of all the essential amino acids. Obtaining a good balance of amino acids is therefore no problem for meat-eating populations. On the other hand, proteins from plants vary widely in their amino acid content. Most plant proteins are deficient in one or more of the essential amino acids.

The most common diet deficiency is inadequate supply of the amino acid lysine, which is missing from corn, wheat and potatoes. Legume crops such as peas and soya beans contain sufficient lysine but lack other amino acids, notably methionine.

Figure 4.41 Wheat – a valuable food crop, but deficient in one essential amino acid

The diet of a European person is normally sufficiently rich in total protein and individual essential amino acids. However, for many people in Africa and Asia, living on a restricted vegetarian diet (through poverty rather than choice), the supply of protein and essential amino acids is inadequate. The disease **kwashiorkor** affects millions of children in Third World countries. A child no longer able to feed on its mother's milk has to depend on protein from rice, wheat, etc., and may suffer the various disorders caused by amino acid deficiency. The most visible symptom of kwashiorkor is a bloated belly.

Providing most of the world's population with a complete protein diet is a major health problem. The solution to this health problem is to produce, cheaply, high quality protein with all the essential amino acids present. Growing plants (crops) and feeding them to animals is a common method of converting incomplete edible protein to more nutritious edible protein. Unfortunately it is a grossly inefficient process.

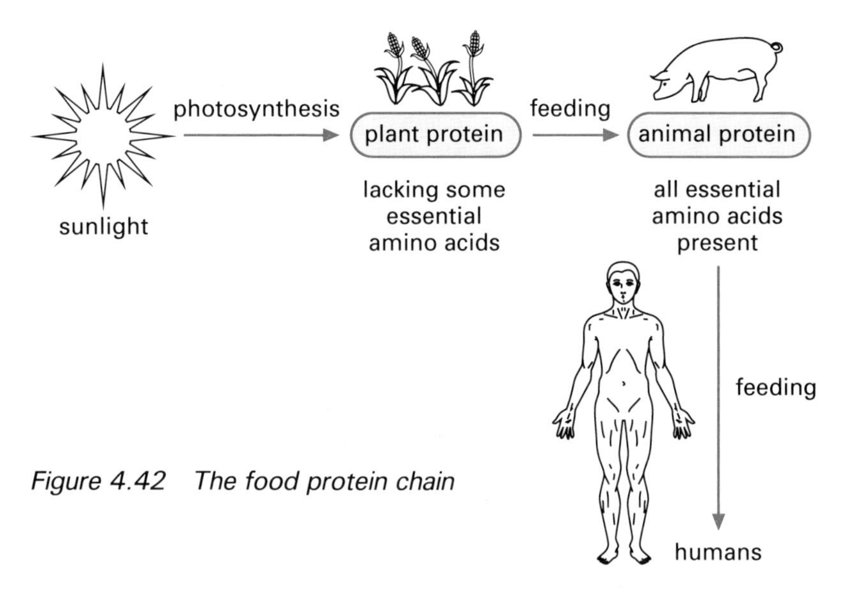

Figure 4.42 The food protein chain

Other methods include adding supplements, such as fish meal, which contain complete protein, to the basic diet, or developing plant strains (e.g. a type of corn which is rich in lysine) that provide the amino acids needed.

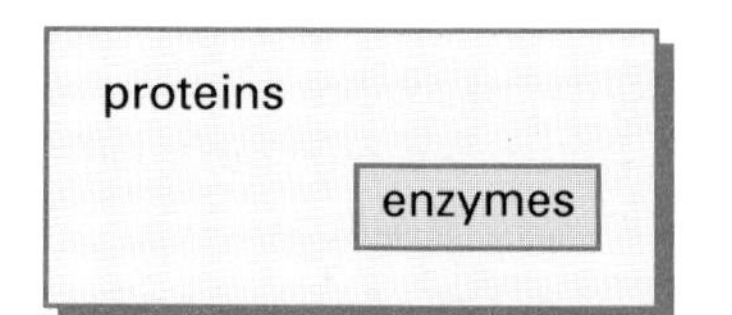

Figure 4.43 Enzymes: a subset of proteins

Figure 4.44 Shape of ribonuclease, an enzyme

4.3.6 Enzymes

All proteins consist of amino acids condensed together, but only a fraction of the proteins in any living organism have the ability to catalyse chemical reactions, i.e. act as **enzymes**.

Enzymes have to be soluble in order to act as catalysts in living cells. They are therefore all **globular** (rather than fibrous) proteins, with their molecules delicately coiled into compact snake-like loops. This carefully structured shape (see Figure 4.44) is maintained by the interplay of many weak bonds. It generally exists under mild conditions of temperature (273–313 K) and pH (5.5–9.0).

Because of its unique shape, a given enzyme will deal with only one set of reactants (substrates). It is also very selective about the reaction it will catalyse. This is essential in living organisms, which cannot cope with a surplus of by-products. The small substrate molecule can only be in contact with a small part of the large enzyme molecule – a critical region known as the **active site**. This is where binding of the substrate to enzyme occurs and where catalysis takes place. Most enzymes have one active site per molecule.

At the active site there is a particular molecular arrangement to form a space more or less precisely fitted to the shape of the substrate. Into this space the substrate fits like a key into a lock. It is held by weak forces in such an orientation that the substrate molecule becomes **activated** and ready to react.

4.3.7 Denaturation

The highly ordered protein (enzyme) shape can be unfolded by moderate heating or by adding chemicals such as acids or alkalis. Proteins changed in this way are said to be **denatured**. The precisely folded shape is undone and a randomly coiled chain produced, as shown in Figure 4.45.

Figure 4.45 Shape of a denatured protein (the enzyme ribonuclease)

When an egg is heated, the neatly coiled, water-soluble albumin molecules in the transparent egg white are denatured to form more loosely coiled chains. These mesh together and trap water to form the opaque gel-like structure we know as cooked egg white. Physical beating of egg white also destroys the precisely curled globular protein shape and the protein begins to precipitate. The resultant material, which is quite different from the original protein solution, is used as the basis for meringues.

Fresh milk has a pH of about 6.6. If its pH is lowered below pH5 (by adding acid) the protein in milk, casein, changes from a soluble globular form to unfolded chains, which interwine to form an insoluble form that precipitates out of solution. The milk is said to have **curdled**. When milk curdles naturally, as in the making of cheese, the acid is produced from bacterial action.

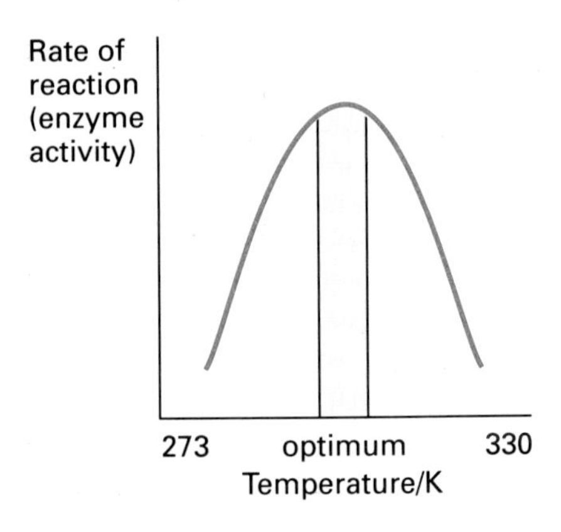

Figure 4.46 Effect of temperature on activity of enzymes

4.3.8 Optimum temperature and pH for enzymes

The effect of denaturation of an enzyme is to rob it of its ability to catalyse. For most enzymes heating to a temperature of around 330 K (57 °C) causes the weak bonds which hold the folded structure together to start to break. The enzyme protein is denatured and catalytic activity is lost. Enzymes therefore operate best within a narrow range of temperature known as the **optimum temperature**.

Most enzymes are at their most active around the pH value of the solution within living cells. Outside these normal pH limits the enzyme's catalytic power drops off, as groups on the side chains ionise. At the extremes of pH (very acidic or very alkaline) the enzyme is denatured.

Since denaturation is usually an irreversible process it has to be avoided in biological systems. It is prevented, in the living body, by maintaining temperature and pH within prescribed limits. Chemical (non-biological) catalysts operate very often in conditions of high temperatures and extreme pH. Enzymes, to their credit, achieve their remarkable catalytic efficiency in conditions which are relatively cool and bland.

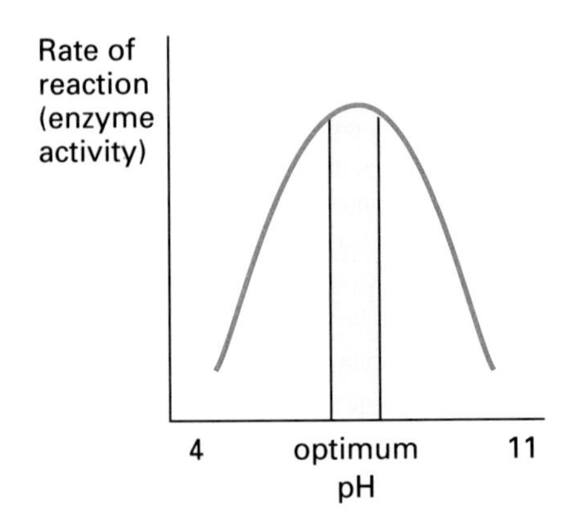

Figure 4.47 Effect of pH on activity of enzymes

Summary of Unit 4

Having read and understood the information and ideas given in this unit, you should now be able to:

classify natural oils and fats according to source

relate differences in melting point to differences in the level of saturation/unsaturation

describe the conversion of oils to fats in terms of hydrogenation of double bonds

explain the process and purpose of digesting edible fats and oils

state that fats and oils are esters composed from fatty acids and glycerol, and can be hydrolysed to form those substances

summarise the arguments linking heart disease with fat intake

identify the ester group and draw structural formulae for simple esters

name an ester systematically and, given its structural formula, identify its component acid and alcohol

describe the main uses of esters and polyesters

outline the part played by amino acids in the structure and shape of protein molecules

identify the peptide link and describe how it is made and broken during the formation and hydrolysis of proteins

- ▲ explain why unsaturation helps to reduce the melting point of a fat or oil
- ▲ outline the composition of fats and oils, in terms of triglyceride mixtures
- ▲ identify, by names and formulae, the products from hydrolysing simple esters
- ▲ explain the concept of reversibility, as applied to ester formation and hydrolysis
- ▲ distinguish polyester fibres from resins, in terms of molecular structure
- ▲ draw the structural formula for a tripeptide, given the structures and sequence of the constituent amino acids
- ▲ explain what is meant by 'essential' amino acids
- ▲ relate the catalytic action of an enzyme to its shape
- ▲ explain how proteins are denatured and the effect of such denaturation.

PROBLEM SOLVING EXERCISES

1. An amino acid will travel up paper at a given rate. This rate depends on how soluble the amino acid is in the solvent compared to how soluble it is in the paper.

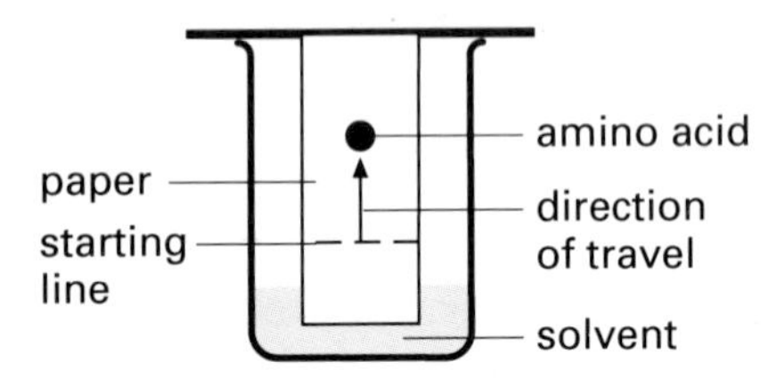

Figure 4.48

Which two of the variables below could be changed without affecting the position reached by the amino acid in a given time?

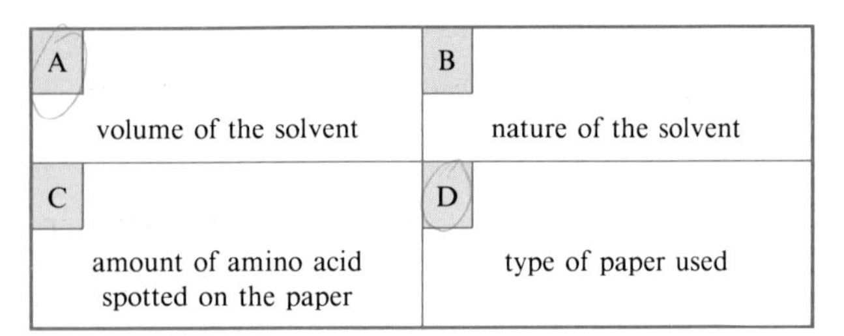

A volume of the solvent	B nature of the solvent
C amount of amino acid spotted on the paper	D type of paper used

(PS skill 6)

2. 'Biopol' is a biodegradable polyester developed by ICI. It has a wide range of applications in medicine and agriculture. Its monomer has the structure $CH_3CHOHCH_2COOH$.
Draw part of the Biopol polymer chain, showing three monomer units linked together.

(PS skill 10)

3. A tripeptide was hydrolysed and the product mixture spotted on paper beside eight amino acids. The resulting chromatogram is shown below.

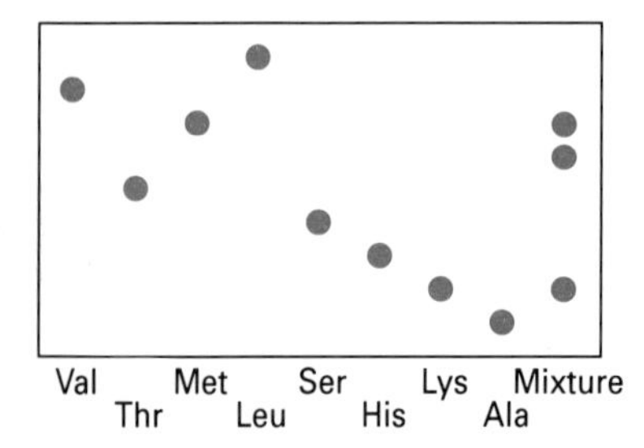

Figure 4.49

What can you conclude about the composition of the mixture?

(PS skill 8)

4. Although fatty acids are stored in the body as triglycerides, it is the fatty acid part which is the main source of energy.
A fatty acid releases on combustion 623 J of energy per mole of CO_2 produced.

If the molar mass of stearic acid, $C_{17}H_{35}COOH$, is 284 g mol^{-1}, calculate the energy released by the combustion of 1 g stearic acid.

(PS skill 4)

5. Octan-2-ol, $CH_3(CH_2)_5CHOHCH_3$, is used in perfumery because of its smell.
Octanoic acid, $CH_3(CH_2)_6COOH$, is obtained from goat fat and has a much less agreeable smell.

Write a shortened structural formula for the ester which is formed from octan-2-ol and octanoic acid.

(PS skill 10)

6. Most fats and oils undergo partial decomposition, on storage, to release free fatty acids. The proportion of 'free acid' in a sample is a general indication of the extent of deterioration of the fat or oil.

Suggest a practical method of measuring the amount (in moles) of free fatty acid in any fat or oil sample.

(PS skill 5)

7. Fats are hydrolysed in the body to form glycerol and fatty acids. To obtain energy both products must be oxidised. The glycerol reacts with ATP to form glyceryl phosphate which is oxidised stepwise to CO_2 and H_2O. The fatty acids are oxidised in steps to ethanoic acid and a smaller fatty acid (two less carbon atoms), e.g. $C_{18} \rightarrow C_{16} \rightarrow C_{14} \ldots$, until C_4 fatty acid, butanoic acid, is obtained. The ethanoic and butanoic acids are then oxidised to CO_2 and H_2O by a complex process.

Present the essential information above by means of a flow diagram.

(PS skill 2)

PROBLEM SOLVING EXERCISES

8. An experiment was carried out with a group of baby chicks.
Group 1 was given a basic diet which lacked the amino acids glycine and methionine.
Group 2 was given a basic diet with glycine added.
Group 3 was given a basic diet with methionine added.
The growth of the chicks was recorded and results obtained as showed in the graph below.

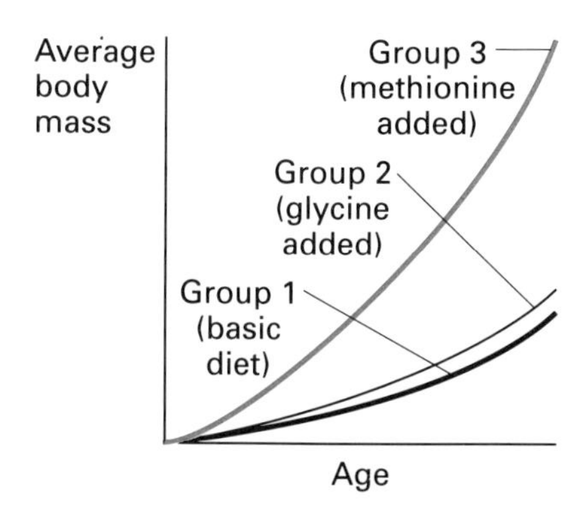

Figure 4.50

Explain why the addition of methionine has a marked effect on the chicks' growth, whereas the addition of glycine has no obvious effect.

(PS skill 9)

9. You wish to show the following items of information.
(i) The relative molecular masses of three different proteins.
(ii) The rate of hydrolysis of a specific protein by an acid.
(iii) The relative proportions of the component amino acids in a specific protein.
Four possible formats for presenting information are given in the grid below.

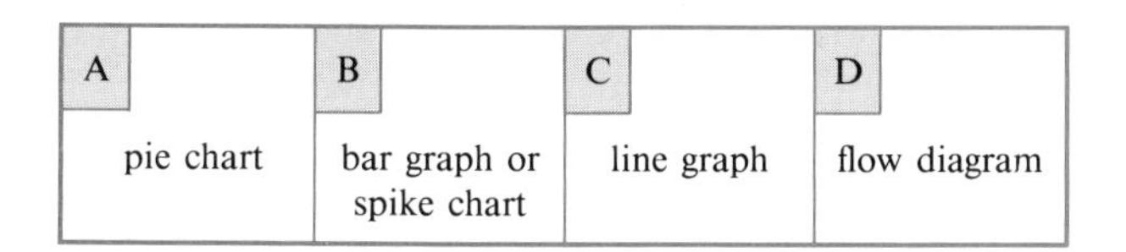

A	B	C	D
pie chart	bar graph or spike chart	line graph	flow diagram

State which would be the best format for displaying the data from:
(a) item (i)
(b) item (ii)
(c) item (iii).

(PS skill 3)

10. The table below lists activation energies for reactions catalysed by different substances.

Reaction	Catalyst	Activation energy/ kJ mol^{-1}
hydrolysis of sucrose	HCl acid	109
	invertase (enzyme)	46
hydrolysis of casein	HCl acid	86
	trypsin (enzyme)	50
hydrolysis of ethyl butanoate	HCl acid	55
	lipase (enzyme)	18

What general statement (generalisation) can you make after considering the information given in the table?

(PS skill 8)

11. The boxes in the grid below contain the names and formulae of some fatty acids.

A	B	C
arachidonic $C_{19}H_{39}COOH$	linoleic $C_{17}H_{31}COOH$	myristic $C_{13}H_{27}COOH$
D	**E**	**F**
stearic $C_{17}H_{35}COOH$	behenic $C_{21}H_{43}COOH$	palmitoleic $C_{15}H_{29}COOH$

A teacher wishes to illustrate the generalisation that melting point increases as chain length increases.
Which two fatty acids should **not** be included with the others if a fair and valid comparison is to be made?

(PS skill 7)

Unit 5
From Bonds to Behaviour

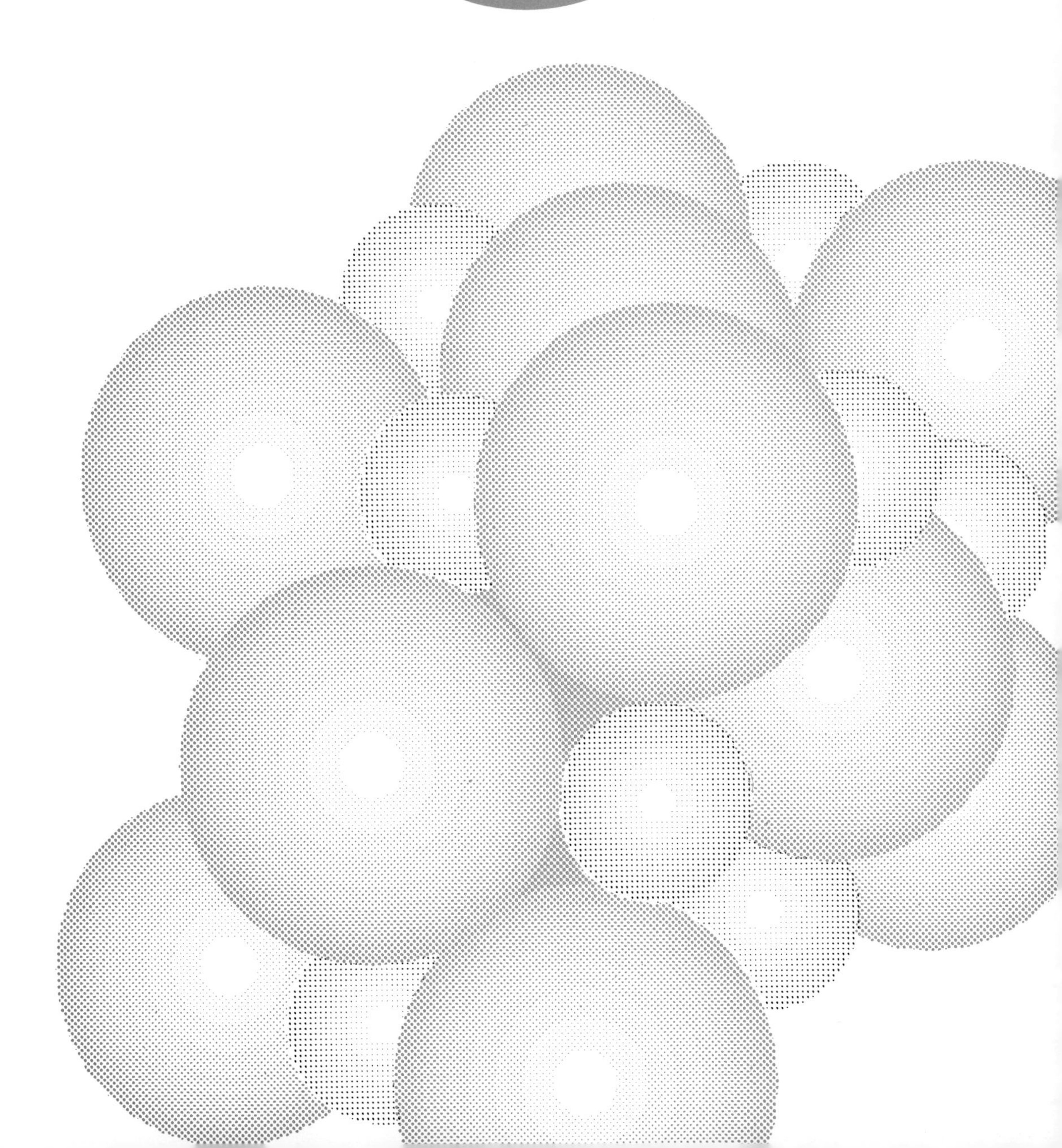

ASSUMED KNOWLEDGE AND UNDERSTANDING

Before starting on Unit 5 you should know and understand:

the reason for elements in the periodic table being arranged as they are

the structure of the atom

covalent bonding

ionic bonding

the reason for all ionic substances being solid at room temperature while covalent substances may be solid, liquid or gas

reactions of elements

reactions of oxides

the molar volume of gases.

OUTLINE OF THE CORE AND EXTENSION MATERIAL

As you progress through Unit 5 you should, at least, try the fundamental content listed under **core** and, if possible, pick up the extra and/or more difficult content listed under **extension**.

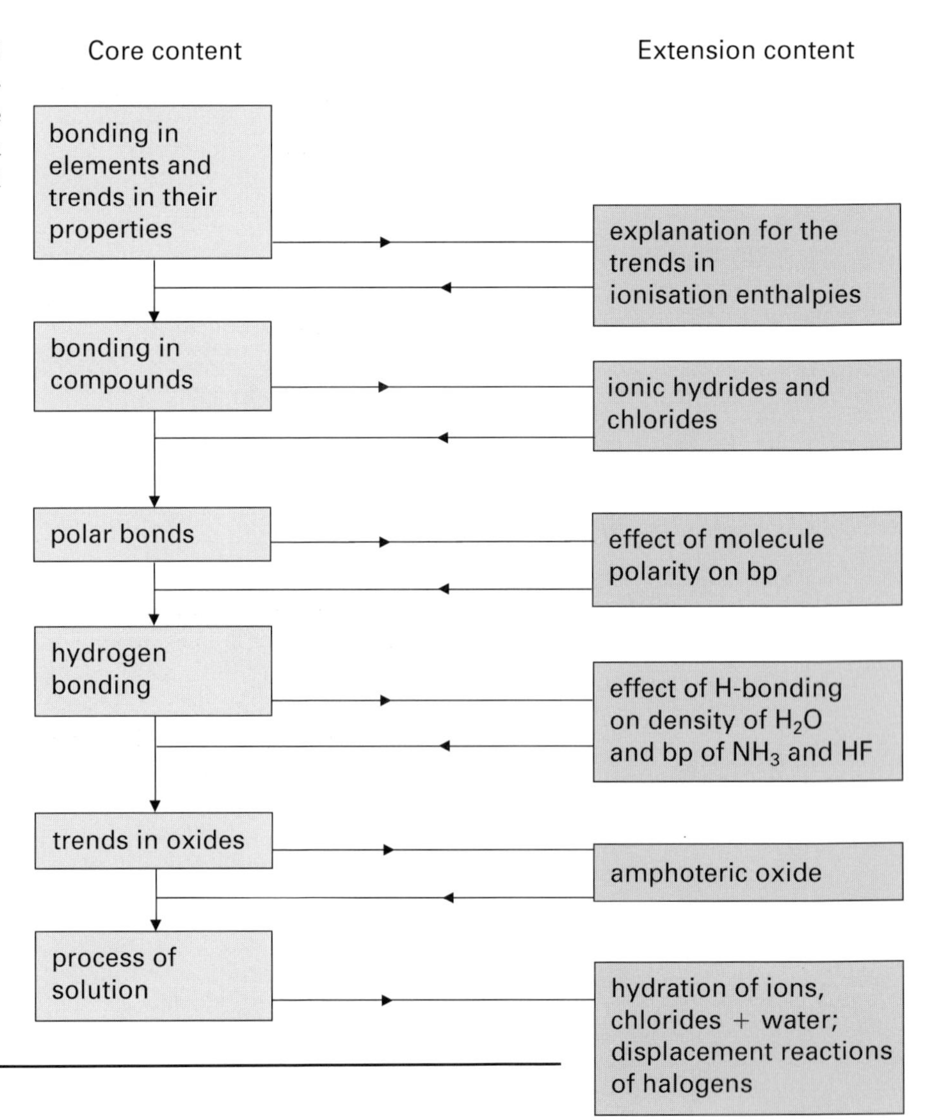

5 FROM BONDS TO BEHAVIOUR

5.1 The periodic table

The essential feature of the periodic table is that, when elements are arranged in order of **increasing atomic number**, similarities in physical and chemical properties are repeated at regular (periodic) intervals.

The elements in a vertical column of the table form a family or **group**. They are identified as belonging to groups 1, 2, 3, etc., as shown at the top of the table.

The horizontal rows across the table are called **periods**. The periods are labelled 1, 2, 3, etc., on the left side of the table.

Figure 5.1 shows a shortened form of the periodic table of elements.

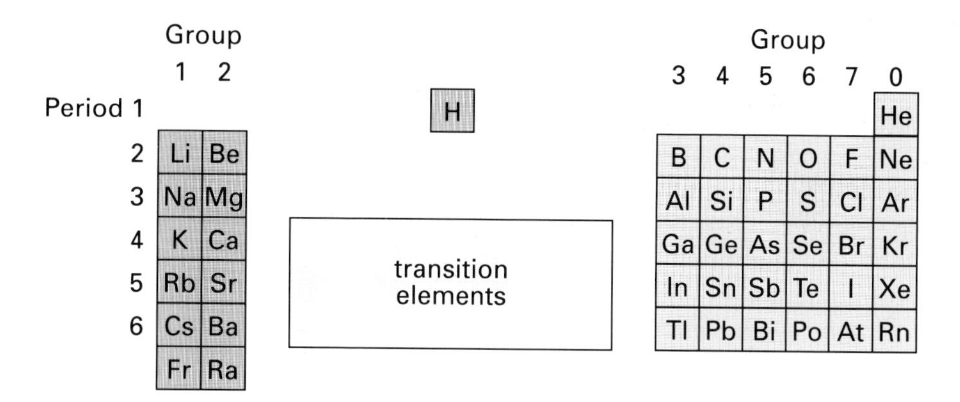

Figure 5.1 A shortened version of the periodic table

5.1.1 Electron arrangements

The atomic number is the number of protons (+) in the nucleus of an atom. Atoms are electrically neutral so the atomic number is also equal to the number of electrons (−) which fill the relatively large space around the nucleus. These electrons are arranged in successive shells of increasing distance from the nucleus and at increasing levels of energy.

Table 5.1 lists the electron arrangements for the first 20 elements in the periodic table.

Atomic number	Element	Electron arrangement
1	H	**1**
2	He	**2**
3	Li	2, **1**
4	Be	2, **2**
5	B	2, **3**
6	C	2, **4**
7	N	2, **5**
8	O	2, **6**
9	F	2, 7
10	Ne	2, **8**
11	Na	2, 8, **1**
12	Mg	2, 8, **2**
13	Al	2, 8, **3**
14	Si	2, 8, **4**
15	P	2, 8, **5**
16	S	2, 8, **6**
17	Cl	2, 8, **7**
18	Ar	2, 8, **8**
19	K	2, 8, 8, **1**
20	Ca	2, 8, 8, **2**

Table 5.1 Electron arrangements

H He

Hydrogen Helium

Figure 5.2 Electrons in hydrogen and helium (Period 1)

From the table we can see that, as one moves along a period, the outer shells (or levels) of atoms are being gradually filled up with electrons. The electrons may be thought of as occupying regions of space called **orbitals**. Each orbital can contain up to two electrons. An orbital containing a single electron is half full. The electrons in hydrogen and helium atoms may be considered to occupy a single orbital as shown. This orbital is the first shell. Atoms of elements 3–10 form Period 2. These atoms may be thought of as having four outer orbitals (shell 2) which can accommodate up to eight electrons. Atoms of elements in Period 3 have a third shell of four orbitals which gradually fill up with electrons to a maximum of eight.

The accommodation of outer electrons in orbitals can be represented as in Figure 5.3. Orbitals vary in shape; for simplicity, they are shown here as circles. Electrons tend to occupy separate orbitals before pressure of numbers forces them to pair up.

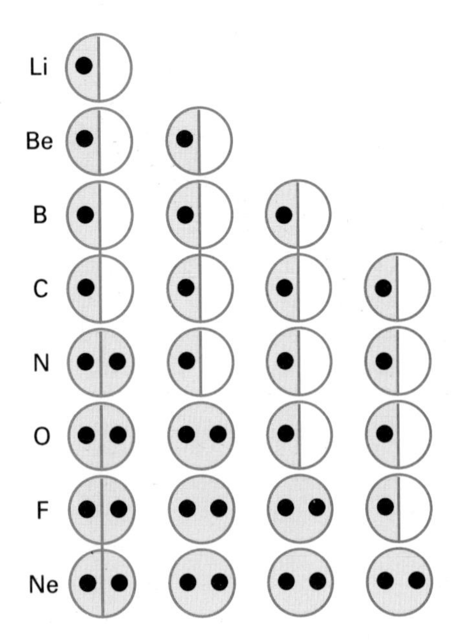

Figure 5.3 Distribution of outer electrons in elements of Period 2

5.2 Bonding in elements

Having considered the outer electron structure of the individual atom, we can now look at what happens when two or more atoms come together. When atoms moving at high speed collide, they bounce apart through repulsion.

However, if the kinetic energy of the atoms is small, attractive forces prevail and the atoms tend to cling together weakly. Such forces are called **van der Waals forces**, after the Dutch scientist who first suggested their existence.

How can one neutral atom attract another neutral atom? Atoms, even those with little energy, are in constant vibration and their electron clouds oscillate to and fro. This causes a **temporary** distortion of the symmetrical shape and creates a small imbalance in the distribution of electric charge. Some slight electrostatic attraction, between similarly deformed atoms, then occurs.

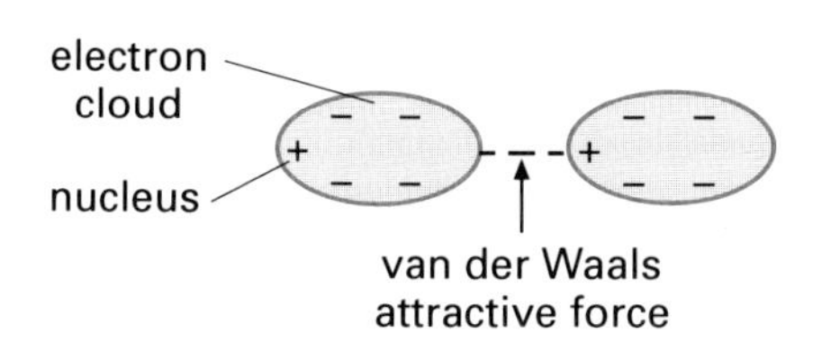

Figure 5.4 Attraction of atoms

Free and isolated atoms are rarely found in nature. Instead atoms of most elements are combined with one another, at normal temperatures, to form larger structural units. A notable exception to this rule is the family of elements known as the noble gases.

5.2.1 The noble gases

Of the 90 or so elements found in the earth's crust only eleven are gases at normal temperature and pressure. Six of these gases are the elements of Group 0 of the periodic table, collectively known as the noble gases.

helium	He	krypton	Kr
neon	Ne	xenon	Xe
argon	Ar	radon	Rn

The noble gas family

Helium is a very light gas and is used to fill balloons and airships. It is safer than hydrogen, which is lighter but flammable.

Figure 5.5 A helium-filled airship; helium, though not as light as hydrogen, is far safer in use

Neon glows reddish orange in electric discharge tubes and gives its name to the advertising signs which brighten the night scene of cities. **Argon** is used as an unreactive gas inside electric light bulbs, to stop oxidation of the metal filament. It also provides an unreactive atmosphere for arc welding, where metals have to be joined, and for growing crystals of silicon and germanium to use in electronic equipment. **Krypton** and **xenon** have been found to react, with fluorine and oxygen, to form compounds. The noble gases are therefore not as totally inert as once believed. **Radon** is a radioactive gas and is dealt with in Unit 8.

5.2.2 Cohesive forces in the noble gases

The atoms of noble gases do not combine with each other to form larger entities such as diatomic molecules. We say the noble gases are **monatomic**.

This reluctance to combine must be associated with the fact that the outermost shells of all noble gas elements contain eight electrons except for helium in which there are only two outer electrons. Two electrons in shell 1, or eight electrons in shell 2 or 3, therefore, appear to represent chemical stability.

The noble gases do not form covalent or ionic bonds between their atoms and yet all of the noble gases will condense to liquids and eventually form solids, provided the temperature is sufficiently reduced. Helium in fact has the lowest bp (4 K) and mp (2 K) of any element.

The fact that it is possible, although difficult, to liquefy and solidify the noble gases implies that forces exist between atoms to hold them together in the liquid and solid state. These weak cohesive forces are the van der Waals attractions.

All other elements (in the atomic number range 1–20) have fewer than eight electrons in their outer shells, and hydrogen of course has fewer than two. To put it another way, all elements other than noble gases have **orbital vacancies** in their outer shells. What effect do these vacancies have on the ability of atoms to become strongly attracted to other atoms of the same element?

atom of noble gas

van der Waals force

Figure 5.6 Forces between atoms in a noble gas

5.3 Covalent bonding

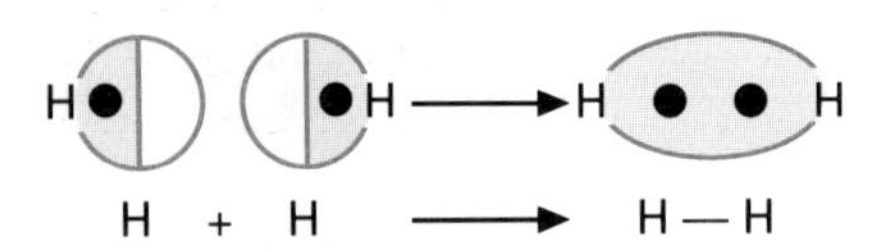

Figure 5.7 Formation of covalent bond between hydrogen atoms

The smallest atom is the hydrogen atom. Each H atom has one electron, which half-fills its single orbital. When two H atoms come together, the nucleus of each is able to attract the electron of the other, so that the two electrons are mutually shared by the two nuclei. This simultaneous attraction of two electrons to two nuclei lowers the energy of the system. We say a **covalent** bond has been formed.

Hydrogen atoms therefore join in pairs to form diatomic H_2 molecules. The requirement for a covalent bond appears to be an unpaired electron in a half-filled outer orbital. Fluorine is similar to hydrogen in this respect. It differs from hydrogen in having three filled outer orbitals, in addition to the half-filled orbital. These orbitals are arranged in space in such a way as to minimise the repulsion between the electron pairs. Fluorine atoms can combine, by mutual sharing of an electron pair, to form an F_2 molecule in which the F atoms have no capability for further covalent bonding.

Oxygen and nitrogen, with six and five outer electrons respectively, have two and three half-filled orbitals. Overlapping of these half-filled orbitals allows the sharing of four and six electrons respectively. Double and triple covalent bonds are produced between the pairs of atoms.

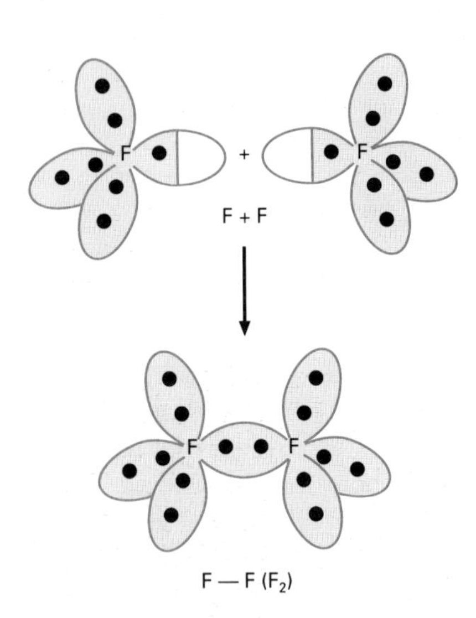

Figure 5.8 Formation of the fluorine molecule

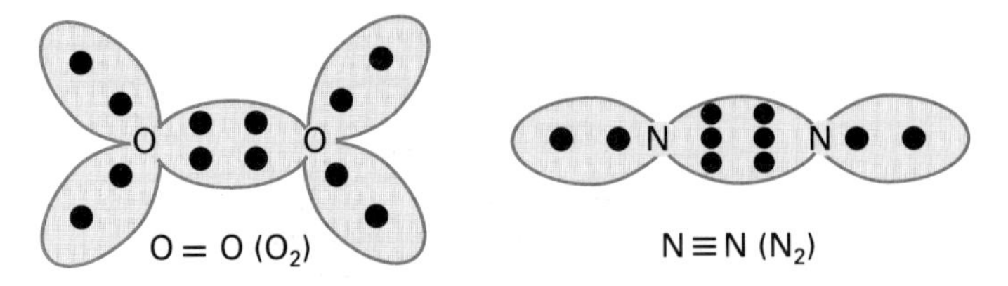

Figure 5.9 Formation of oxygen and nitrogen molecules

Figure 5.10 Polyatomic molecules

Other elements to form diatomic molecules are chlorine (Cl_2), bromine (Br_2) and iodine (I_2). Phosphorus and sulphur atoms combine to form **polyatomic** (more than two atoms) molecules. Two forms of phosphorus exist – the stable form, red phosphorus, is a tetratomic P_4 molecule in which each phosphorus atom is covalently bonded to three other atoms. The molecule is tetrahedral in shape. Sulphur, in the solid state, exists as octatomic S_8 molecules which have a ring structure as shown in Figure 5.10.

All these diatomic and polyatomic molecular elements have strong covalent **intra**molecular (within molecule) bonding but weak **inter**molecular (between molecules) attraction in the form of van der Waals forces. The van der Waals forces arise due to the unsymmetrical distribution of charge in the molecules. At relatively low temperatures molecules move more slowly and are subject to less disruption through collisions. Van der Waals forces can therefore produce weak binding between molecules in the liquid or solid state.

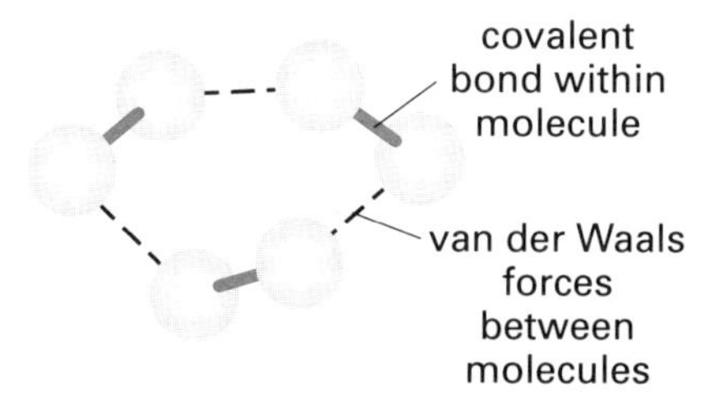

Figure 5.11 Bonding in molecular substances

5.4 Covalent networks

Elements whose atoms contain more outer electrons than orbital vacancies, i.e. whose atoms have outer shells which are more than half full, form small discrete (independent) molecules such as O_2, P_4, S_8, etc. Covalent bonding need not and often does not lead to the formation of such small structural units. **Carbon** has four outer electrons and generally four half-filled orbitals. These orbitals take up a **tetrahedral** arrangement when engaged in bonding within the element.

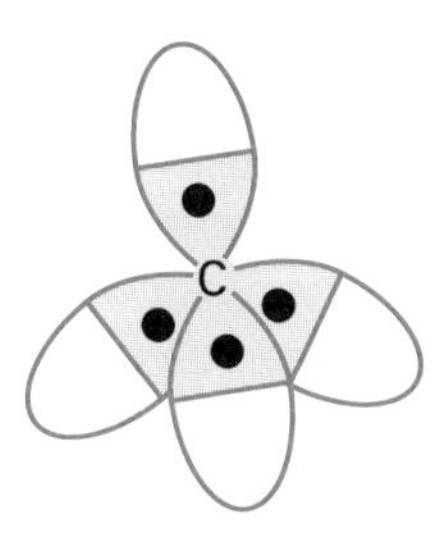

Figure 5.12 Outer orbital arrangement for carbon

Carbon element is found in two different crystalline forms, **diamond** and **graphite**.

Diamond

The structure of diamond is shown in Figure 5.13.

Diamond is a three-dimensional network of carbon atoms. Each atom is surrounded tetrahedrally by four neighbours and attached by single covalent bonds. The covalent network is infinite; diamond is simply **one** giant molecule. The tremendous rigidity which arises from the lattice of strong interlocking covalent bonds makes diamond one of **the hardest substances known**. Each carbon–carbon bond is an interconnected part of a giant network. To rupture a few bonds on the surface you have to break, theoretically, the whole piece of diamond. Consequently the most demanding cutting or drilling jobs are achieved by edging or tipping the tool with diamond. The material being cut or drilled cracks up, not the diamond.

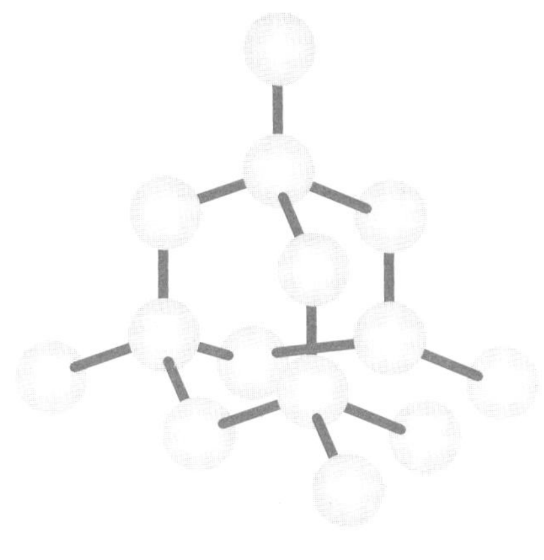

Figure 5.13 Arrangement of atoms in diamond

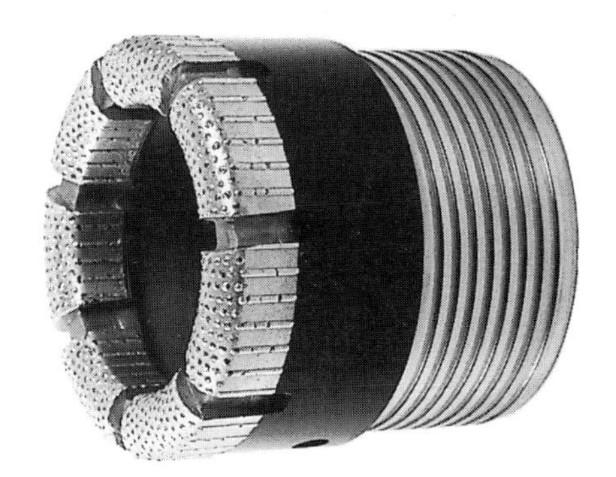

Figure 5.14 A diamond drill bit; the small diamonds are set into the surface of a tungsten carbide matrix which is only slightly less hard than the diamonds themselves

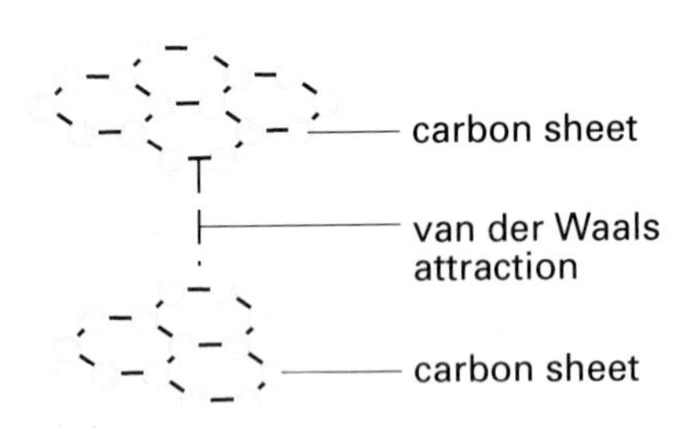

Figure 5.15 Arrangement of atoms in graphite

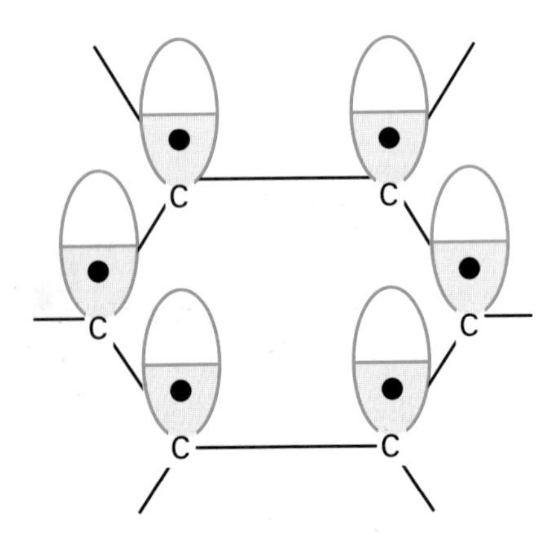

Figure 5.16 Non-bonded electrons in graphite

When diamond is heated, in the absence of air, to about 2070 K it converts to its other form, graphite.

Graphite

Graphite is a two-dimensional covalent network of carbon atoms. Each atom is surrounded in the same plane by three neighbours. Sheets of carbon atoms, arranged in interlocking hexagonal rings, are created. These sheets of carbon atoms are stacked in layers like a pack of cards and the layers held together merely by weak van der Waals forces.

Graphite is as remarkable for its softness as diamond is for its hardness. It is easily crumbled, due to the ability of the sheets to overcome van der Waals forces and slide over one another (as in the pack of cards). This capacity to provide slippiness enables graphite to be used as a lubricant between moving metallic parts. On a more familiar note, writing with a pencil involves continuously rubbing graphite layers from the pencil on to the surface of the paper.

In graphite, only three of carbon's four outer electrons are used in localised (fixed) single covalent bonds. The fourth electron is delocalised and free to move. As a result graphite, unlike diamond, is able to conduct an electric current in the direction of the layers. It is used in dry cells and as electrodes in industrial electrolysis, such as the extraction of aluminium.

Graphite can be converted into diamond but only at enormous temperatures and pressures. These 'synthetic' diamonds are not acceptable as gemstones but are used for industrial work.

Scientists have recently discovered a *third* stable form of carbon. It is an assembly of sixty carbon atoms bonded to one another to form a perfectly symmetrical, hollow sphere. Its structure can be compared to a football: the surface is divided into twenty hexagons and twelve pentagons. The 'seams' represent the C—C bonds, and the sixty intersections indicate the positions of the carbon atoms. This new C_{60} form of carbon has been named 'fullerene' after Fuller, the architect who designed the Expo Exhibition dome with the same shape.

Other elements with covalent networks

Silicon, like carbon, has four outer electrons and four half-filled orbitals. It therefore has the capacity to form three-dimensional networks, using covalent bonds. Silicon is a hard grey shiny solid with a structure similar to diamond.

Boron, another hard shiny solid, has three outer electrons. It is, however, a small atom and its nucleus is able to exert a strong pull on outer electrons. These electrons are used to form three single covalent bonds to other boron atoms. This produces an extended network structure.

5.5 Metallic elements

The vast majority of elements belong to the set called 'metals'. Metals are characterised by qualities such as good conductivity of heat and electricity, shininess, mechanical strength, malleability (able to be shaped) and ductility (able to be drawn into wire form).

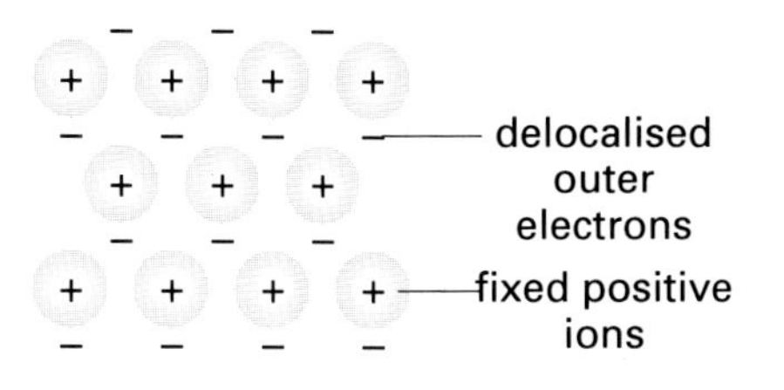

Figure 5.17 Bonding in a metal

The outer shells of 'metal' atoms are less than half-full, i.e. there are more outer orbital vacancies than outer electrons. For example, Na is $\frac{1}{8}$ full, Mg is $\frac{2}{8}$ full, and so on.

In a large cluster of metal atoms there are, therefore, sufficient empty orbitals to allow outer electrons to move freely from one atom to another and throughout the whole cluster. Outer electrons are, therefore, not fixed to any particular atom but **delocalised** throughout the structure. Each 'atom', having lost control of its outer electron(s), is now effectively a positive ion.

We can think of a metal as a lattice of regularly spaced positive ions surrounded by a 'sea' of mobile electrons. Metallic bonding is the **attraction** between the lattice of **positive** ions and the sea of **negative** electrons.

5.6 Variations in structure among elements

Elements can be assigned to one of four classes on the basis of chemical structure. The classes are as follows:

(a) metallic lattice
(b) covalent network
(c) molecules of two or more atoms
(d) atoms

(a) Elements with outer electron shells which are **less than half full** show (with the exception of boron) **metallic** properties. Their loosely held delocalised outer electrons hold the lattice of ions together. Metals make up about two-thirds of the 105 elements tabled.

(b) Elements whose outer electron shells are **half full** form **networks of atoms** held together by covalent bonds. The covalent network is sometimes referred to as a giant molecule of n atoms, where n is an indefinitely large number.

(c) Elements whose outer electron shells are **more than half full** bond covalently in pairs or occasionally larger groups of atoms to form **molecules**. Attraction between molecules is caused by van der Waals forces.

(d) Elements with **full** outer electron shells do not combine with each other and so remain as **single atoms**. Van der Waals forces are the only form of attraction between them. They form the group known as noble gases.

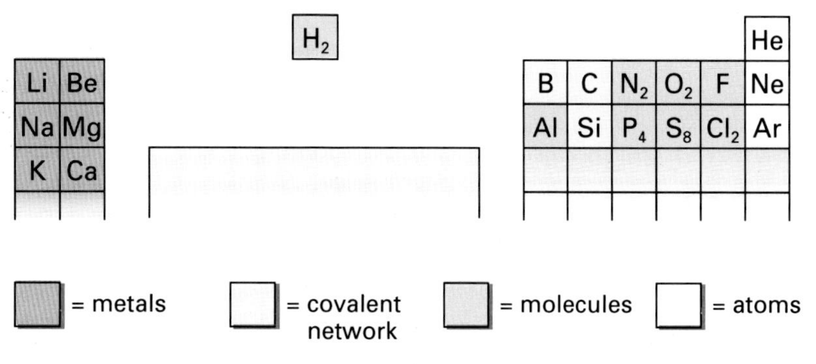

Figure 5.18 Variation in bonding for elements 1–20

5.7 Variation in physical properties for elements 1–20

Many of the properties of the elements vary, in a more or less regular fashion, as we proceed from left to right across a period or from top to bottom down a group. Most of these variations can be explained in terms of the electron structure of the elements.

5.7.1 Covalent radius

Atoms, unlike snooker balls, do not have a definite size. A free, uncombined atom has no sharp boundary to its mobile electron cloud. The density of an electron cloud drops off with increasing distance from the nucleus and approaches zero at large distances. There is therefore no definite edge to any atom. However, there are techniques which make it possible to measure the distances between atoms which are covalently bonded.

For example, the distance between the centres of chlorine nuclei in a Cl_2 molecule is measured as 198 pm (1 pm = one picometre = 10^{-12} metre). This distance is known as the Cl—Cl **bond length**.

Half of this distance is then defined as the radius of a chlorine atom. Because the radius has been calculated from measuring the covalent bond length it is called the **covalent radius**.

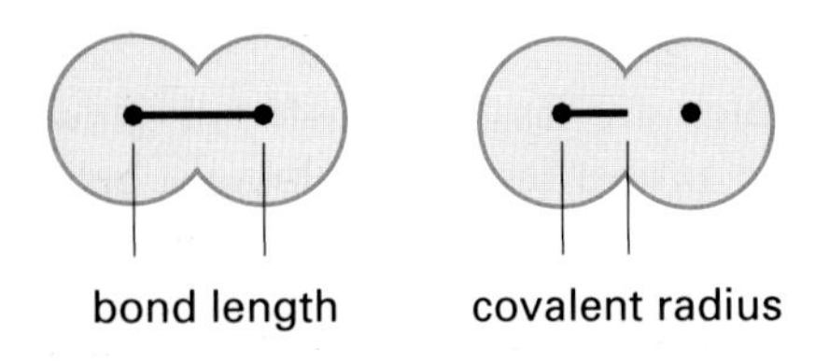

Figure 5.19 Bond length and covalent radius

The covalent radius of an element is half the distance between the nuclei of two of its covalently bonded atoms.

$$\text{The covalent radius of a Cl atom} = \frac{\text{Cl—Cl bond length}}{2} = \frac{198}{2}\text{ pm} = 99\text{ pm}$$

A covalent radius is therefore the effective radius of an atom in the direction of its covalent bond. It has been found that the single covalent bond length between two atoms of **different** elements is approximately equal to the sum of their individual covalent radii.

As noble gases do not generally form covalent bonds their constant radii cannot be measured. For the noble gases the inter-nuclear distance between adjacent (neighbouring) atoms in the solid state is halved to provide a radius called the van der Waals radius. This gives some measure of the noble atom's size but, for other elements, a van der Waals radius is always larger than the corresponding covalent radius.

The effective size of an atom depends on how it is bonded to other atoms. The observed bond lengths for a given pair of atoms may vary, due to multiple bonding. The more multiple the bonding, the shorter is the inter-nuclear distance and the lower the covalent radius. This is illustrated in Table 5.2.

Bond	Bond length/pm
O—O	148
O=O	114
N—N	148
N=N	122
N≡N	111

Table 5.2 Bond lengths of oxygen–oxygen and nitrogen–nitrogen bonds

In general, the covalent radius of an element corresponds to half the length of a **single** covalent bond. Thus the covalent radius of oxygen and nitrogen is not half the bond lengths in the diatomic molecules, O_2 and N_2, since these bonds are multiple bonds (O=O and N≡N).

Covalent radii for O and N are calculated from molecules where the bonds to be measured are single, e.g. in molecules of hydrogen peroxide (a bleach) and hydrazine (a rocket fuel).

H—O—O—H

H_2O_2

hydrogen peroxide

(H)(H)N—N(H)(H)

N_2H_4

hydrazine

The covalent radii of the elements in any period decrease with increasing atomic number. This is illustrated, for elements of Period 3, in Figure 5.20.

	Na	Mg	Al	Si	P	S	Cl
covalent radius/pm	157	136	125	117	110	104	99

Figure 5.20 Variation in covalent radius across Period 3

How does one account for this trend of diminishing size? As we move from one element to the next **across a period**, electrons are being added to the same outer shell and protons are being added to the nucleus. The outer electrons are attracted more strongly to the more positive nucleus. The net effect is a pulling in (contraction) of the electron cloud and a **reduction** in size.

As we proceed **down any group** the size of the atom generally **increases**, as shown for Group 1 elements in Figure 5.21.

The increase in size going down the group is caused by each succeeding element having **one more shell** of electrons than the preceding element. The outer electrons are shielded from the positive attraction of the nucleus by the screening effect of inner shells. The additional space occupied by these loosely held outer electrons more than offsets the effect of the extra nuclear charge (more protons in nucleus). Each atom (electron cloud) is therefore bigger than one with a smaller number of shells.

	Electron arrangement	Covalent radius/pm
Li	2, 1	123
Na	2, 8, 1	157
K	2, 8, 8, 1	203
Rb	2, 8, 18, 8, 1	216
Cs	2, 8, 18, 18, 8, 1	235

Figure 5.21 Covalent radii of Group 1 elements

5.7.2 Ionisation enthalpy

The formation of positive ions is not the result of atoms willingly and spontaneously shedding electrons. Energy is always required to pull electrons completely away from the attraction of the nucleus.

A lithium atom contains three electrons. The energy required to remove the least tightly held electron is called the **first ionisation enthalpy** (enthalpy is a more precise label for this type of energy and will be dealt with more fully in Unit 6).

First ionisation enthalpy of an element is the energy required to remove one electron from each atom in a mole of isolated gaseous atoms.

The equation for the removal of the first electron of lithium is:

$$Li(g) \longrightarrow Li^{+}(g) + e^{-}$$

Ionisation enthalpies are measured in units of kilojoules per mole of atoms. For lithium, first ionisation enthalpy = 526 kJ mol^{-1}.

In any **period** of the periodic table there is a gradual **increase** in first ionisation enthalpy with increasing atomic number. Figure 5.22 charts values for the elements which make up the first three rows (Periods 1, 2 and 3).

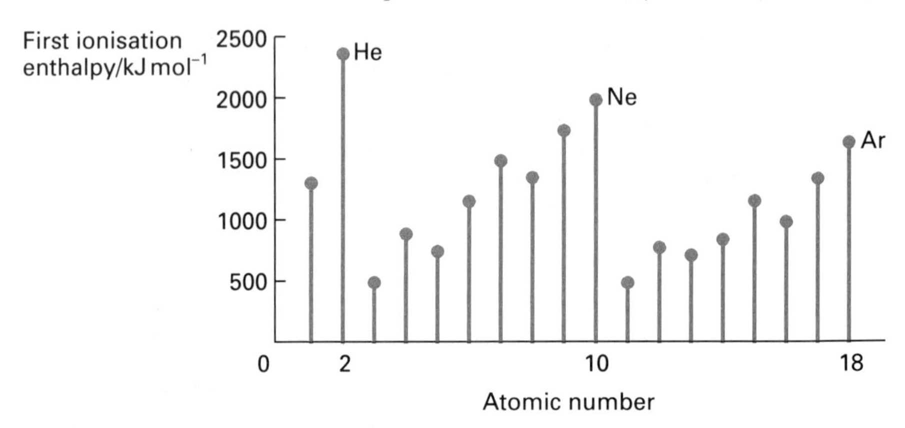

Figure 5.22 Ionisation enthalpies for elements 1–18

In Period 2, for instance, lithium has the lowest ionisation enthalpy and neon the highest. How can we explain this general upward trend? As we proceed along a row, electrons are being added to the same outer shell at the same time as protons are added to the nucleus. The extra electrons, being in the same shell, do not effectively shield one another from the nucleus and consequently experience a stronger pull towards the nucleus. Just as this has the effect of reducing atomic size, so it also causes electrons to become harder to remove.

For elements in any column or **group** there is a gradual **decrease** in first ionisation enthalpy with increasing atomic number, as Figure 5.23 illustrates for Group 1.

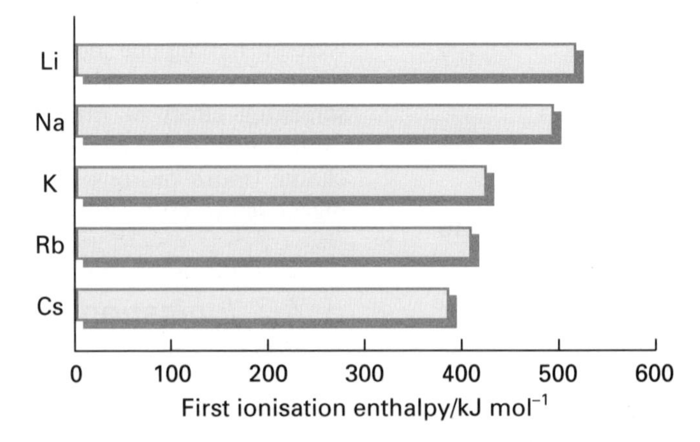

Figure 5.23 Ionisation enthalpies of Group 1 elements

When we compare elements of a vertical group, we are comparing elements with the same number of outer electrons but an increasing number of shells. As we go down the group, the atoms become larger and the outer electrons are further away from the nucleus. The electrons become easier to remove from the atom.

The size of an atom indicates the tightness with which electrons are held to the nucleus. Elements with larger covalent radii have lower ionisation enthalpies and vice versa.

Figure 5.24 The doors of a sports centre, controlled by photocells, open automatically for a laden visitor

Caesium and potassium have low first ionisation enthalpies. When these elements absorb light energy, the outer electrons are readily released and produce a detectable current. Caesium and potassium are used in photocells which control the opening of automatic doors.

5.8 Metallic character in elements

Elements which display properties such as shininess, good conductivity of heat and electricity, malleability and ductility are said to be metals. Elements which do not show these characteristics are described as **non-metals**. The metallic elements tend to be found to the left and bottom of the periodic table, whereas the non-metals are to be found to the right and top of the table. In several cases it is hard to assign an element to either category. We sometimes use the term **metalloid** to describe those elements which are difficult to classify one way or the other.

What feature in an element's structure determines whether it will behave like a metal or a non-metal? Elements which are metallic appear to have two characteristics in common.

(a) Their atoms have **empty orbitals** available to their outer electrons.
(b) Their outer electrons have **low ionisation enthalpies**.

There seem to be the two conditions which lead to metallic bonding. The picture of a solid metal is that of an array of positive ions embedded in a pool of mobile electrons. Such electron delocalisation is most easily achieved by atoms having empty available orbitals and weakly bound outer electrons.

These free-ranging electrons are responsible for the properties which metals display and which were mentioned earlier. Reduce the easy movement of electrons from one atom to another and elements begin to lose some of their metallic character.

5.8.1 Variation in metallic character across a period

Figure 5.25 indicates the classification of elements in Periods 2 and 3.

Period								
2	Li	Be	B	C	N	O	F	Ne
3	Na	Mg	Al	Si	P	S	Cl	Ar

= mainly metal

= metalloid

= mainly non-metal

Figure 5.25 Metallic and non-metallic character of elements 3–18

As we move across a row metallic character **decreases**. Why? There are two reasons. Firstly, as the effective nuclear charge on the atom increases, the **ionisation enthalpy** for outer electrons rises rapidly. It becomes more and more difficult for atoms to release the electrons needed for delocalised metallic bonding. Secondly, as outer orbitals are gradually filled up, there are fewer empty orbitals available. This restricts the free movement of outer electrons and the opportunity for metallic bonding.

There is no abrupt changeover from metallic to covalent bonding. This is shown by the presence of the metalloids which show both metal and non-metal behaviour.

To sum up, as we cross the periodic table,

—— outer e^- become harder to remove ——→

—— outer orbitals fill up ——→

—— delocalisation of e^- becomes harder to achieve ——→

—— **metallic character decreases** ——→

5.8.2 Variation in metallic character down a group

Figure 5.26 outlines the classification of elements in Groups 4, 5 and 6.

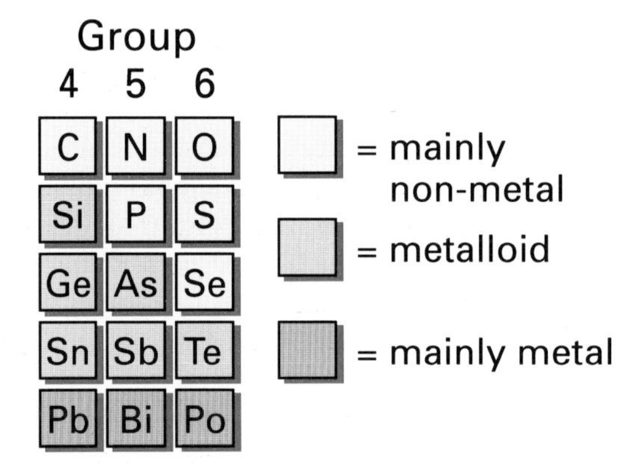

Figure 5.26 Metallic and non-metallic character of elements in Groups 4–6

As we descend families or groups in the table, metallic character **increases.** This can be explained again in terms of ionisation enthalpies and vacant orbitals. The size of the atom increases down the group and its outer electrons become more loosely held. In addition, larger atoms have a wider selection of vacant outer orbitals available to accommodate electrons. Both factors contribute to the increasing delocalisation of outer electrons and the increasing opportunity for metallic bonding.

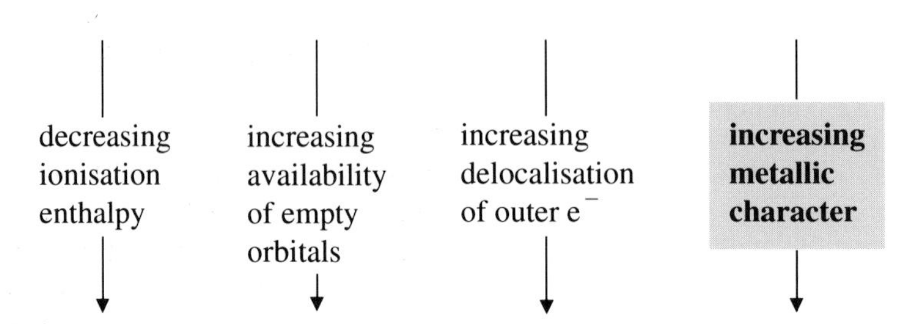

5.9 Variation in densities

The density of a substance is its **mass per unit volume**. Elements with heavy atoms which are packed closely together will have high densities. The closeness of the packing depends on the strength of binding forces which may be metallic, covalent or van der Waals.

Variation in the densities of the elements across Period 3 is outlined in Figure 5.27.

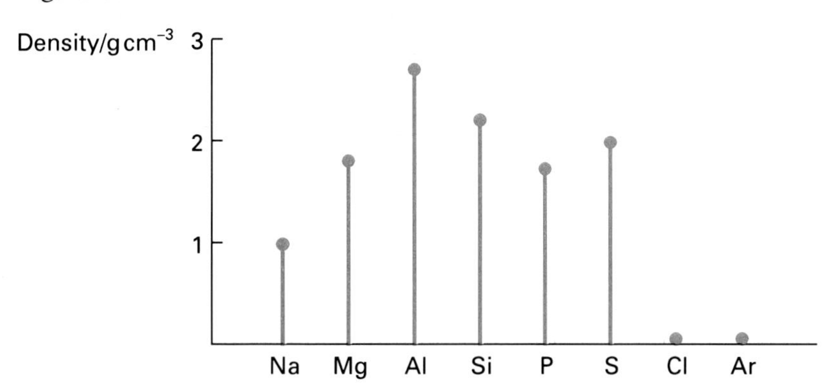

Figure 5.27 Densities of elements in Period 3

Na, Mg and Al are all metals. The metallic bonding within each is due to the attraction between the delocalised outer electrons and the fixed positive ions. The larger the number of outer electrons, the larger is the charge on each ion. This produces stronger attraction.

As we move from Na to Mg to Al the bonding become stronger. The ions are pulled more closely together and this leads to an increase in density.

P and S have heavier and smaller atoms than Si so we might expect their densities to be higher. However, Si is a covalent network of tightly packed atoms, whereas P and S form P_4 and S_8 molecules which are only loosely held by van der Waals forces.

The van der Waals attraction between the Cl_2 molecules and between Ar atoms is very slight. These elements are gases at NTP and occupy a large amount of space (about 24 litres per mole). Cl_2 and Ar, therefore, have very low densities.

There is a general **increase** in density as one descends a group of elements. Along with the rise in atomic mass there is an increase in metallic bonding or van der Waals attraction. These effects lead to closer packing and higher density.

5.10 Variation in melting and boiling points

When a solid element melts, the attraction between its particles is sufficiently reduced to let it take up the less tightly bound liquid arrangement. When the liquid boils, attraction between particles is more or less removed.

By comparing the melting points and boiling points of elements, we obtain some idea of how strong the attractive forces are between their particles in solid or liquid states. The stronger this binding, the greater the energy that must be applied to overcome it. This means that higher temperatures must be reached before change of state takes place.

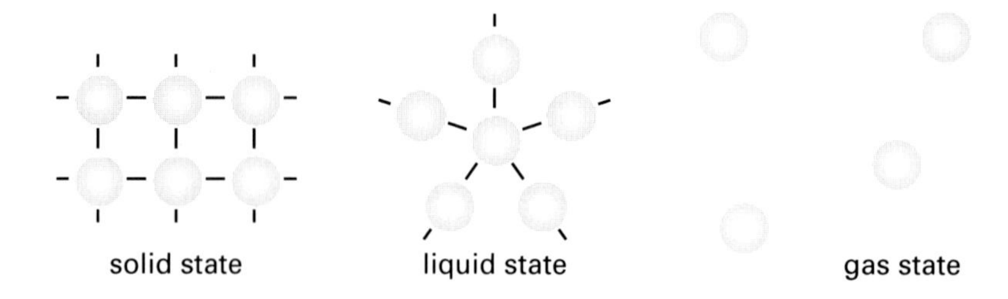

Figure 5.28 Attraction between particles

5.10.1 Melting and boiling points across a period

Figure 5.29 outlines the variation in melting and boiling points for elements in Period 3.

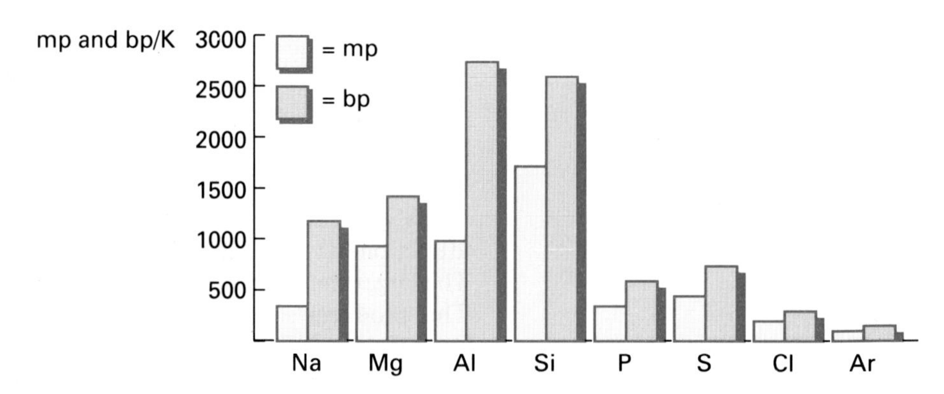

Figure 5.29 Melting and boiling points of elements in Period 3

Melting points

Whether a melting point is high or low is determined by the strength of the attractive forces holding the particles of the solid together. The rise in melting point across the first three elements, all of them bound by **metallic bonding**, can be put down to the increasing strength of the bonding Na → Mg → Al, as explained (for densities) in section 5.9.

The melting point reaches a maximum at silicon and this is not surprising. Silicon is held together by **strong covalent bonds** in a giant **network** of silicon atoms. It cannot be melted without breaking these covalent bonds.

After silicon the melting point drops rapidly. The last four elements form solids in which the molecules (single atoms in the case of argon) are held together by **van der Waals forces**. These weak intermolecular forces are usually easily overcome. The variation in melting point between phosphorus (P_4), sulphur (S_8) and chlorine (Cl_2) implies that van der Waals forces increase as the size of the molecules increases.

Boiling points

The boiling point depends on the strength of the bonds which are broken in converting liquid to gas. The variation in boiling point across the row follows that for melting point as first metallic bonding, then covalent bonding and finally van der Waals forces are overcome, to bring about separation of atoms or molecules.

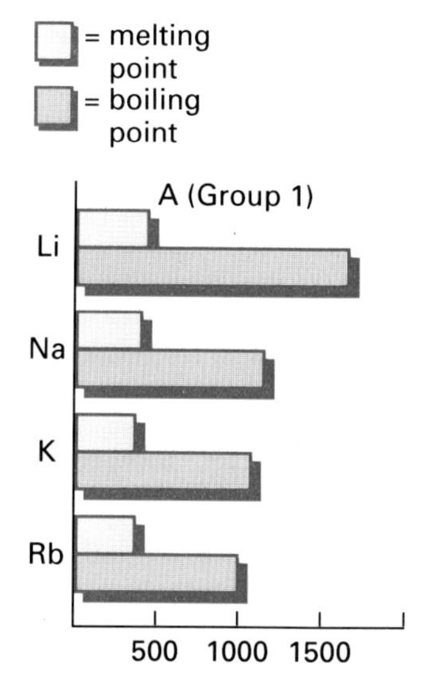

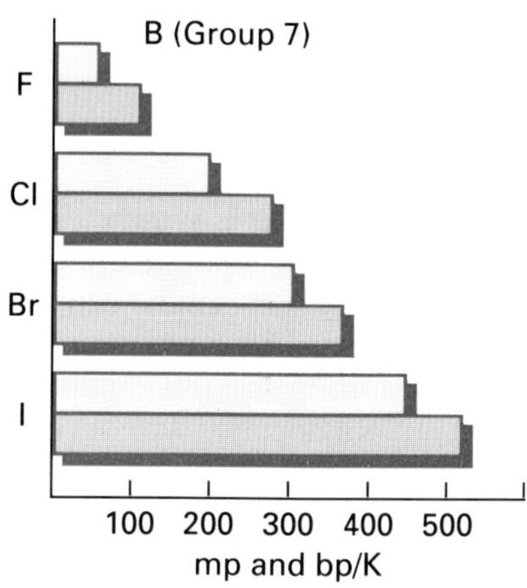

Figure 5.30 Variations in mp and bp

One must be careful to distinguish between the attractive forces which act **between** molecules and the attractive forces **within** molecules of the same substance. Table 5.3 indicates the weakness of van der Waals forces between Cl_2 molecules and the strength of covalent bonds within Cl_2 molecules.

Change	Bonds broken	Energy needed/kJ mol^{-1}
$Cl_2(l) \longrightarrow Cl_2(g)$	van der Waals	23
$Cl_2(g) \longrightarrow 2Cl(g)$	covalent	243

Table 5.3

5.10.2 Melting and boiling points down groups

The trend in melting and boiling points down a group depends on the nature of bonding within the elements of the group. Group 1 contains elements whose atoms are held by metallic bonding. Group 7 contains non-metals which exist as diatomic molecules. The attraction between these molecules is due to van der Waals forces. The variations within these groups, on opposite sides of the periodic table, are shown in Figure 5.30.

Figure A implies that Li is more strongly bonded than Rb. The smaller Li^+ ions in the metallic lattice of Li exert a greater pull on their delocalised electrons than do the larger Rb^+ ions in Rb.

Figure B implies that iodine is more strongly bound than fluorine. This is because the van der Waals forces between the larger I_2 molecules are stronger than those between the smaller F_2 molecules.

5.11 Bonding in compounds

Two electrons shared between two atoms produce a covalent bond between these atoms. When two **like** atoms (atoms of the same element) share an electron pair as in the fluorine molecule, F_2, the electrons are distributed **equally**. We describe such a bond as **non-polar**. At the other extreme there may be no sharing of electrons. In lithium fluoride the bonding electron originally belonging to lithium is effectively **transferred** to the outer orbital of fluorine. The ions formed, Li^+ and F^-, are of opposite charge and are attracted to each other. The attraction is referred to as an **ionic** bond.

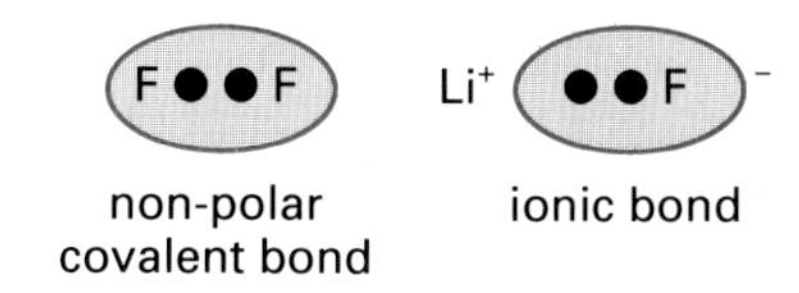

Figure 5.31 Extremes of bonding

Most covalent bonds fall somewhere between the two extremes of non-polar and ionic, i.e. most bonds are **polar**. In a polar bond one of the atoms has a stronger attraction for the bonding electrons than the other atom. This unequal pull on the shared electrons causes an imbalance in the distribution of charge within the bond. The more electron-greedy atom acquires more than its half share of the electrons and takes on a **partial negative charge**, δ-. (δ is the lower case Greek letter 'delta' and means 'a small amount'.) The other atom, with a less than half share of the electrons, acquires a **partial positive charge**, δ+. These signs are placed in formulae over the atoms in the covalent bond to show the bond's polarity or partial ionic character.

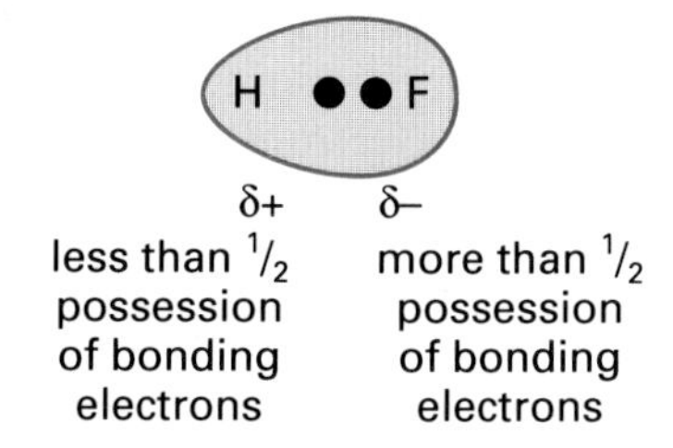

Figure 5.32 Polarisation of H—F bond

The bond in hydrogen fluoride is an example of a polar bond and is written $\overset{\delta+}{H}—\overset{\delta-}{F}$.

The capacity of an atom to attract electrons to itself in a chemical bond

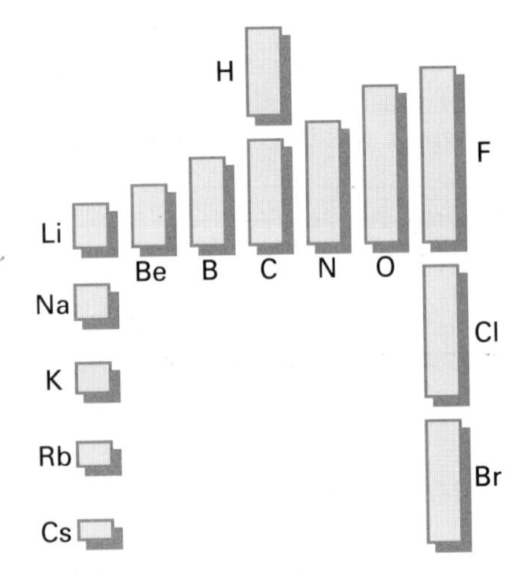

Figure 5.33 Comparison of electron-attracting abilities

varies from element to element. We cannot measure, directly, any single property that would give us a reliable numerical value for this electron-attracting ability. However, atoms that have **high ionisation enthalpies** also show **strong attraction** for electrons in bonds. Conversely, atoms with low ionisation enthalpies exert relatively little pull on bonding electrons.

In Figure 5.33, relative electron-attracting abilities are outlined for hydrogen, elements across Period 2 and elements down Groups 1 and 7.

Across the period, moving from left to right, there is a steady **increase** in attraction for bonding electrons. Electron attractiveness is lowest for the most metallic elements and highest for the most non-metallic elements. It can also be seen that electron attraction falls away with the increasing atomic size down any group. Taking both observations into account, we can say ability to attract bonding electrons is lowest for large Group 1 metals such as caesium and greatest for small non-metals such as fluorine. This is what we would expect, given that caesium has a low first ionisation enthalpy, and fluorine a high ionisation enthalpy. We can thus predict that elements which are **widely separated** in the periodic table, and so widely different in their electron attractiveness, will form compounds which are predominantly **ionic.**

5.12 Ionic compounds

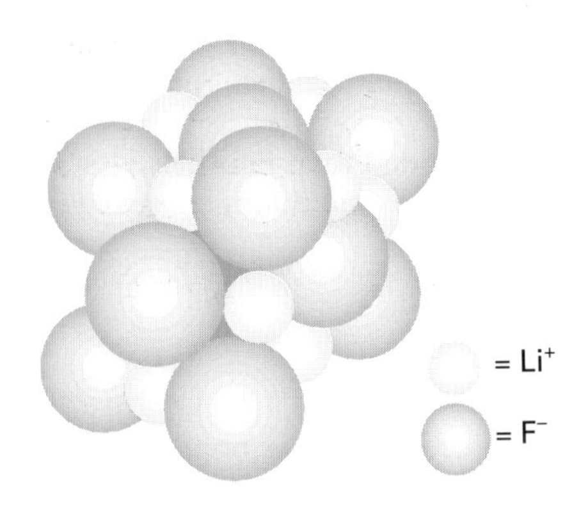

Figure 5.34 Arrangement of ions in lithium fluoride

In a pure covalent bond the distribution of bonding electrons is symmetrical. In an ionic bond the distribution is skewed or biased so completely that separate positive and negative ions are produced. Ionic solids contain positive ions surrounded by negative ions and vice versa. This maximises the attraction between oppositely charged ions, and minimises the replusion between similarly charged ions. As a result, each ion is associated with several (often six) ions of opposite sign and not with any one other ion in particular. The ions pack together to form an ionic lattice of indefinite size.

In **lithium fluoride**, for instance, each Li^+ ion is surrounded by six F^- ions, and each F^- ion is surrounded by six Li^+ ions. This system of packing is fairly common. However, the packing method used in other ionic compounds may have to be different, as it all depends on the relative sizes of positive and negative ions and their relative numbers. Different packing arrangements lead to a different crystalline shape for the ionic solid.

The formation of an ionic compound such as Li^+F^- arises from a large difference between the electron attracting powers of Li and F, so large in fact that it leads to complete charge transfer. The range of polarity in bonds between Period 2 elements and fluorine is outlined in Figure 5.35.

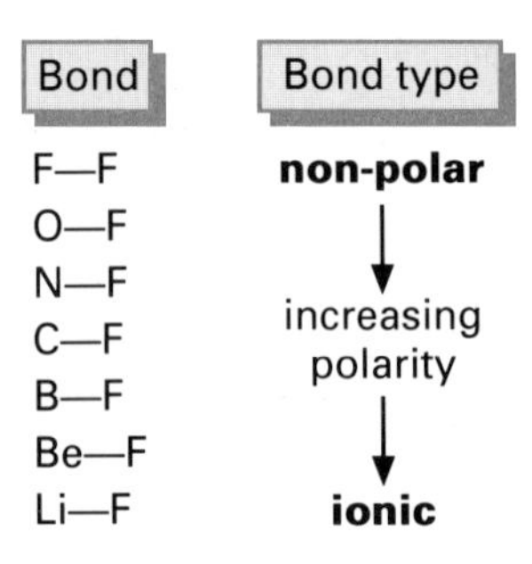

Figure 5.35 Polarity of bonds to fluorine

5.12.1 Ionic hydrides

A range of polarity is found in compounds formed between hydrogen and Period 2 elements. Hydrogen has average attraction for bonding electrons. Depending on how strongly the other element pulls electrons towards it, hydrogen may range from having a full negative charge to a fairly pronounced positive charge. The trend in the polarity of Period 2 hydrides is described in Figure 5.36.

Figure 5.36 Polarisation of hydrogen in Period 2 hydrides

The hydride ion H^- is found only in hydrides of metals with low attraction for bonding electrons. All Group 1 metals form **ionic** hydrides containing the **negative** hydrogen ion, e.g. Li^+H^-, Na^+H^-, K^+H^-, etc. In Group 2, beryllium and magnesium hydrides are partially ionic but predominantly covalent. The other metals in Group 2, calcium, strontium and barium, have larger atomic sizes and lower ionisation enthalpies. This makes it possible for transfer of the outer electrons to hydrogen and the formation of ionic hydrides, e.g. $Ca^{2+}(H^-)_2$.

Ionic hydrides are white solids. When they are melted and electrolysed, hydrogen gas is given off at the **positive** electrode. This confirms the presence of the H^- ion.

5.13 Covalent network compounds

Boron, carbon and silicon have a covalent network structure. These elements combine with each other and with other suitable elements to form compounds which are also covalent network solids.

Carbon and silicon both have four half-filled orbitals and can bond to each other indefinitely. When the two elements are heated together in an electric furnace, the compound **silicon carbide** (carborundum) is formed. It is a covalent network with a structure like diamond; each carbon bonds tetrahedrally to four silicon atoms and vice versa, as shown in Figure 5.37. To bring about physical disintegration of a covalent network solid is very difficult, as each atom is locked into a giant construction of covalent bonds. Silicon carbide is therefore an **extremely hard** material and finds wide use as an abrasive.

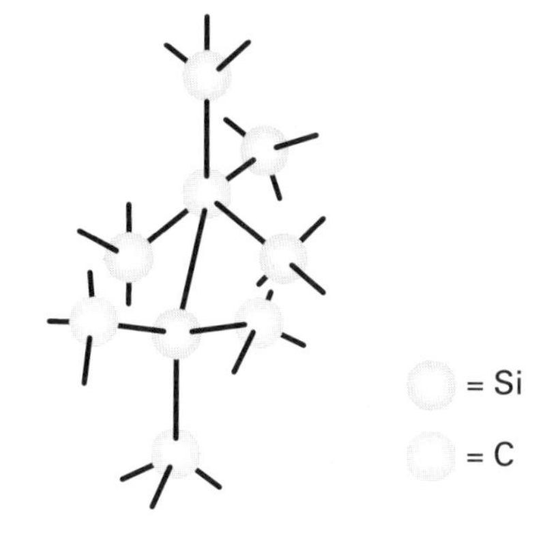

Figure 5.37 Part of the SiC network

A list of covalent network compounds manufactured for industrial and commercial use is given in Table 5.4.

Name of compound	Formula
silicon carbide	SiC
boron nitride	BN
boron carbide	B_4C
aluminium silicide	Al_4Si_3
silicon nitride	Si_3N_4
aluminium carbide	Al_4C_3
aluminium nitride	AlN

Table 5.4 A selection of covalent network compounds

Figure 5.38 Silicon carbide grit is used to coat abrasive discs and papers

A covalent network need not be restricted to two elements in combination. Aluminium borosilicide (AlBSi) is an example of a three-element network, while very strong ceramic materials have been produced from Si-Al-O-N networks called sialons.

5.14 Discrete molecular compounds

Elements having outer shells which are **half full** or **more than half full** of electrons can share electron pairs to form discrete (separate and independent) molecules: molecules of a finite (limited) size containing atoms linked by covalent bonds.

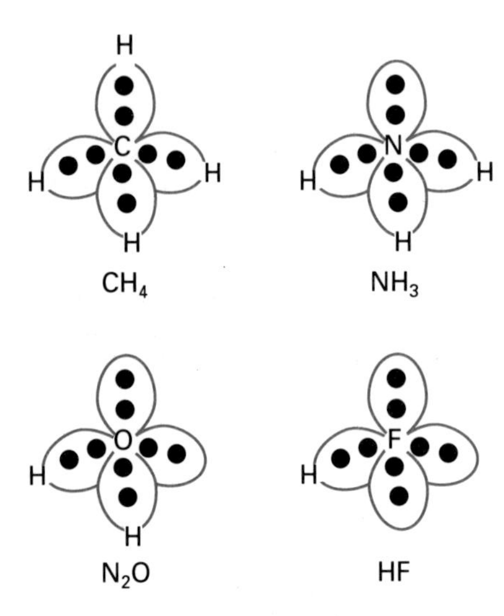

Figure 5.39 Molecular hydrides from Period 2

A hydrogen atom with its single half-filled orbital can overlap and share an electron pair with each of the four half-filled orbitals on carbon, the three half-filled orbitals on nitrogen, the two half-filled orbitals on oxygen and the single half-filled orbital on fluorine. Covalent bonds are formed and small discrete molecules produced (Figure 5.39).

5.14.1 Polarity of molecules

The covalent bonds between unlike atoms are all polarised ($\delta+/\delta-$) to a greater or lesser extent because of unequal sharing of electrons. The bonds formed between hydrogen and carbon, nitrogen, oxygen and fluorine are all polar bonds. Does this mean that CH_4, NH_3, H_2O and HF are all polar molecules? Not necessarily so.

A molecule has to be **unsymmetrical** in its distribution of charge in order to be polar. A polar molecule contains polar bonds which are not symmetrically opposed to each other. If the polar bonds are in fact symmetrically opposed, they cancel each other out and the net polarity of the molecule is zero.

The polarity of a molecule can be determined by measuring a quantity known as its 'dipole moment'. All we need know is that **the larger the dipole moment, the greater the polarity of the molecule**.

Table 5.5 outlines the shapes and polarities (dipole moments) of some selected molecules.

Table 5.5 Polar and non-polar molecules

Molecule	Bond arrangement	Shape	Dipole moment/ cm × 10^{-30}
CH_4	H above C; C—H; H (hashed), H (wedge)	tetrahedral	0
NH_3	N with H (hashed), H (wedge), H	pyramidal	4.9
HCl	H—Cl	linear	3.6
HF	H—F	linear	6.1
CO	C≡O	linear	0.4
CO_2	O=C=O	linear	0
SO_2	O=S=O (bent)	bent	5.5
H_2O	H—O—H (bent)	bent	6.2

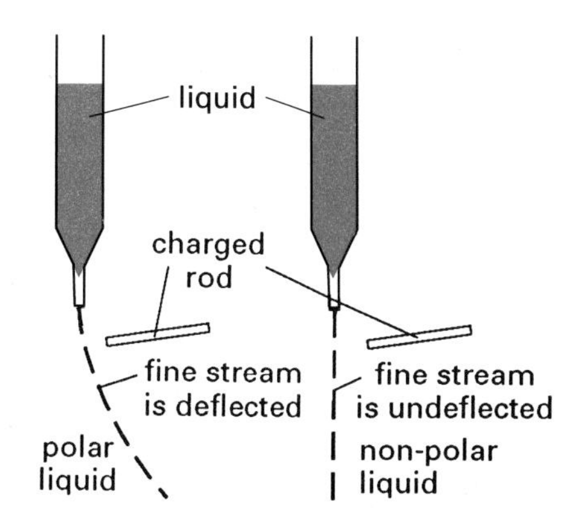

Figure 5.40 Experiment to detect polar liquids

Liquids whose molecules are polar can be distinguished from liquids containing non-polar molecules by a simple experiment. A plastic rod can be charged by rubbing it with a dry cloth. The charged rod is brought up close to a fine stream of test liquid. The stream will be **deflected** towards the rod if the liquid is **polar.** It does not matter whether the rod is charged positively or negatively. In every case the stream of polar liquid will be deflected by attraction (not by repulsion).

A fine stream of the liquid trichloromethane, $CHCl_3$, is deflected by a charged rod. A similar stream of tetrachloromethane, CCl_4, remains undeflected. The reason for the difference in behaviour can be traced to the arrangement of bonds in each molecule.

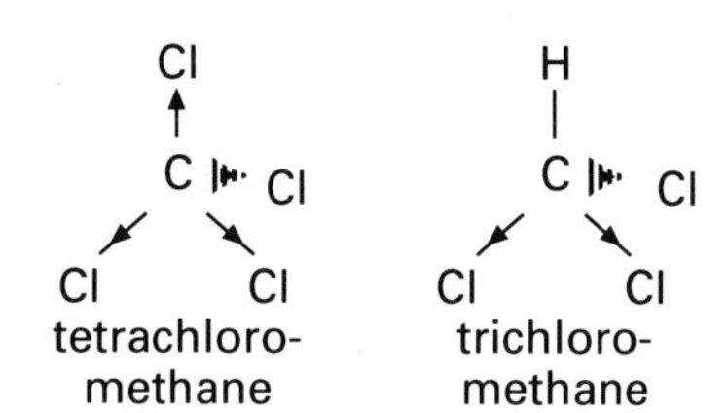

Figure 5.41 Symmetrical and unsymmetrical arrangements of polar bonds

The four polar bonds in CCl_4 are symmetrically arranged in space: the molecule is therefore non-polar. In $CHCl_3$, the single C—H bond throws the distribution of charge within the molecule out of balance. $CHCl_3$ is polar and its molecules are therefore deflected by a charged rod.

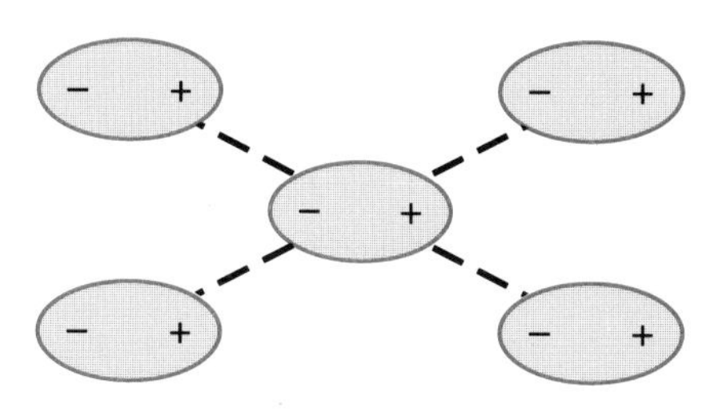

Figure 5.42 Attraction between polar molecules

5.14.2 Attraction between polar molecules

When discussing the intermolecular forces in **elements**, we referred to van der Waals forces as attractions arising from **temporary** polarisation of atoms and non-polar molecules. Polar molecules in compounds, however, possess separation of charge, i.e. **permanent dipoles**. They can align themselves so that electrostatic attraction will arise between the positive end of one molecule and the negative end of another, as shown in Figure 5.42.

The dipole–dipole attractions are included in the category of van der Waals forces. The extra intermolecular attraction between polar molecules makes them more difficult to separate than non-polar molecules of similar size and shape. As a result polar compounds tend to have higher boiling points.

CH_3

CH_3

1,4-dimethylbenzene
(*para*-xylene)
non-polar, bp = 411 K

CH_3 CH_3

1,2-dimethylbenzene
(*ortho*-xylene)
polar, bp = 417 K

Figure 5.43 Non-polar and polar molecules

5.15 Hydrogen bonding

Van der Waals forces are not the only type of intermolecular attraction. An extra strong attraction occurs between molecules in which hydrogen is bonded to the very small, highly electron-attracting atoms of **oxygen** or **nitrogen** or **fluorine**. These bonds are quite polar, with hydrogen at the positive end.

$$\overset{\delta-}{\mathrm{O}}—\overset{\delta+}{\mathrm{H}} \qquad \overset{\delta-}{\mathrm{N}}—\overset{\delta+}{\mathrm{H}} \qquad \overset{\delta-}{\mathrm{F}}—\overset{\delta+}{\mathrm{H}}$$

This positively charged hydrogen is attracted to the unshared electron pair on the O, N or F atom of the **adjacent molecule**. Being so small, the hydrogen atom can place itself between the two neighbouring O, N or F atoms, attract each of them and thus 'bond' them together.

This type of electrostatic attraction is called a **hydrogen bond** and is usually indicated by a dotted or dashed line. It is found, for example, in H_2O, NH_3 and HF. Here there is an unshared pair of electrons on the O, N or F atoms.

Figure 5.44 Hydrogen bonds (shown by broken lines)

The strengths (energies) of hydrogen bonds vary but they are, generally, about ten times as strong as van der Waals forces and about one tenth as strong as covalent bonds. Experimental evidence of hydrogen bonding comes from comparisons of the physical properties of hydrogen compounds. Figure 5.45 shows the boiling points of hydrides of elements in Groups 5, 6 and 7.

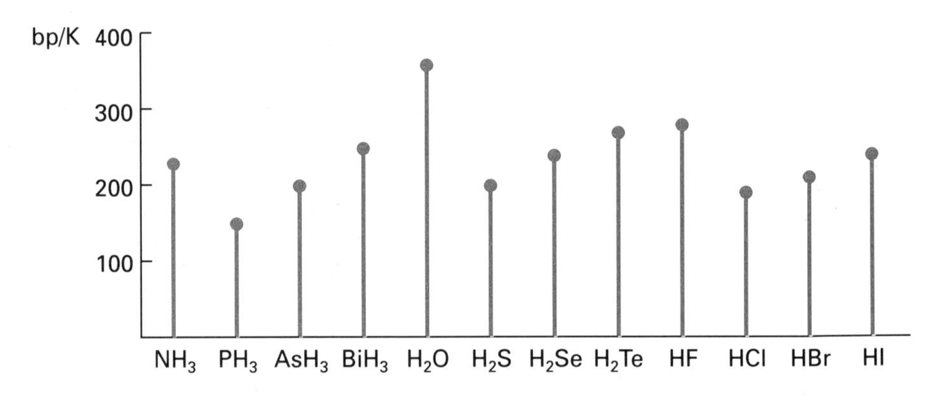

Figure 5.45 Boiling points of hydrides

Our understanding of van der Waals forces would lead us to predict that, in a series of similar molecules, boiling point (and melting point) should increase with increasing molecular mass. Clearly NH_3, H_2O and HF are out of step in their own groups. Their boiling points are much higher than would be predicted on the basis of their size (Table 5.6).

Table 5.6 Predicted and actual bp

Compound	Predicted bp	Actual bp
NH_3	about 143 K	240 K
H_2O	about 293 K	373 K
HF	about 273 K	293 K

These anomalously (abnormally) high boiling points can be attributed to the extra cohesion between molecules caused by hydrogen bonding.

5.15.1 *Hydrogen bonding in compounds with O—H or N—H bonds*

Hydrogen bonds are very common and have an important structural function in many molecules of biological importance, particularly carbohydrates and proteins.

Figure 5.46 Cornflour is pure starch and acts as a thickening agent

The polyglucose chains in **cellulose** interact through hydrogen bonding. These hydrogen bonds hold the ribbon-like cellulose molecules firmly together to form the strong cellulose fibres which act as scaffolding in the cell walls of plants and trees.

Hydrogen bonding is also the reason why **starches** are used in cooking. In natural starch the molecules are twisted by hydrogen bonding into a compact solid which is almost impenetrable to water. When the starch is heated in water, its internal hydrogen bonding breaks down. Water molecules flood in, to form hydrogen bonds with the many OH groups on starch molecules. Swelling then occurs and the suspension becomes very viscous. Starch is therefore used to thicken watery solutions such as gravy and sauces.

In the polypeptide chains of proteins, the CO group of each amino acid is hydrogen bonded to the NH_2 group of the amino acid situated four units ahead in the chain.

C=O---H—N
↑
hydrogen bond

All the CO and NH groups are linked in this way. As a result the polypeptide chain is twisted into a spiral shape. Hydrogen bonds may also be found between CO and NH groups in **different** polypeptide chains. Silk fibre, for example, consists of polypeptide chains cross-linked by means of hydrogen bonds.

Hydrogen bonding is present in small alcohols and carboxylic acids. As a result, their molecules cling together more strongly than molecules of similar mass which are not hydrogen bonded. This causes them to have relatively high boiling points, as Table 5.7 suggests.

Compound	Relative molecular mass	Bonding between molecules	bp/K
CH_3OCH_3*	46	vdW	149
CH_3CH_2OH	46	vdW+H-bonds	351
HCOOH	46	vdW+H-bonds	373

Table 5.7 Bps of dimethyl ether, ethanol and methanoic acid

* The compound CH_3OCH_3 is called dimethyl ether and is marketed as an ozone-friendly aerosol propellant.

5.15.2 Hydrogen bonding in water

The structure of water is very important as water is the medium in which so much chemistry, particularly the processes of life, takes place. The structure of **ice** is of interest since it provides clues about the structure of water. In ice, each H_2O molecule is surrounded, tetrahedrally, by four other H_2O molecules which are bound to the central molecule by means of inter-molecule hydrogen bonds. This arrangement creates a very **open** structure. Ice is therefore, unusually, **less dense** than its liquid form, water.

When ice melts, its cage-like lattice of H_2O molecules disintegrates and molecules can pack more closely together, thus raising the density. The density of water is a maximum at 277 K (4 °C). Strong hydrogen bonding

Figure 5.47 Hydrogen bonded H_2O molecules in ice

still persists in liquid water, with hydrogen bonds being continuously broken and reformed.

Figure 5.48 Icebergs floating in the Arctic Ocean – a spectacular illustration of the fact that ice is less dense than water

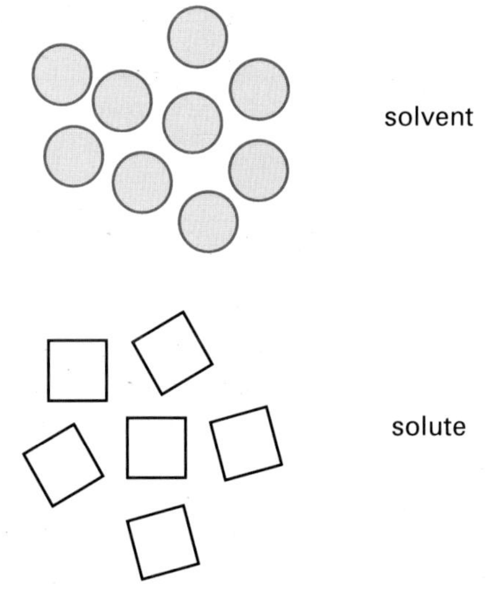

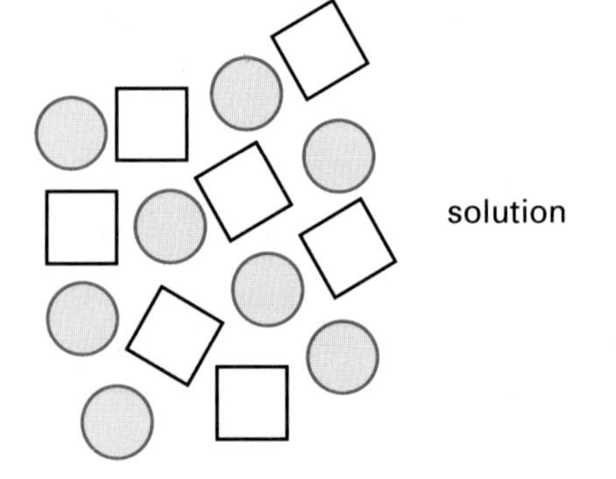

Figure 5.49 Mixing of solvent and solute particles

5.16 The process of solution

The process of solution, i.e. dissolving, involves three processes which occur simultaneously rather than one after another. These three processes are:

(a) separation of the solute particles (ions or molecules) from one another
(b) separation of the solvent molecules from one another
(c) intermixing and association of solute and solvent particles.

(a) and (b) are energy-absorbing (endothermic) processes, since the bonds between solute particles or solvent molecules have to be broken. Step (c) is an exothermic, energy-releasing process since bonds are formed between solute and solvent.

Dissolving will occur when solute–solvent bonding is a more attractive option (mainly in terms of energy) than maintaining separate solute–solute and solvent–solvent bonding.

Solvents may be divided into three broad groups:

(a) non-polar solvents
(b) moderately polar solvents
(c) highly polar solvents.

5.16.1 Non-polar solvents

Non-polar solvents include benzene, tetrachloromethane and white spirit. Such liquids cannot generally dissolve ionic solids or highly polar molecular substances. They can offer only van der Waals attraction, and this cannot compensate for strong ionic or polar interaction in the solute. Non-polar solvents will, however, readily dissolve 'like' substances, i.e. those with little or no polarity, such as oils, waxes and other compounds with substantial hydrocarbon content.

5.16.2 Polar solvents

A polar solvent is one whose molecules possess strong permanent dipoles. Polar solvents can dissolve polar solutes. **Strongly polar** solvents like water, hydrogen fluoride, liquid ammonia and methanoic acid are also capable of dissolving ionic compounds. They are therefore classed as **ionising solvents**.

5.17 Water, the universal solvent

Water is an excellent solvent. It mixes readily with ethanol, as in beers, wines and spirits, because it is able to form hydrogen bonds with molecules of the alcohol. Sugars and amino acids are also soluble in water – a result of their capacity to form hydrogen bonds.

Water is essential for life as it can provide a liquid medium, inside and outside cells, through which ions and molecules can move. Its properties as a solvent relate to the **polarity** of water molecules. The H_2O molecule is bent, with the shared electrons in the O—H bonds biased towards the O atom. The oxygen end (side) of the molecule takes on a partial negative charge, leaving a partial positive charge on the other (H) side.

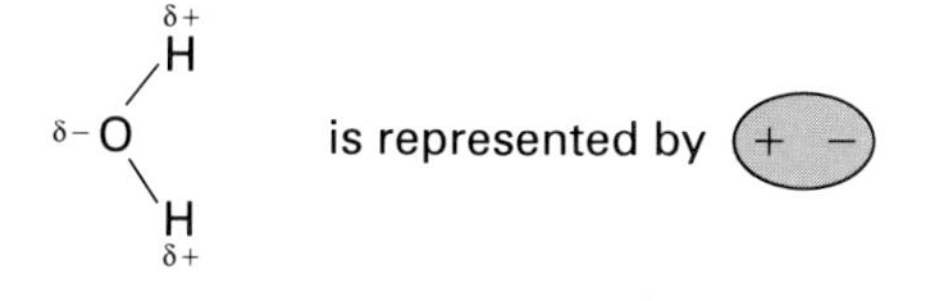

Figure 5.50 Polarisation of H_2O molecule

Ionic solids melt only at high temperatures. Considerable energy has to be applied in order to dismantle the rigid ionic lattice. When an ionic solid dissolves in water, the ionic lattice disintegrates and the ions are freed to move about the aqueous solution. Where does the energy to separate the ions come from?

At the surface of the ionic crystal, ions become attracted to and surrounded by the polar water molecules. The ions separate because the charged ends of the water molecules attract the ions more than the ions attract each other. As ions are dislodged from the lattice other ions are exposed to the attraction of water molecules and the process continues.

The water molecules align themselves around positive ions with their negative poles pointing in towards the ion. Around a negative ion the orientation of the water molecules is reversed. An ion enclosed within a cluster of solvent molecules is said to be solvated. When the solvent is water, which is usually the case, the ion is said to be **hydrated** (Figure 5.51).

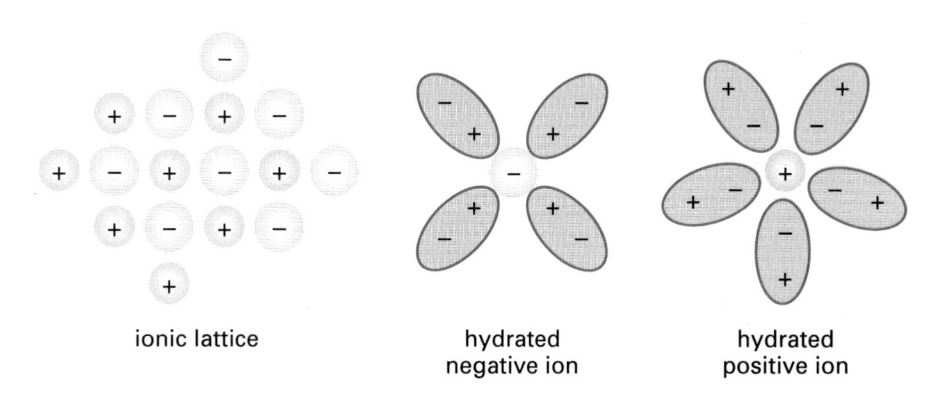

Figure 5.51 Solution process

So much for the **process** of solution. What are the energy implications? The attraction between ions and water dipoles releases energy. This output of energy compensates for the energy which has to be put into separating the ions fixed in the crystal lattice. In other words **the exothermic hydration process supplies the energy needed to continue the dismantling of the ionic lattice**.

The process of solution may be represented by a simple equation. The dissolving of sodium chloride in water can be shown as follows:

$$Na^+Cl^-(s) \xrightarrow{H_2O(l)} Na^+(aq) + Cl^-(aq)$$

The symbol **(aq)** indicates that the ion is surrounded by a small group of water molecules (the exact number varies and is often difficult to pin down). Because of their fixed orientation around ions, these water molecules cannot participate so readily in forming ice. This explains why the higher the salt content of water, the more resistant it is to freezing. Conversely, when salt is sprinkled on icy roads and pavements, hydration of the sodium and chloride ions by H_2O molecules in the ice brings about the demolition of the ice structure: the ice 'melts'.

5.17.1 The ionisation of some molecular compounds in water

Certain substances remain as molecules when they dissolve in a non-ionising solvent (e.g. a hydrocarbon liquid) and yet when they dissolve in water they form ions, to some extent at least. These substances have molecules containing strongly polar bonds. Interaction with polar H_2O molecules can increase the separation of charge within the bonds to a point where the atoms break apart as separate, individual ions.

Take **hydrogen chloride**, for example. Neither liquid HCl nor HCl gas, dissolved in a non-polar solvent such as benzene, conducts electricity. We can infer that the hydrogen chloride is present as HCl **molecules**. Aqueous solutions of hydrogen chloride, on the contrary, strongly conduct an electric current. This suggests the presence of **ions**: $H^+(aq)$ and $Cl^-(aq)$.

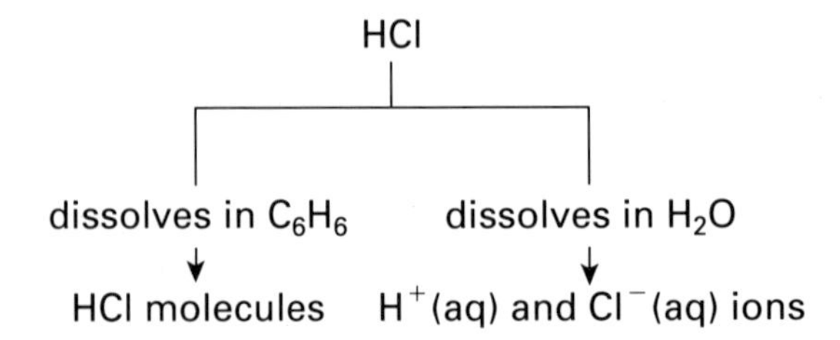

A very large amount of energy is required to split an HCl molecule to give H^+ and Cl^- ions. In the gaseous state, temperatures well above 1300 K are required, and yet this ionisation proceeds readily in water at room temperature.

The energy to break the H—Cl bonds comes from the reaction between the polar H_2O molecules and the polar HCl molecules.

$$H-\overset{\delta-}{O}(H)\cdots\overset{\delta+}{H}-\overset{\delta-}{Cl}\cdots\overset{\delta+}{H}-O(H)$$

The ionisation of polar HCl molecules, as they dissolve in water to form hydrochloric acid, can then be represented by the simple equation:

$$HCl(aq) \longrightarrow H^+(aq) + Cl^-(aq)$$

and the simple drawing in Figure 5.52.

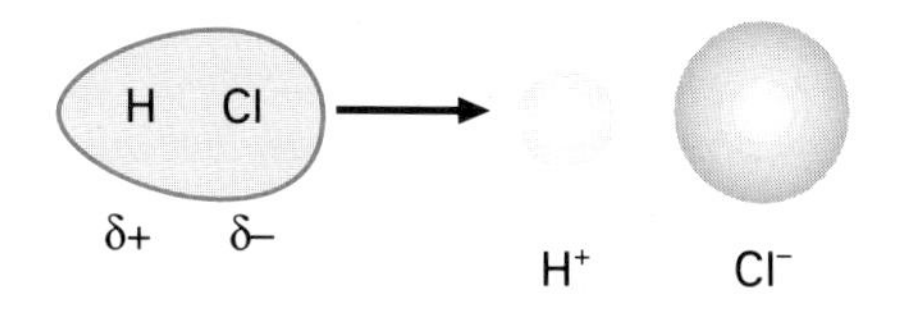

Figure 5.52 Ionisation of HCl

In dilute solutions of HCl(aq) there is normally 100 per cent dissociation of molecules into ions. In concentrated solutions of HCl(aq), where the water molecules are in more limited supply, then ionisation is less complete. A small proportion of molecules may remain undissociated.

All pure acids are molecular compounds. Only in the presence of an ionising solvent such as water do H^+ ions form.

5.18 Variation in oxides

There are obvious trends in the structures and properties of oxides of elements across Periods 2 and 3 of the periodic table.

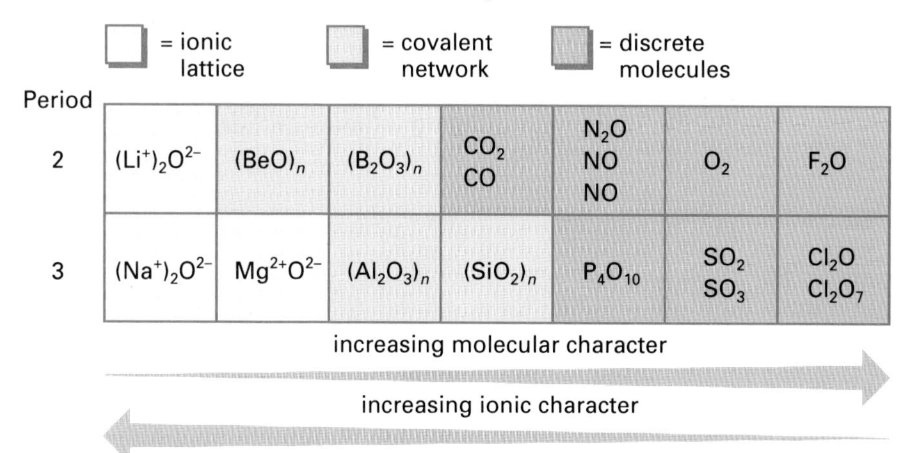

Period							
2	$(Li^+)_2O^{2-}$	$(BeO)_n$	$(B_2O_3)_n$	CO_2 CO	N_2O NO NO	O_2	F_2O
3	$(Na^+)_2O^{2-}$	$Mg^{2+}O^{2-}$	$(Al_2O_3)_n$	$(SiO_2)_n$	P_4O_{10}	SO_2 SO_3	Cl_2O Cl_2O_7

Figure 5.53 Trends in the structures of oxides

The oxides of metals or metalloids have giant structures which are either ionic or covalent. The oxides of Li, Na, and Mg form ionic lattices. They are solids at room temperature, with high melting and boiling points. When melted these ionic oxides will **conduct electricity**.

Aluminium oxide (alumina) occurs in more than one form. It seems unsure whether to exist as a giant structure of ions or a giant structure of atoms. It has a very high melting point and is used as a support for catalysts in the cracking and reforming of hydrocarbons.

The metalloids B and Si form oxides which are covalent networks of atoms. Their melting and boiling points are high but, unlike ionic oxides, covalent network oxides do not conduct electricity in the molten (liquid) state.

The oxides of the non-metals C, N, P, S and Cl consist of small discrete molecules. These oxide molecules are fairly easily separated and as a result show low melting and boiling points. The oxide of phosphorus is solid at room temperature, while both sulphur trioxide, SO_3, and dichlorine heptoxide, Cl_2O_7, are liquids. The other non-metal oxides are normally gases. Being uncharged, molecular oxides are **non-conductors** in the molten state.

5.18.1 The acid/base character of oxides

As we pass along a period, the structure of the oxide changes from ionic lattice to covalent network to discrete molecules. The nature of the oxygen differs in each type of structure. As the electron-attracting power of the element increases, the polarity of its bond to oxygen gradually decreases. The element–oxygen bond becomes increasingly non-polar as electrons become more evenly shared.

The amount of negative charge carried by the oxygen determines how the oxide will react with water or with acids and alkalis. Oxides which do react can be classed as **basic**, **acidic** or **amphoteric**. Oxides which do not react with water, acids or alkalis are described as neutral. CO, N_2O and NO are all neutral oxides.

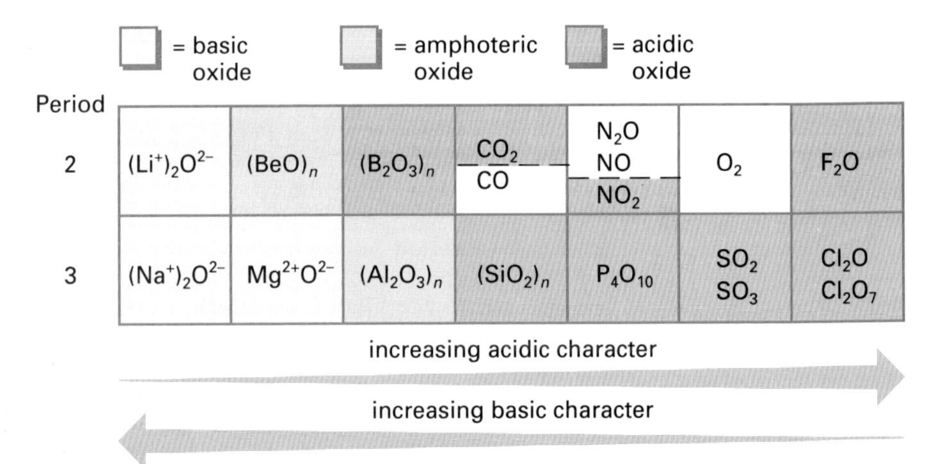

Figure 5.54 Trends in the base/acid nature of oxides

5.18.2 Basic oxides

If a metal oxide is soluble in water, its oxide ion, O^{2-}, reacts with the water to form the hydroxide ion. The reaction involves the transfer of a proton (H^+) from H_2O to the O^{2-} ion.

$$O^{2-} + H_2O \longrightarrow 2OH^-$$

Ionic oxides such as Li_2O, Na_2O and MgO therefore dissolve in water to form **alkaline** solutions, e.g.

$$(Na^+)_2O^{2-}(s) + H_2O(l) \longrightarrow \underset{\text{sodium hydroxide solution (an alkali)}}{2Na^+(aq) + 2OH^-(aq)}$$

All three oxides react with acids (to form salts) and are therefore classed as **basic** oxides, e.g.

$$Mg^{2+}O^{2-}(s) + 2H^{+}(aq) + 2Cl^{-}(aq) \longrightarrow Mg^{2+}(aq) + 2Cl^{-}(aq) + H_2O(l)$$

It should be noted that many metal oxides, e.g. CuO, PbO, FeO, NiO, ZnO, do not dissolve in water but do react with acids to form salts. These metal oxides are basic but not alkaline. Only the soluble metal oxides form alkalis.

5.18.3 Amphoteric oxides

The oxides of beryllium and aluminium are in the form of a lattice which is neither clearly ionic nor clearly atomic. The 'atoms' have significant partial positive or negative charge. Both BeO and Al_2O_3 are insoluble in water. However, they react with **both** acids and bases (alkalis) and are therefore described as **amphoteric**.

An oxide which can behave like an acidic oxide or like a basic oxide is called an amphoteric oxide.

Aluminium oxide can react with acid and thus act the part of a basic oxide. On the other hand, it is also capable of behaving like an acidic oxide by reacting with alkali.

(a) $Al_2O_3 + 6H^{+} \longrightarrow 3H_2O + 2Al^{3+}$
acid

Al: the **positive** ion in the salt formed

(b) $Al_2O_3 + 2OH^{-} + 3H_2O \longrightarrow 2Al(OH)_4^{-}$
alkali

Al: part of the **negative** ion (aluminate) in the salt formed

The purification of aluminium ore in the manufacture of aluminium metal depends on the amphoteric nature of Al_2O_3. The major source of aluminium is the ore **bauxite**, Al_2O_3. This is contaminated with Fe_2O_3 which gives it its reddish-brown colour. When bauxite is added to a strongly basic solution such as NaOH(aq), the Al_2O_3 dissolves but the Fe_2O_3 impurity, not being amphoteric, remains as a solid. The solution is filtered and white Al_2O_3 (alumina) obtained. The alumina is eventually electrolytically smelted to give aluminium.

Figure 5.55 Pouring off molten aluminium inside British Alcan's Lochaber smelter, Fort William

5.18.4 *Acidic oxides*

Non-metals with a fairly strong attraction for bonding electrons are able to stop oxygen monopolising these electrons and forming O^{2-} ions. These elements form oxides which generally dissolve in water to form **acidic** solutions or react with basic solutions to form salts.

Non-metals form acidic oxides.

Table 5.8 lists some oxides and the acids which they form.

Oxide	Acid formed
B_2O_3	H_3BO_3 (boric)
SiO_2	†
NO_2	HNO_2/HNO_3 (nitrous/nitric)
P_4O_{10}	H_3PO_4 (phosphoric)
SO_2	H_3SO_3 (sulphurous)
SO_3	H_2SO_4 (sulphuric)
Cl_2O	HClO (hypochlorous)
Cl_2O_7	$HClO_4$ (perchloric)

Table 5.8 Acidic oxides

* B_2O_3 is insoluble but H_3BO_3 is formed by reaction of water with many boron compounds.
† SiO_2 is insoluble but reacts with alkalis to form the silicate ion.

$$SiO_2 + 2OH^- \longrightarrow SiO_3^{2-} + H_2O$$

5.19 Variation in chlorides

Chlorine is a very reactive element and combines with most other elements to form compounds called chlorides. The **structure** of chlorides of elements in Periods 2 and 3 changes as we move across each row. The pattern appears to be for metals to form ionic lattices and non-metals to form discrete molecules, as illustrated in Figure 5.56.

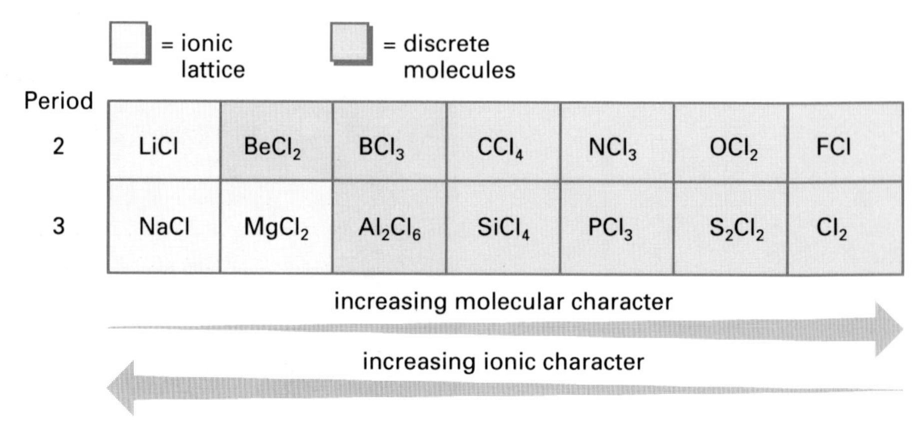

Figure 5.56 Trends in the structures of chlorides

The change from ionic chloride to molecular chloride is gradual rather than abrupt. Beryllium chloride is predominantly molecular since the small Be^{2+}

atoms hold on to their bonding electrons and prevent the formation of Cl^- ions (and therefore Be^{2+} ions). Another chloride which is partly ionic but mainly molecular is aluminium chloride, Al_2Cl_6. Its structure is as shown.

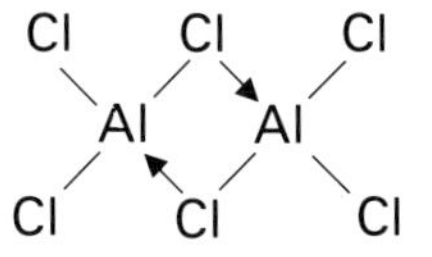

One of the three Cl atoms on each Al atom donates an electron pair to the empty orbital on Al. This special type of bond is shown with an arrow at one end.

Al_2Cl_6 is a poor conductor of electricity at its melting point, but conductivity increases with a further rise in temperature. This suggests a structure which has a degree of both covalent and ionic bonding.

The structure of the chloride also determines the type of reaction it undergoes with water. **Ionic chlorides dissolve in water** without further reaction. The ionic lattice is taken apart by the hydrating action of polar H_2O molecules. A neutral solution containing separate and free-moving hydrated ions is produced, e.g.

$$Mg^{2+}(Cl^-)_2(s) \longrightarrow Mg^{2+}(aq) + 2Cl^-(aq)$$

With most **molecular chlorides** interaction with water molecules goes a lot farther than simple hydration: new chemical species are formed. Such reaction with water is called **hydrolysis**. Of the molecular chlorides in Periods 2 and 3, only tetrachloromethane, CCl_4, is unreactive to water. The other molecular chlorides are liquids or gases, and hydrolyse readily in the presence of water, generally to form acidic solutions and hydrogen chloride gas.

$$BCl_3 + 3H_2O \longrightarrow H_3BO_3 + 3HCl$$
$$SiCl_4 + 4H_2O \longrightarrow SiO_2.2H_2O + 4HCl$$
$$PCl_3 + 3H_2O \longrightarrow H_3PO_3 + 3HCl$$
$$OCl_2 + H_2O \longrightarrow 2HClO$$

In each case, polar H_2O molecules attack the central atom and bring about replacement of Cl atoms by OH groups. The mechanism for the hydrolysis of $SiCl_4$ can be outlined as follows.

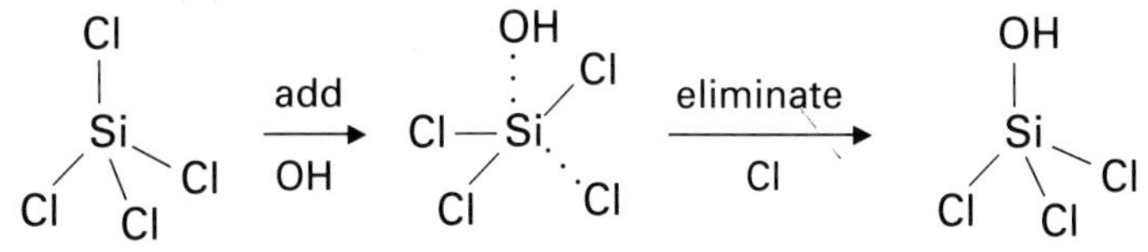

add OH → unstable complex → eliminate Cl →

unstable complex

Repetition of this process eventually yields $Si(OH)_4$, which rearranges to give $SiO_2.2H_2O$.

The hydrolysis of silicon chloride is an important step in the manufacture of silicone polymers. Silicone polymers are water repellent and chemically inert. They have many commercial uses.

5.20 Variations in the halogens

The large number of chloride compounds is a reflection of the reactivity of the chlorine element. Chlorine is a member of the Group 7 family of elements called the **halogens**. Each halogen readily forms compounds –

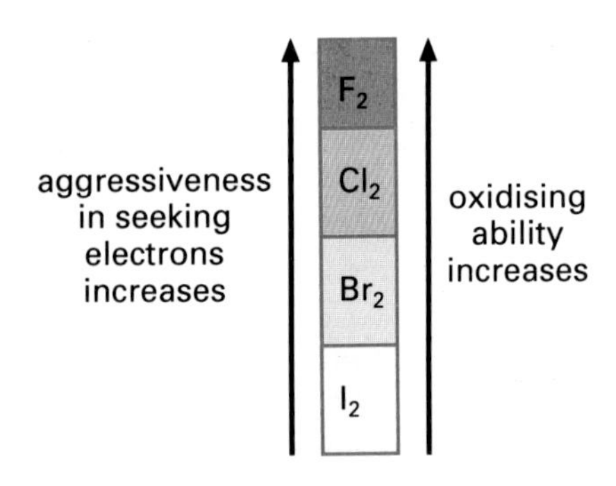

Figure 5.57 Relative oxidising ability of halogens

fluorides, bromides and iodides as well as chlorides. Their reactivity with metals is due to the ability of halogen atoms to attract electrons and form negative (halide) ions. The halogen oxidises the metal and is itself reduced.

The oxidising ability of the halogens increases up Group 7 as the atomic size decreases. Fluorine is the most powerful oxidising agent and the most reactive halogen. Iodine is the weakest oxidiser and least reactive.

One result of this trend in oxidising power is that each halogen will displace any halogen below it in the periodic group from a solution of its ions. For example, if either aqueous chlorine or bromine solution is added to aqueous iodide ions, iodine molecules are produced – the result of oxidation of iodide ions.

The equation for the reaction between chlorine and iodide is:

$$2I^-(aq) + Cl_2(aq) \longrightarrow I_2(aq) + 2Cl^-(aq)$$

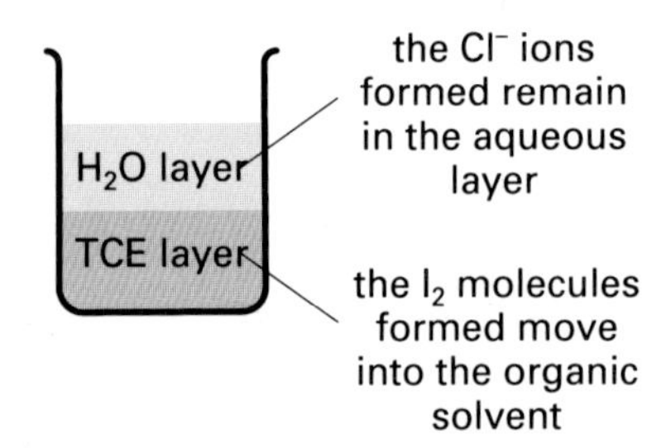

Figure 5.58 Displacement of iodine by chlorine

The production of I_2 molecules can be shown by adding trichloroethane (TCE) to the aqueous reaction mixture. The I_2 formed dissolves in the lower TCE layer and gives it a violet colour.

If we neglect the part played by the solvent (water) this halogen displacement reaction can be broken down into four key changes:

bond breaking: $Cl_2 \longrightarrow 2Cl$
oxidation: $2I^- \longrightarrow 2I + 2e^-$
reduction: $2e^- + 2Cl \longrightarrow 2Cl^-$
bond forming: $2I \longrightarrow I_2$

The energy needed to bring about bond breaking and oxidation is supplied from the energy released by reduction and bond forming.

Halogen displacement is not reversible. Iodine cannot displace chlorine from solution: it is not strong enough to oxidise chloride ions. Electron transfer is always from the less reactive halogen to the more reactive halogen.

Bromine is used for making herbicides, disinfectants, flame retardants and photographic emulsions (silver bromide). It is obtained commercially by using chlorine to oxidise the small concentration of bromide ions present in seawater.

Figure 5.59 A factory in Anglesey, North Wales, for the extraction of bromine from seawater

The extraction is carried out by passing chlorine gas into the seawater solution. A redox reaction occurs between the dissolved chlorine and the bromide ions.

$$Cl_2(aq) + 2Br^-(aq) \longrightarrow 2Cl^-(aq) + Br_2(aq)$$

The free bromine is swept out of solution by a current of air and subsequently removed.

Summary of Unit 5

Having read and understood the information and ideas given in this unit, you should now be able to:

describe the origin and strength of van der Waals forces

state that noble gases exist as atoms

distinguish between discrete molecules and covalent networks

describe metallic bonding

identify the structural type of each of the first 20 elements

relate the different properties of diamond and graphite to the differences in their structures

define covalent radius and explain its gradual decrease on crossing a period, or gradual increase on descending a group of the periodic table

define first ionisation enthalpy

relate trends in metallic character, density, melting and boiling points of elements 1–20 to changes in the type of bonding and structure

relate the strength of an element's attraction for bonding electrons to its position in the periodic table

use the position of two elements in the periodic table to predict the type of bonding in the compound which they form

describe the ionic lattice structure

describe the structure of covalent network compounds and explain why they are generally hard

indicate how polar bonding arises

explain hydrogen bonding and its effect on the boiling point of water

explain how some polar covalent bonds split to give ions in aqueous solution

distinguish between intramolecular and intermolecular bonding

relate trends in the structural type of oxides across Periods 2 and 3 to trends in their melting and boiling points, states at room temperature and abilities to conduct electricity

describe how polar solvents dissolve polar and ionic compounds

explain why non-polar solvents can dissolve non-polar compounds, but not highly polar or ionic compounds

▲ explain trends in the first ionisation enthalpies of elements across a period or down a group in the periodic table

▲ state that some metals form ionic hydrides in which the hydrogen is negatively charged

▲ state that chlorine combines directly with metals to form ionic chlorides

▲ explain how molecules with polar bonds may be, overall, polar or non-polar

▲ account for the difference in boiling point between polar and non-polar compounds of similar mass

▲ explain why ice is less dense than water

▲ account for the boiling points of NH_3 and HF being much higher than expected from comparison with other hydrides in Groups 5 and 7

▲ define amphoteric oxide and relate its chemical properties to the structure of the oxide

▲ describe the process of hydration and the solution of an ionic compound in terms of energy input and output

▲ state that ionic chlorides dissolve in water without reaction, but that most molecular chlorides undergo a hydrolysis reaction which generally produces HCl

▲ explain the displacement reactions of halogens in aqueous solutions in terms of redox and covalent bond breaking/making.

PROBLEM SOLVING EXERCISES

1. Information about two aluminium compounds is given in the table below.

Compound	Conductivity of molten compound	Melting point/K
aluminium fluoride	conducts	1473
aluminium bromide	does not conduct	371

From the above data state what you can infer about the bonding in:

(a) aluminium fluoride
(b) aluminium bromide.

(PS skill 8)

2. You wish to show that the presence of hydrogen bonding between molecules of a compound causes its boiling point to be higher than would be expected otherwise.
You choose two compounds to illustrate the difference:
X is hydrogen bonded.
Y is not hydrogen bonded.
For comparison to be fair and valid two characteristics of X and Y ought to be similar.
Which two boxes in the grid below contain these characteristics?

A X and Y should have similar molecular masses	B X and Y should be in the same compound series
C X and Y should contain the same functional group	D X and Y should have similar molecular shapes

(PS skill 7)

3. Bond lengths are given in picometres (pm). The lengths of single C—C and Si—Si bonds are 154 pm and 232 pm respectively. Calculate the length of the C—Si bond.

(PS skill 4)

4. Explain why Mg^{2+} ions are smaller than S^{2-} ions, even though Mg atoms are larger than S atoms.

(PS skill 9)

5. You are given a sample of white solid.
Describe some simple experiments that you could do to help you decide whether the solid was held together by covalent bonds, van der Waals bonds or ionic bonds.

(PS skill 5)

6. Draw the structure and indicate the shape you would predict for the molecules listed in the table.

Molecule	Polarity
BeH_2	non-polar
NO_2	polar
BF_3	non-polar
NCl_3	polar

(PS skill 10)

7. Fluorine is manufactured by electrolysis of potassium hydrogen difluoride as shown.

$$KHF_2(l) \longrightarrow H_2(g) + F_2(g) + 2KF(l)$$

The negative ion in the salt is HF_2^-.
Can you predict the arrangement of atoms and offer an explanation for the stability of this three-atom ion?

(PS skill 9)

8. 'The first ionisation enthalpy decreases as the covalent radius increases.'
In each of the boxes in the grid below, the first ionisation enthalpy drops as the covalent radius rises.

A	B	C	D
Si K Cs	Be Al Rb	Ca Sr Ba	H C S

Which box contains the most valid evidence for relating energy needed to remove an outer electron (ionisation enthalpy) to size of atom (covalent radius)?

(PS skill 7)

PROBLEM SOLVING EXERCISES

9. The solubilities of some gases, at 298 K, are listed in the table.

Gas	Solubility g gas/kg water
CO_2	1.45
Cl_2	6.41
NH_3	480
SO_2	94.1

How might you measure the solubility of a gas in water? Suggest a possible experimental method, using diagrams for assistance.

(PS skill 5)

10. The melting and boiling points of some silicon compounds are listed in the table.

Compound	Melting point /K	Boiling point /K
SiH_4	88	161
$SiBr_4$	278	426
$SiCl_4$	205	330
SiF_4	183	188
SiI_4	393	563

If we take normal room temperature to be 298K, which of these compounds are liquid at this temperature?

(PS skill 8)

11. 'As you go down a group of column of metals in the periodic table, the melting point decreases as the covalent radius increases.'
Use the databook to find a piece of evidence to support the general statement given above.

(PS skill 1)

12. Butenedioic acid contains two carboxyl groups per molecule.
These groups may be on the same side of the double bond or on opposite sides of the double bond.
This gives rise to two different acids, maleic and fumaric.

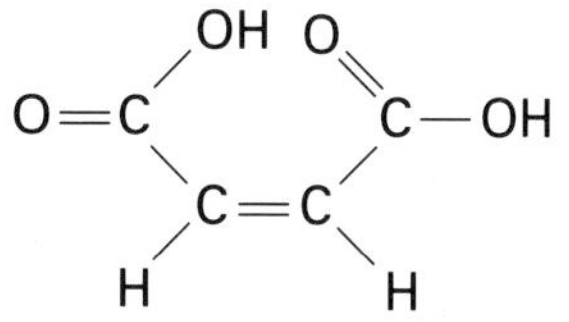

maleic acid
mp = 430 K

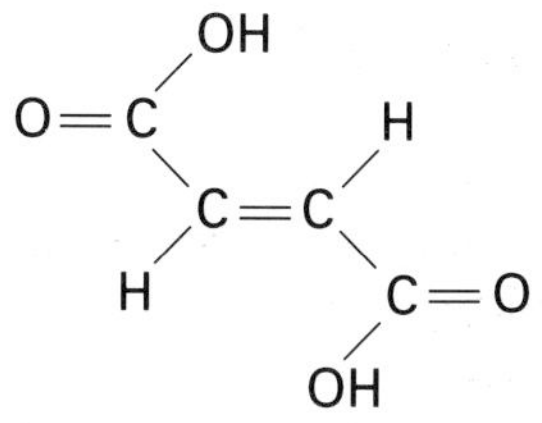

fumaric acid
mp = 560 K

Explain, in terms of hydrogen bonding, why fumaric acid has a much higher melting point than maleic acid.

(PS skill 9)

Unit 6
Thermochemistry

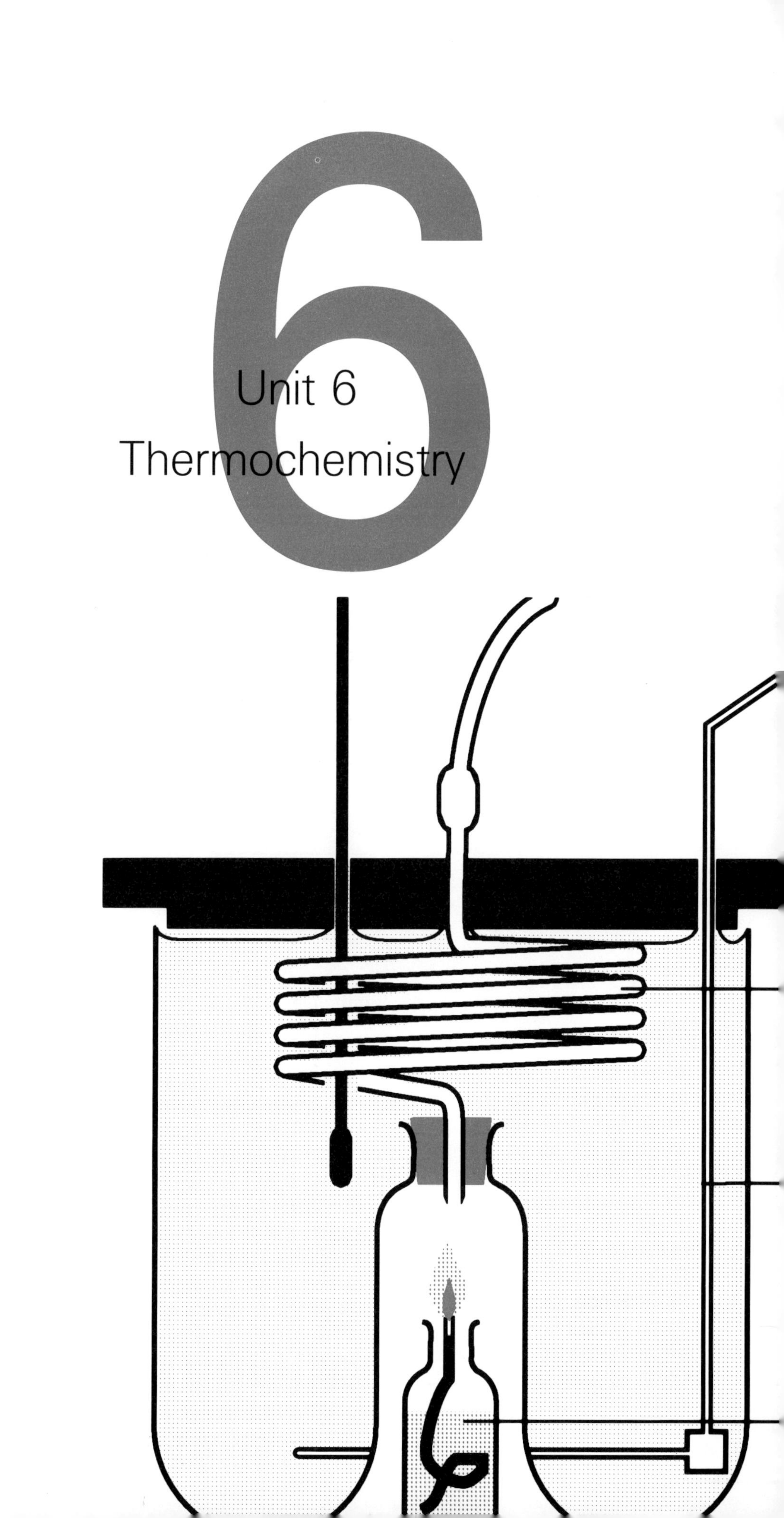

ASSUMED KNOWLEDGE AND UNDERSTANDING

Before starting on Unit 6 you should know and understand:

what is involved in the combustion of substances

what is meant by an exothermic reaction

the relationship between bond breaking, bond making and the heat energy evolved from a reaction

the process of acid–base neutralisation

diagrams which describe changes in potential energy during the course of a reaction

how to carry out calculations involving amounts of substances in moles

the polar nature of the water molecules and the process of hydration of ions

what is meant by first ionisation enthalpy

the process by which ionic compounds dissolve in water and the energies involved.

OUTLINE OF THE CORE AND EXTENSION MATERIAL

As you progress through Unit 6 you should, at least, try to grasp the fundamental content listed under **core** and, if possible, pick up the extra and/or more difficult content listed under **extension**.

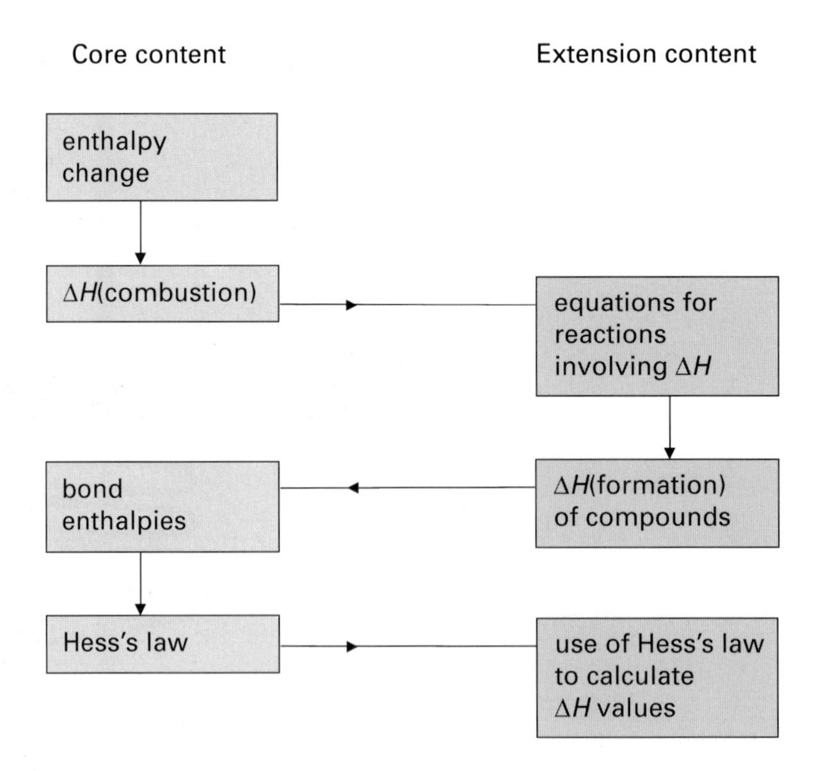

6 THERMOCHEMISTRY

6.1 Enthalpy

Thermochemistry is the study of the heat energy taken in or given out by chemical reactions. This heat, absorbed or released, can be related to the internal heat content of the substances involved. Such internal heat content is called **enthalpy** and is given the symbol H.

Every substance is assumed to have an enthalpy. A set of reactants will have a definite total enthalpy, H_r. Similarly, a set of products will have a definite total enthalpy, H_p.

During a chemical change, reactants convert to products. This brings about a **change in enthalpy**, ΔH. (The upper case Greek symbol Δ is used to denote 'change in'.)

$$\Delta H = \text{enthalpy(products)} - \text{enthalpy(reactants)}$$

$$\Delta H = H_p - H_r$$

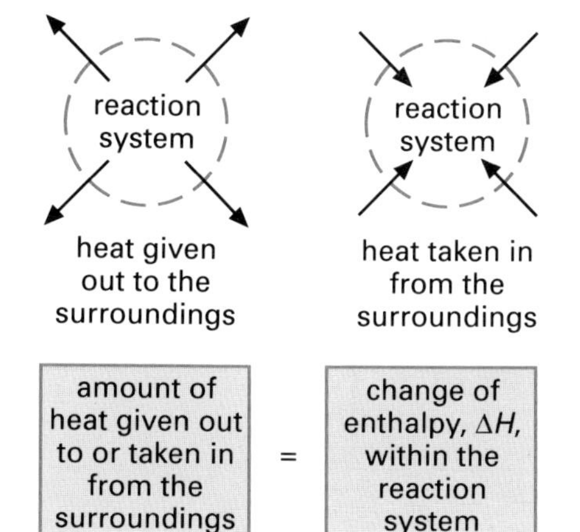

Figure 6.1

6.1.1 Measuring enthalpy change

We can measure a property such as the mass or temperature of a substance. We have no way of measuring its enthalpy directly. We can only measure the **differences** in enthalpy between the reactants and products in a chemical change. We are able to do this by measuring the **amount of heat taken in** or **given out** by the reaction.

If we describe any reaction open to the atmosphere as the 'reaction system' and the materials around it as the 'surroundings', we can form the equation given in Figure 6.1.

This equation is based on the principle of the **conservation of energy**. Energy can be converted from one form to another or shifted from one place to another: it cannot, however, be created or destroyed. Using this concept, we assume that the surroundings gain whatever heat the reaction system loses. Similarly, if the reaction system gains heat we assume all of this heat is taken from the surroundings.

The total energy of the system + surroundings remains constant.

The direction of the heat transfer, between the reaction system and the surroundings, indicates whether there has been a rise or a fall in enthalpy within the system.

The sign of ΔH depends on the relative sizes of H_p and H_r. If the reaction is **exothermic**, heat is given out from the reaction system and H_p must be smaller than H_r. ΔH will therefore be **negative** in sign. If the reaction is **endothermic**, heat is taken into the reaction system and H_p must be larger than H_r. ΔH will therefore be **positive** in sign.

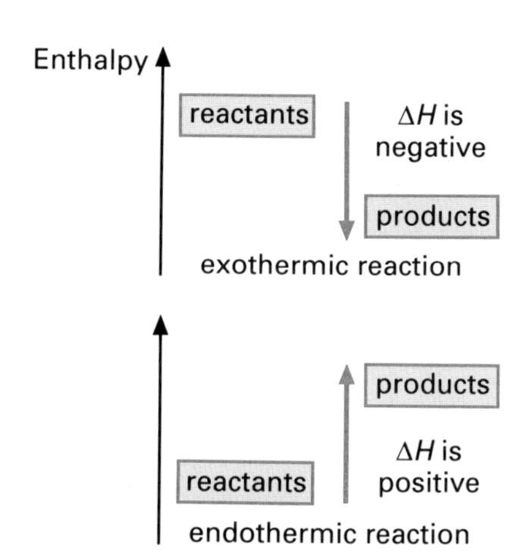

Figure 6.2 Enthalpy changes

6.1.2 Activation energy and enthalpy change

We must be careful not to confuse **activation energy** and **enthalpy change** for an **endothermic** reaction. Both represent amounts of heat absorbed by the reaction system. All reactions, be they endothermic or exothermic, require the input of a minimum amount of energy to get them going and this is known as activation energy. When we apply an electric spark, a lighted spill or a roaring Bunsen flame to a reaction mixture we are supplying activation energy, in the form of heat. The reaction system will absorb what it needs of this heat to get started.

Whether or not there is a net absorption of heat from the surroundings at the end of the reaction depends on the enthalpy level of the products. If this final state is above that of the reactants, not all the activation energy absorbed will be returned to the reaction system and a shortfall in heat energy is created.

As the reaction mixture cools to its original temperature, it makes up this shortfall by absorbing heat energy from the surroundings. This net amount of heat absorbed will be a measure of the difference in enthalpy (ΔH) between products and reactants and will always be less than the activation energy (E_a).

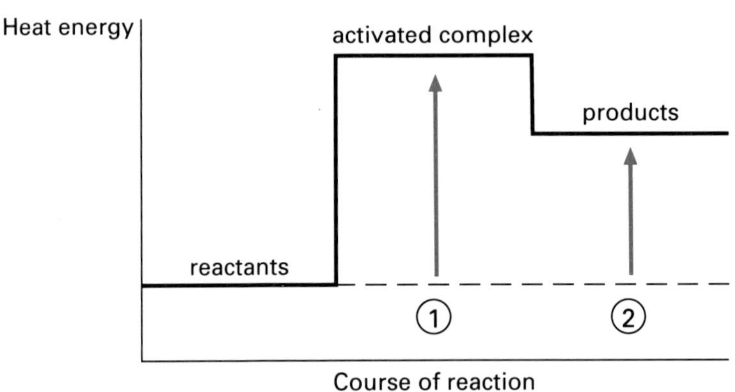

Figure 6.3 Amounts of heat absorbed

6.1.3 Factors affecting ΔH

In giving enthalpy changes for a reaction it is important to specify:

(a) amounts of substances
(b) direction of reaction
(c) the states of the substances.

Amounts of substances

The formation of one mole of water molecules from hydrogen and oxygen molecules is an exothermic reaction. The amount of heat given out to surroundings, and therefore the change in enthalpy, is 286 kilojoules. This can be shown by the equation:

$$H_2(g) + \tfrac{1}{2}O_2(g) \longrightarrow H_2O(l) \qquad \Delta H = -286 \text{ kJ mol}^{-1}$$

If **two** moles of water molecules are formed, then the amount of heat given out and the enthalpy change are **doubled**.

$$2H_2(g) + O_2(g) \longrightarrow 2H_2O(l) \qquad \Delta H = -572 \text{ kJ mol}^{-1}$$

Change in enthalpy, ΔH, is **proportional to the amount of substance** which is converted or produced. In thermochemistry, 'amount of substance' is always measured in moles. We therefore use the unit **kilojoules per mole** when giving ΔH values.

The enthalpy change for the formation of water as listed in a data book refers to the formation of **one mole of water molecules** and is given as $\Delta H = -286 \text{ kJ mol}^{-1}$.

Direction of reaction

It is also important to be clear about the sign of the ΔH value. The enthalpy change for the **same amount of reverse reaction** will be equal in size but **opposite in sign** to the original reaction.

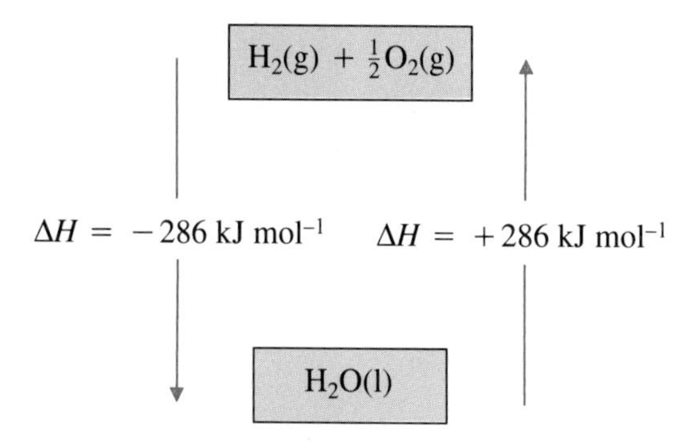

ΔH values for formation and decomposition of water

The states of the substances

When a substance undergoes a change in state its enthalpy changes even though its temperature remains constant. The changes

solid $\longrightarrow$ liquid

and

liquid $\longrightarrow$ gas

are endothermic changes, while the reverse changes are exothermic.

It is therefore important that a state symbol is attached to all the formulae in an equation, e.g.

$$H_2(g) + \tfrac{1}{2}O_2(g) \longrightarrow H_2O(\mathbf{l}) : \Delta H = -286 \text{ kJ mol}^{-1}$$

$$H_2(g) + \tfrac{1}{2}O_2(g) \longrightarrow H_2O(\mathbf{g}) : \Delta H = -242 \text{ kJ mol}^{-1}$$

The difference of 44 kJ mol^{-1} in the ΔH values represents the enthalpy change involved in the change from **liquid** water to water **vapour**.

6.2 Enthalpy of combustion

To understand the action of fuels and the metabolism of organisms, we need to understand the energy changes involved in **combustion** reactions.

> ***The enthalpy of combustion of a substance, ΔH_c, is the enthalpy change that occurs when one mole of the substance undergoes complete combustion.***

For hydrocarbons and other organic compounds, complete combustion means the products must be carbon **dioxide** and liquid water.

All combustions are exothermic. In some cases a very large amount of heat is released. Oxyacetylene burners, used for melting and welding metals, depend on the heat output from the combustion of C_2H_2 (acetylene) and O_2 to provide the very high temperatures required.

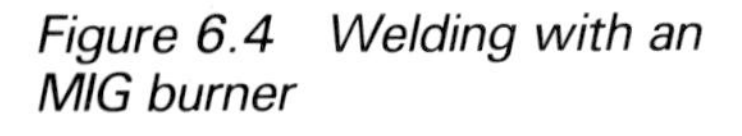

Figure 6.4 Welding with an MIG burner

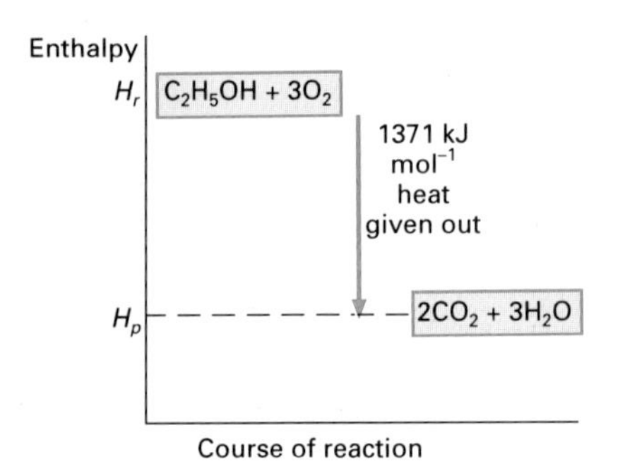

Figure 6.5 Enthalpy diagram for combustion of ethanol

An important nonhydrocarbon fuel is ethanol. Its combustion can be represented by the equation below.

$$C_2H_5OH(l) + 3O_2(g) \rightarrow 2CO_2(g) + 3H_2O(l) \quad \Delta H_c = -1371 \text{ kJ mol}^{-1}$$

The energy change brought about by the combustion can be shown by means of an enthalpy diagram (Figure 6.5).

6.2.1 *Measuring enthalpy of combustion by experiment*

Figure 6.6 shows a very simple apparatus that can be used in a school lab to determine the enthalpy of combustion of ethanol.

All the heat evolved from the burning ethanol is assumed to be absorbed by the water in the beaker. By measuring the quantity of heat taken in by the water, we are attempting to measure the heat released by the reaction.

The heat absorbed by the water is obtained by knowing the mass of the water (m), its specific heat capacity (c) and the temperature rise (ΔT), as read on the thermometer. (Specific heat capacity is the amount of heat required to raise the temperature of 1 kg of the substance through 1 K.)

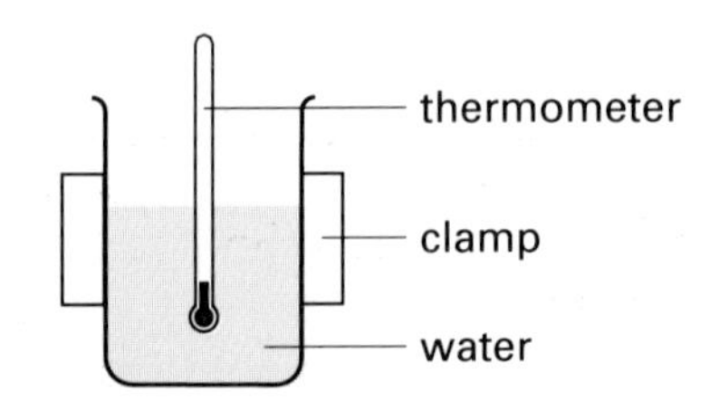

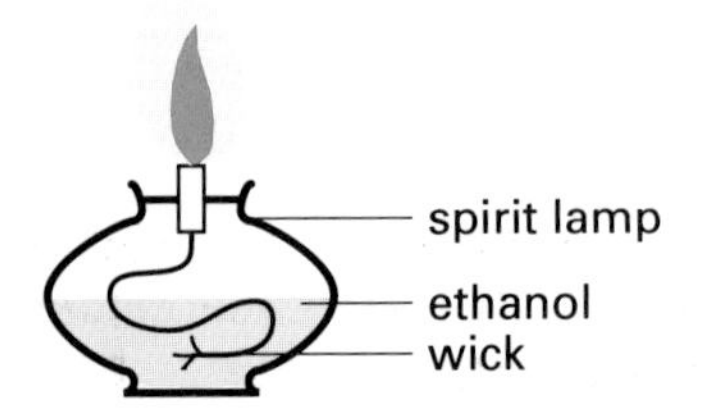

Figure 6.6 Combustion apparatus

Any value for the enthalpy of combustion obtained by this simple but relatively crude apparatus is likely to be significantly **less** than the standard figure, determined by more sophisticated apparatus. A certain amount of the heat produced by the combustion is absorbed by the beaker, the thermometer itself and most importantly the surrounding air.

heat given out by combustion reaction	=	heat absorbed by the water	← measured
		heat 'lost'	← not measured

Sample exercise

Ethanol is burned in a spirit lamp until 150 g of water in a can rises 10 K in temperature. The flame is snuffed and the mass of ethanol used up found (by subtraction) to be 0.253 g.
What figure for the enthalpy of combustion of ethanol can be obtained using such data?

Method

Molar mass of $C_2H_5OH = 46\ g\ mol^{-1}$

Therefore amount of ethanol used $= \frac{0.253}{46}$ mol

$= 0.0055$ mol

Heat absorbed by water $= cm\,\Delta T$

$= (4.18 \times 0.15 \times 10)$ kJ

$= 6.27$ kJ

Assuming all the heat produced by the combustion was absorbed by the water, we can say

0.0055 mol ethanol burns to give 6.27 kJ heat.

Therefore 1 mol ethanol should burn to give $(6.27 \times \frac{1}{0.0055})$ kJ

$= 1140$ kJ heat

Estimated enthalpy of combustion $= 1140\ kJ\ mol^{-1}$

An apparatus designed to give more accurate measures of enthalpies of combustion is the **flame calorimeter** shown in Figure 6.7. Oxygen gas is passed through the combustion chamber and the heat released by the reaction (burning) raises the temperature of the surrounding water bath.

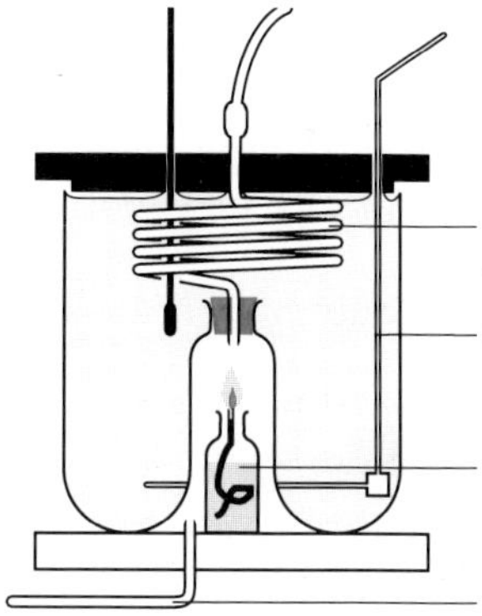

Figure 6.7 A flame calorimeter

6.3 Enthalpy of neutralisation

The neutralisation of an acid with a base, like combustion, is invariably exothermic. Unlike combustion, neutralisation usually takes place in solution. Again, the enthalpy change must be precisely defined.

The enthalpy of neutralisation is the enthalpy change when one mole of water molecules is formed by the neutralisation of an acid with a base.

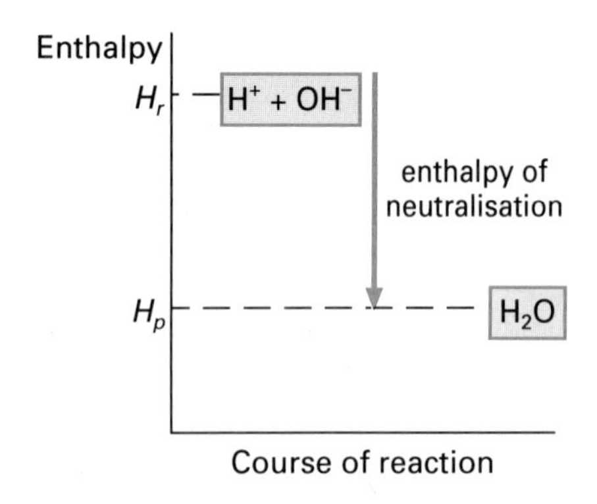

Figure 6.8 Enthalpy diagram for neutralisation

In neutralisation reactions hydrogen ions, from the acid, combine with hydroxide ions, from the base, to form molecules of water. The other ions present in the reaction mixture are merely spectators.

$$\mathbf{H^+(aq)} + Cl^-(aq) + Na^+(aq) + \mathbf{OH^-(aq)} \rightarrow \mathbf{H_2O(l)} + Na^+(aq) + Cl^-(aq)$$

hydrochloric acid + sodium hydroxide → water + sodium chloride

The heat released to the surroundings is the result of a fall in enthalpy in going from ions to molecule. The enthalpies of neutralisation for selected acid–base reactions are given in Table 6.1.

Acid	Base	ΔH(neutralisation)/kJ mol^{-1}
HCl(aq)	KOH(aq)	−57.3
HCl(aq)	NaOH(aq)	−57.3
HNO_3(aq)	NaOH(aq)	−57.3

Table 6.1 Enthalpies of neutralisation

The three neutralisation reactions release the same amount of heat. This is not surprising if we consider that each reaction is, effectively, the combination of the same pair of ions.

$$H^+(aq) + OH^-(aq) \longrightarrow H_2O(l)$$

A similar enthalpy change is therefore to be expected.

6.4 Enthalpy of solution

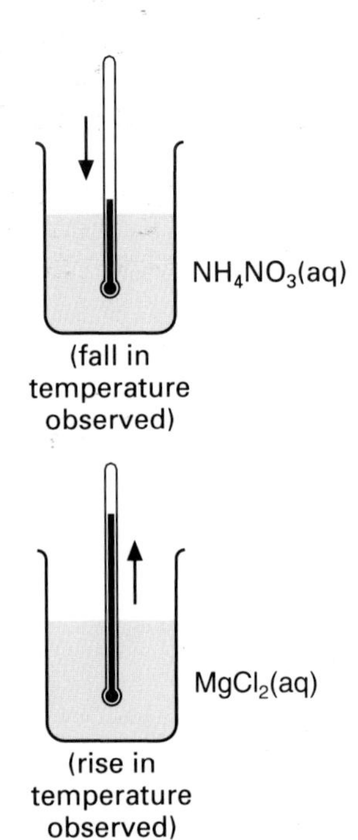

Figure 6.9 Dissolving endothermically and exothermically

The dissolving or solution process nearly always occurs with either an absorption or a release of heat. This indicates a change in enthalpy.

> ***The enthalpy of solution is the enthalpy change which occurs when one mole of solute is dissolved completely in a solvent.***

If ammonium nitrate, for example, is dissolved in water the mixture becomes cool indicating that, for ammonium nitrate, the solution process is **endothermic**. On the other hand, when magnesium chloride is dissolved in water the mixture becomes warm, suggesting that, in this case, the solution process is **exothermic**.

Some hot and cold packs, used by athletes to treat minor injuries, contain inorganic salts which dissolve in water, either exothermically or endothermically, to produce the heat or coolness required.

Sodium chloride dissolves in water with only a tiny output of heat. The equation for its solution in water is as follows.

$$Na^+Cl^-(s) \longrightarrow Na^+(aq) + Cl^-(aq) \qquad \Delta H = +5 \text{ kJ mol}^{-1}$$

In order to understand the energy changes involved, the above solution process can be divided into two distinct processes which occur simultaneously:

(a) lattice-breaking
(b) hydration.

Lattice-breaking

The first process involves separating the ions, packed firmly within the solid, so that they are infinitely far apart. To do this, it is necessary to break apart the lattice, by pulling the ions away from one another. Thus the process

$$Na^{+}Cl^{-}(s) \longrightarrow Na^{+}(g) + Cl^{-}(g)$$

requires an input of energy, i.e. it is endothermic. The separated ions are assigned the 'gaseous' label (g) to indicate their complete freedom

The amount of heat absorbed when one mole of an ionic compound is decomposed into isolated ions is termed lattice enthalpy or, more accurately, enthalpy of lattice-breaking.

Ionic substances, because of the very strong electrostatic attraction between oppositely charged ions, have quite large enthalpies of lattice-breaking.

Hydration

In the second step of the solution process, the 'free' ions enter the water where they become hydrated.

$$Na^{+}(g) + Cl^{-}(g) \longrightarrow Na^{+}(aq) + Cl^{-}(aq)$$

Each ion is attracted to the oppositely charged sides of the water molecules and this attraction results in a release of heat.

The amount of heat released when one mole of gaseous ions is completely hydrated is described as the enthalpy of hydration.

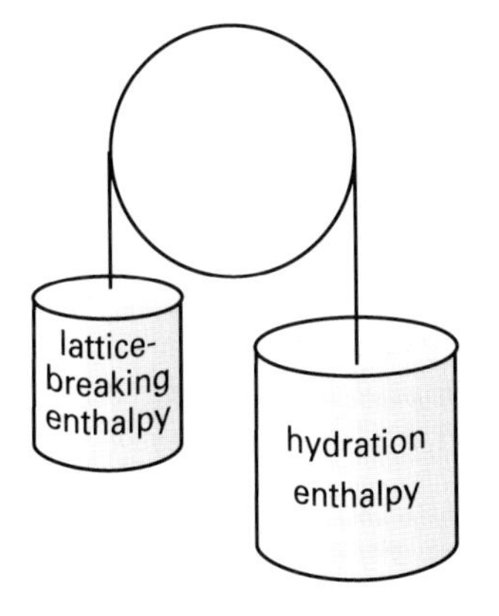

Figure 6.10 Enthalpy of solution: balance of lattice-breaking and hydration

The overall change that takes place when sodium chloride dissolves in water can be represented by the two steps that we have just considered.

The enthalpy of solution corresponds to the net energy change that occurs, and is equal to the **difference** between the amount of heat supplied during the first process and the amount of heat evolved during the second process.

If lattice enthalpy is greater than hydration enthalpy, a net input of heat is required when the substance dissolves and the process will be endothermic. On the other hand, if hydration enthalpy is larger than lattice enthalpy, more heat is released when the ions become hydrated than is required to break up the lattice, and an exothermic change will be observed.

The enthalpy diagram for the solution of sodium chloride

$$Na^{+}Cl^{-}(s) \longrightarrow Na^{+}(aq) + Cl^{-}(aq)$$

is given in Figure 6.11.

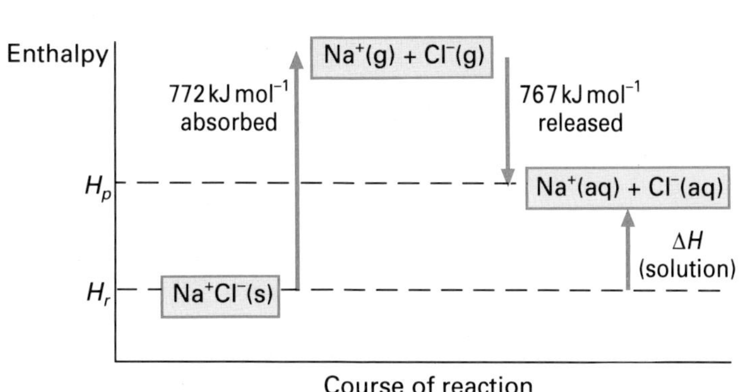

Figure 6.11 Enthalpy diagram for solution of NaCl (y axis not to scale)

Energy data for sodium chloride's solution in water can be compared to those of lithium and potassium chloride (Table 6.2).

Compound	ΔH(lattice-breaking)/ kJ mol^{-1}	ΔH(hydration)/ kJ mol^{-1}	Calculated ΔH(solution)/ kJ mol^{-1}
LiCl	+832	−882	−50
NaCl	+772	−769	+3
KCl	+690	−685	+5

Table 6.2 Enthalpies of solution

6.5 Hess's law

In calculating enthalpy of solution, we assumed that the overall enthalpy change for the solution will be equal to the **sum** of the enthalpy changes for each step, i.e. we have assumed that enthalpy changes are additive. This assumption is based on the law of conservation of energy which says that no energy is ever lost or gained during changes in the **form** of energy. There may, of course, be several steps to an overall conversion but the principle still applies.

If a chemical conversion can be carried out in a series of steps, the enthalpy change for the conversion will be equal to the sum of the enthalpy changes for each step.

If we extend this reasoning further, then it follows that the enthalpy change in converting R to P will be the same **regardless of the route** by which the chemical change R → P occurs. This principle was established, in 1840, through the work of a professor of chemistry in Leningrad, Germain Henri Hess. It is known as **Hess's law of constant heat summation**.

Hess's law states that the total enthalpy change accompanying a chemical conversion is independent of the route by which the conversion takes place.

For example, the conversion of sodium hydroxide solid to sodium chloride solution can be achieved by two possible routes. One is a direct single-step process, the other an indirect two-step process. All steps are exothermic.

Route 1. $NaOH(s) \xrightarrow{+HCl(aq)} NaCl(aq)$

Route 2. NaOH(s) → $+ H_2O$ → NaOH(aq) → $+HCl(aq)$ → NaCl(aq)

The enthalpy diagrams for each route are shown in Figure 6.12.

If Hess's law holds, then the enthalpy change for route 1 must be equal to the overall enthalpy change for route 2, i.e. $\Delta H_1 = \Delta H_{2A} + \Delta H_{2B}$.

If Hess's law did not hold, then ΔH_1 would be greater or less than $\Delta H_{2B} + \Delta H_{2B}$. This implies that we could convert NaOH(s) to NaCl(aq) by one route, releasing an amount of energy which is greater or less than the

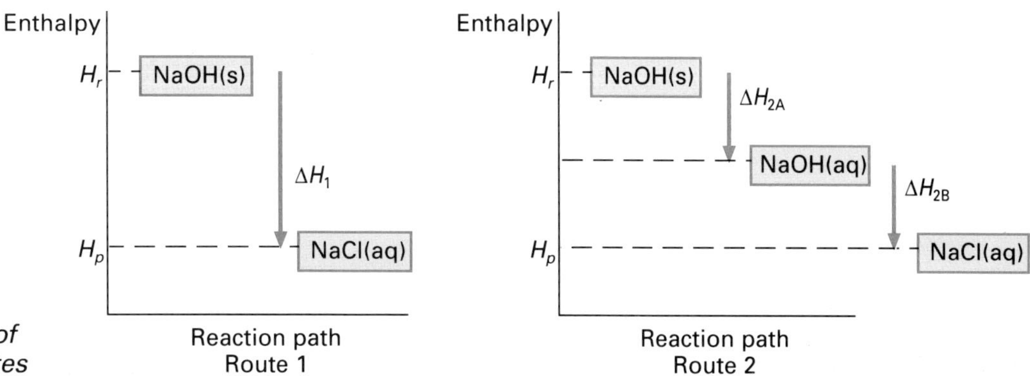

Figure 6.12 Formation of NaCl(aq) by different routes

energy absorbed if we reversed the change, using the other route. In other words, we could recover the NaOH(s) with a gain or loss of energy overall.

In effect this would be a means of creating or destroying energy and as such it goes against the law of conservation of energy.

6.5.1 Experiment to verify Hess's law

The three chemical reactions referred to in the NaOH → NaCl conversion can be carried out in the lab, the amounts of heat released measured and Hess's law confirmed. For simplicity we have to make the following approximation: the density and specific heat capacity of NaOH(aq), and of NaCl(aq), is the same as that of water, i.e. 1 kg l^{-1} and 4.18 kJ kg^{-1} K^{-1}.

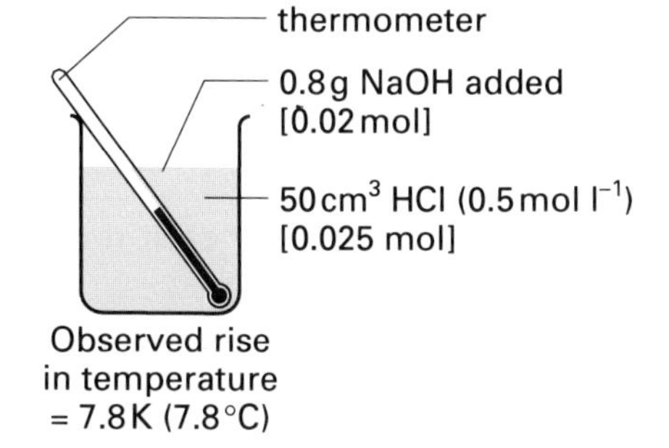

Figure 6.13 Reaction 1

Route 1 **$NaOH(s) + HCl(aq) \longrightarrow NaCl(aq) + H_2O(l)$**

Heat evolved by reaction of 0.02 mol NaOH(s) = amount of heat absorbed by solution

$= cm\,\Delta T$

$= (4.18 \times 0.05 \times 7.8)$ kJ

$= 1.63$ kJ

Amount of heat that would be evolved by reaction of 1 mol NaOH(s)

$= (1.672 \times \frac{1}{0.02})$ kJ

$= 81.51$ kJ

Enthalpy change for route 1 = ΔH_1 = −81.51 kJ mol^{-1}

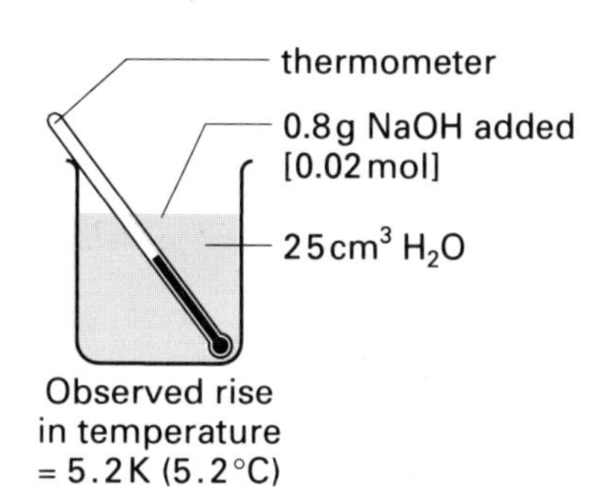

Figure 6.14 Reaction 2A

Route 2 **Reaction 2A** **$NaOH(s) + H_2O(l) \longrightarrow NaOH(aq)$**

Heat evolved by reaction of 0.02 mol NaOH(s) = amount of heat absorbed by solution

$= cm\,\Delta T$

$= (4.18 \times 0.025 \times 5.2)$ kJ

$= 0.543$ kJ

Amount of heat that would be evolved by reaction of 1 mol NaOH(s)

$= (0.54 \times \frac{1}{0.02})$ kJ

$= 27.00$ kJ

Enthalpy change for reaction 2A = ΔH_{2A} = −27.00 kJ mol^{-1}

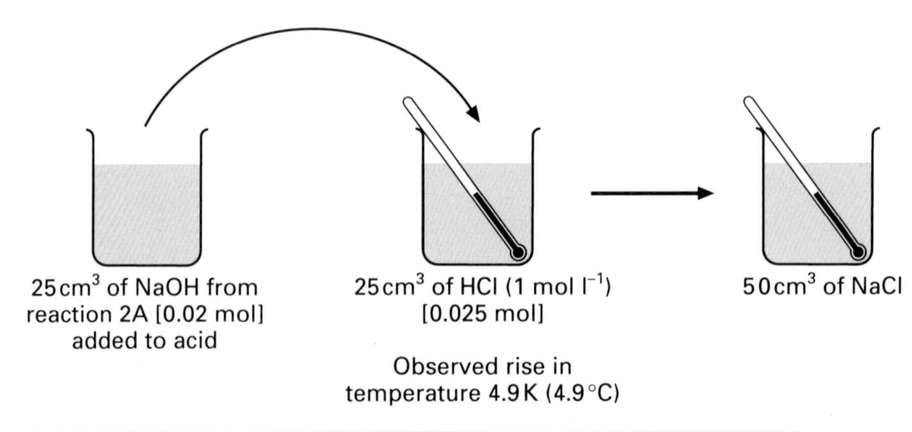

Figure 6.15 Reaction 2B

Reaction 2B $NaOH(aq) + HCl(aq) \longrightarrow NaCl(aq) + H_2O(l)$

Heat evolved by reaction of 0.02 mol NaOH(aq) = amount of heat absorbed by solution

$$= cm\,\Delta T$$
$$= (4.18 \times 0.05 \times 4.9)\text{ kJ}$$
$$= 1.02\text{ kJ}$$

Amount of heat that would be evolved by reaction of 1 mol NaOH(aq)

$$= (1.02 \times \tfrac{1}{0.02})\text{ kJ}$$
$$= 51.12\text{ kJ}$$

Enthalpy change for reaction 2B = ΔH_{2B} = -51.12 kJ mol^{-1}

To summarise the experimental results,

$$\Delta H \text{ for route } 1 = -81.51\text{ kJ mol}^{-1}$$

$$\text{overall } \Delta H \text{ for route } 2 = [-27.00 + (-51.12)]\text{ kJ mol}^{-1} = -78.12\text{ kJ mol}^{-1}$$

Within the limits of experimental error the enthalpy changes for each route are the same. This confirms Hess's law.

6.6 Using Hess's law to determine ΔH values

The great value of Hess's law is that it can be used to calculate enthalpy changes that cannot be measured directly by experiment.

A useful characteristic of any compound is its enthalpy of formation.

The enthalpy of formation is the enthalpy change which occurs when one mole of the compound is formed from its elements, in their normal physical states.

Now for many compounds the enthalpy of formation cannot be determined directly by experiment. Take propene, for example. Its formation from elements is described by the equation:

$$3C(s) + 3H_2(g) \longrightarrow C_3H_6(g)$$

Unfortunately this synthesis cannot be performed in a lab. Carbon and hydrogen will not combine directly and the reverse change is also not possible; propene will not decompose into its elements.

How then can we find the ΔH for the formation of propene? All three substances, carbon, hydrogen and propene, can be burned and their enthalpies of combustion measured accurately.

Substance	Enthalpy of combustion/kJ mol^{-1}
carbon	−394
hydrogen	−286
propene	−2056

Table 6.3 Enthalpies of combustion

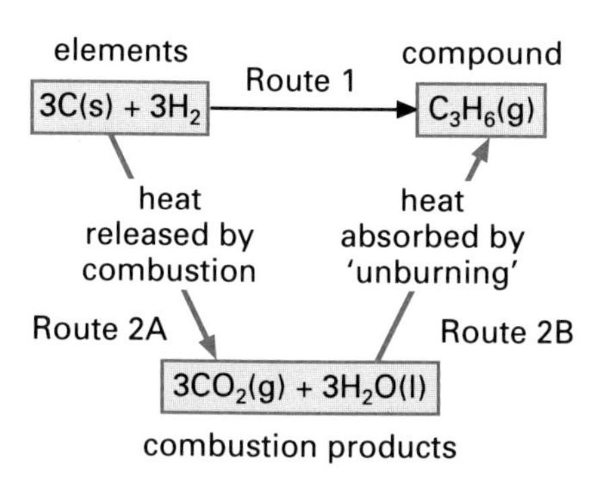

Figure 6.16 Route to propene via combustion

The products of the combustions, carbon dioxide and water, act as the 'stepping stone' which enables us to link carbon and hydrogen (the reactants) with propene (the product).
The equation for the formation of propene can be constructed by:

(a) burning 3 mol C atoms
(b) burning 3 mol H_2 molecules
(c) 'unburning' 1 mol C_3H_6 molecules (reversing the combustion, in theory).

The ΔH values in kJ mol^{-1} are adjusted accordingly (i.e. multiplied or sign reversed) to give ΔH values (in kJ) for the amounts of reaction shown in the equations.

ΔH/kJ mol^{-1}

Route 2A
$3C(s) + 3O_2(g) \longrightarrow 3CO_2(g)$ $\quad -(3 \times 394) = -1182$

Route 2A
$3H_2(g) + 1\frac{1}{2}O_2(g) \longrightarrow 3H_2O(l)$ $\quad -(3 \times 286) = -858$

(total for Route 2A: −2040)

Route 2B
$3CO_2(g) + 3H_2O(l) \longrightarrow C_3H_6(g) + 4\frac{1}{2}O_2(g) + 2056$

Route 1
$3C(s) + 3H_2(g) \longrightarrow C_3H_6(g)$ $\quad -2040 + 2056 = +16$

Therefore ΔH(formation) of propene $= +16$ kJ mol^{-1}

An enthalpy diagram for the formation of propene is given in Figure 6.17.

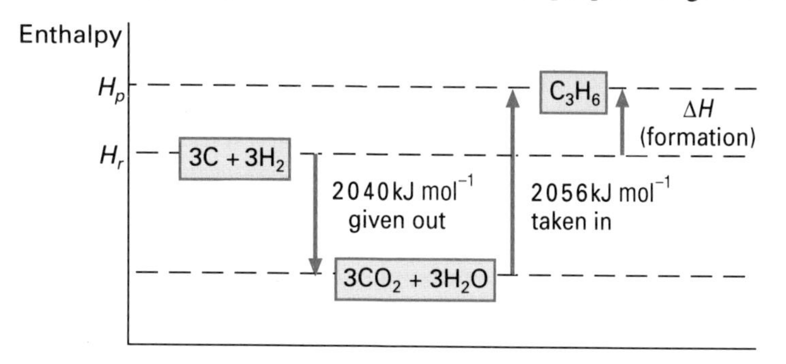

Figure 6.17 Enthalpy diagram for propene formation (y axis not to scale)

Sample exercise

Calculate the enthalpy of formation of benzene, C_6H_6.

Method

We have to calculate ΔH for the reaction:

$$6C(s) + 3H_2(g) \longrightarrow C_6H_6(l)$$

An alternative route, with of course the same overall ΔH , is by burning 6 mol C and 3 mol H_2 to form combustion products which can then be 'unburned' to form 1 mol C_6H_6.

The enthalpies of combustion of carbon, hydrogen and benzene can be obtained from a databook (−394, −286 and −3273 kJ mol^{-1} respectively).

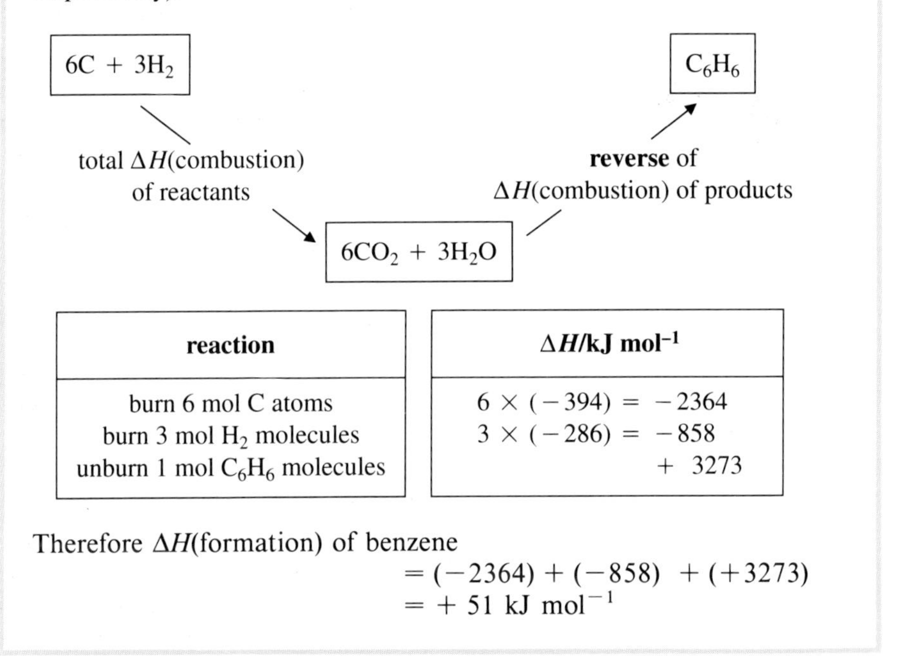

reaction	ΔH/kJ mol^{-1}
burn 6 mol C atoms	$6 \times (-394) = -2364$
burn 3 mol H_2 molecules	$3 \times (-286) = -858$
unburn 1 mol C_6H_6 molecules	$+ 3273$

Therefore ΔH(formation) of benzene

$$= (-2364) + (-858) + (+3273)$$
$$= +51 \text{ kJ mol}^{-1}$$

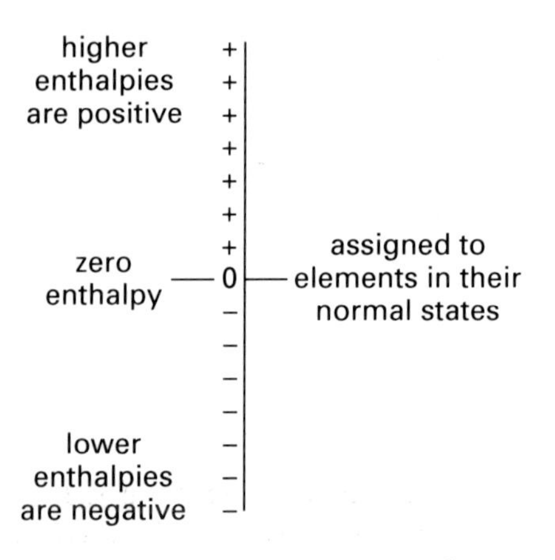

Figure 6.18 Enthalpy scale for elements and their compounds

As has been said already, absolute enthalpies cannot be measured, only **differences** between enthalpies. For the purposes of calculation, however, a value of **zero** is given to the enthalpies of **elements in their normal states**. This does not mean that elements have no enthalpy. In enthalpy diagrams we wish to use elements as the reference line, against which the enthalpies of their compounds can be compared: positive values for compounds with higher enthalpies and negative values for compounds with lower enthalpies.

6.7 Relative stability of compounds

The enthalpies of formation of compounds may be positive or negative. The vast majority of such enthalpies are **negative**: this indicates that the forming of most compounds, from their elements, is an **exothermic** process.

Hydrogen halide compounds show a range of ΔH(formation) values (Figure 6.19). As usual, the elements (hydrogen and halogen) are assigned an enthalpy value of zero and act as a reference mark for other enthalpies.

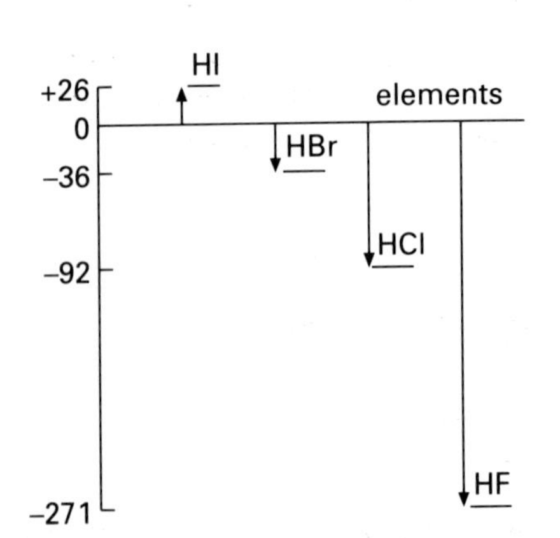

Figure 6.19 Enthalpies of formation of hydrogen halides (in kJ mol^{-1})

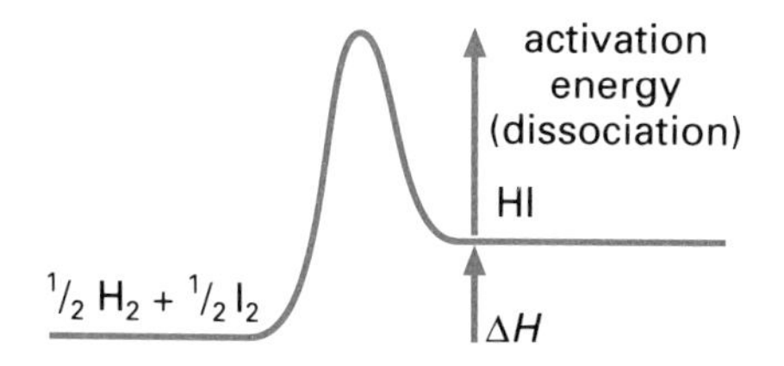

Figure 6.20 Energy diagram for HI

We describe hydrogen iodide, HI, as being an **unstable** compound **with respect to its elements.** This is because HI is at a higher enthalpy level, and dissociation into its elements would be energetically more favourable than combination of the elements to form HI. We must be careful to stress, however, that the **rate** at which unstable HI dissociates depends on the size of the **activation energy** for that reaction, not the enthalpy change, which is the same size for formation and dissociation. The other halides, HBr, HCl and HF, are more stable than their elements. Hydrogen fluoride has the largest negative ΔH(formation) (-271 kJ mol^{-1}) and is the most stable of these compounds with respect to its elements.

6.7.1 Reduction of oxides

Metal oxides show considerable variation in stability. Table 6.4 lists the enthalpies of formation of selected oxides.

Table 6.4 ΔH(formation) for selected metal oxides

Metal oxide	ΔH(formation)/kJ mol^{-1}
Al_2O_3	-1676
Mn_3O_4	-1388
Cr_2O_3	-1140
Co_3O_4	-879
Fe_2O_3	-824
ZnO	-348
PbO	-217
CuO	-157

Few metals are found uncombined. Most metals have to be extracted by industrial process from ores, which are often oxides or easily converted to oxides. Less stable oxides, such as those of zinc, lead and copper, are reduced to the metal by reaction with carbon. More stable oxides, those belonging to metals like manganese, chromium and cobalt, are reduced by reaction with aluminium.

$$\text{aluminium} + \text{metal oxide} \longrightarrow \text{aluminium oxide} + \text{metal}$$

The energy released by the formation of the exceptionally stable aluminium oxide helps to bring about the decomposition of the less stable metal oxide. One example of this 'thermit' process is the extraction of chromium.

		ΔH/kJ mol^{-1}
$2\,Al(s) + 1\frac{1}{2}O_2(g)$	$\longrightarrow Al_2O_3(s)$	-1676
$Cr_2O_3(s)$	$\longrightarrow 2Cr(s) + 1\frac{1}{2}O_2(g)$	$+1140$
$2Al(s) + Cr_2O_3(s)$	$\longrightarrow Al_2O_3(s) + 2Cr(s)$	-536

The surplus heat from the reaction (536 kJ per mole of Al_2O_3 produced) ensures that the process, once started, is self-sustaining. The output of heat is also sufficient to raise the temperature of the products above 2300 K.

At this temperature both Al_2O_3 and Cr are molten (liquids) rather than solid and can be easily separated from one another.

Even more heat is released from the reaction between aluminium and iron(III) oxide, Fe_2O_3. The reaction is used, not to extract iron (there are other cheaper processes), but to provide the high temperatures needed to weld metals together (Figure 6.21).

Figure 6.21 Molten iron, produced in the reaction of iron(III) oxide and aluminium powder (the 'thermit' reaction), being used for welding steel rails

6.8 Lattice enthalpy

As we have seen, the sign and size of its enthalpy of formation indicates how stable a compound is with respect to its elements. For an ionic compound, a major factor in determining how stable the compound will be is lattice enthalpy or, to give it its more precise name, **enthalpy of lattice-breaking.** If we deal with singly charged ions, it is the enthalpy change for the process.

$$M^+X^-(s) \longrightarrow M^+(g) + X^-(g)$$

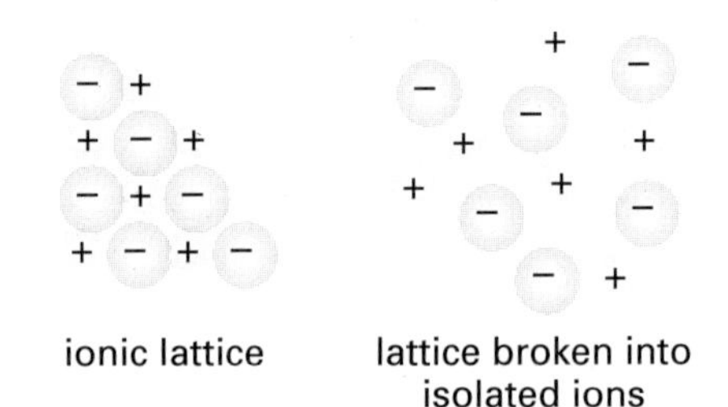

Figure 6.22 Lattice-breaking

This ΔH is always **positive** since heat energy is always required to dismantle an ionic lattice into separate isolated ions.

Unfortunately, lattice-breaking enthalpies cannot be measured directly by experiment. They can, however, be determined indirectly from other experimentally measurable data, linked in an energy sequence known as the **Born Haber cycle.**

The cycle is similar to that for obtaining enthalpies of formation of molecular compounds in that it consists of two routes:

(a) a direct one-step conversion

(b) an indirect two-step pathway.

Figure 6.23 outlines the choice of routes using sodium chloride as an example. The enthalpy of formation of the compound can be measured from the direct reaction of sodium metal with chlorine gas in a calorimeter.

$$\Delta H(\text{formation}) \text{ of NaCl} = -410 \text{ kJ mol}^{-1}$$

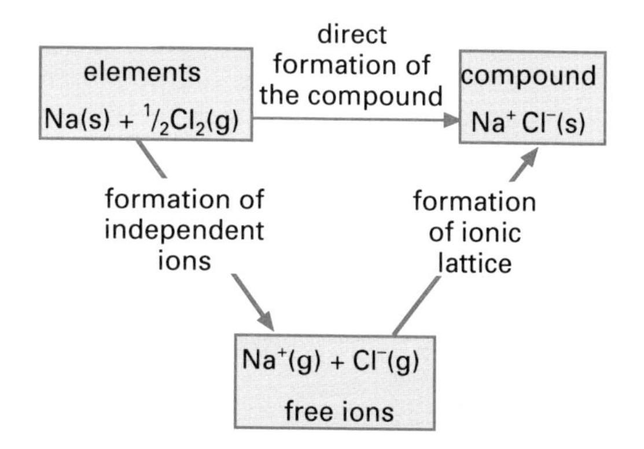

Figure 6.23 Formation of NaCl

The alternative **indirect** route to NaCl will have the same enthalpy change since the same product and reactants are involved.

We can calculate lattice-forming enthalpy by subtraction, if we can first measure the energy involved in creating the free sodium and chloride ions. This latter process can be considered as four individual steps, the first two involving sodium, the last two involving chlorine.

Sublimation and ionisation of sodium

Solid sodium must be converted into gaseous sodium (sublimed) in order to have free atoms. Each of these free atoms can then be stripped of its outer electron (ionised) to form a positive sodium ion. Both steps are **endothermic**.

$$\text{Na(s)} \xrightarrow[+107\ \text{kJ mol}^{-1}]{\text{enthalpy of sublimation } (\Delta H_s)} \text{Na(g)} \xrightarrow[+502\ \text{kJ mol}^{-1}]{\text{ionisation enthalpy } (\Delta H_i)} \text{Na}^+\text{(g)}$$

Atomisation and electron gain of chlorine

Chlorine molecules must be dissociated into single atoms, each of which can then attach an electron to form a negative chloride ion. The first step (atomisation) is **endothermic**, the second step (electron gain) is **exothermic**.

$$\tfrac{1}{2}\text{Cl}_2\text{(g)} \xrightarrow[+121\ \text{kJ mol}^{-1}]{\text{enthalpy of atomisation } (\Delta H_a)} \text{Cl(g)} \xrightarrow[-368\ \text{kJ mol}^{-1}]{\text{enthalpy of electron gain } (\Delta H_{eg})} \text{Cl}^-\text{(g)}$$

The enthalpy diagram for the formation of sodium chloride (showing both routes) is given in Figure 6.24.

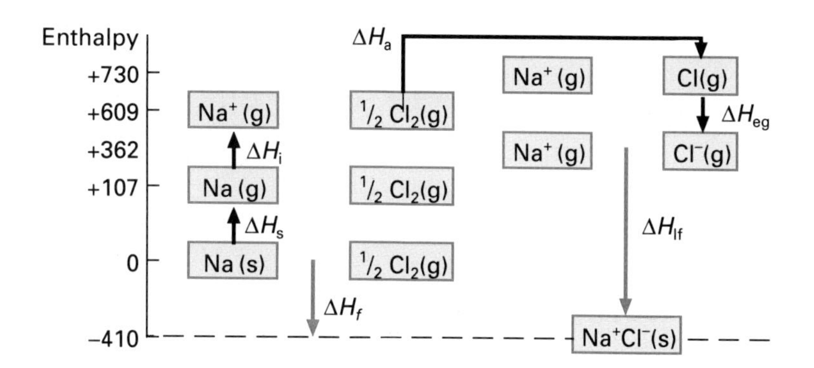

Figure 6.24 Enthalpy diagram: formation of NaCl (y axis not to scale)

The enthalpy of lattice-forming (ΔH_{lf}) is calculated by equating the formation enthalpy (ΔH_f) of the compound (route 1) with the algebraic sum of the enthalpies associated with the five steps in the formation (route 2).

= exothermic change

= endothermic change

ΔH_f −410	=	ΔH_s 107	+	ΔH_i 502	+	ΔH_a 121
Formation		Sublimation		ionisation		atomisation

+	ΔH_{eg} −368	+	ΔH_{lf} ?
	electron gain		lattice-forming

$$-410 = 107 + 502 + 121 - 368 + \Delta H_{lf}$$
$$\Delta H_{lf} = -772$$

Therefore enthalpy of lattice-forming of NaCl(s) = −772 kJ mol^{-1}
and enthalpy of lattice-breaking of NaCl(s) = +772 kJ mol^{-1}

6.9 Bond enthalpies

A chemical reaction is brought about by the breaking of existing bonds in the reactant(s) and the forming of new bonds in the product(s). In **ionic** reactions we are often dealing with **general** electrostatic attraction between oppositely charged particles, where it is difficult to pinpoint individual bonds. In **molecular** reactions, where covalent bonds are being broken and reformed, it is possible to assign a particular enthalpy change to a **specific** bond.

Bond enthalpy is the amount of heat absorbed when one mole of free bonds is broken to form separate atoms.

$$A{-}B(g) \longrightarrow A(g) + B(g)$$

A similar amount of heat would be **given out** when one mole of A—B bonds is formed from the free atoms.

Useful evidence, in support of the argument that you can attach an enthalpy value to a particular covalent bond, comes from the combustion of alkanes. Figure 6.25 shows the relative sizes of **enthalpies** of combustion for four successive members of the alkane family.

ethane CH_3CH_3
propane $CH_3CH_2CH_3$
butane $CH_3CH_2CH_2CH_3$
pentane $CH_3CH_2CH_2CH_2CH_3$

Enthalpy of combustion /kJ mol^{-1}

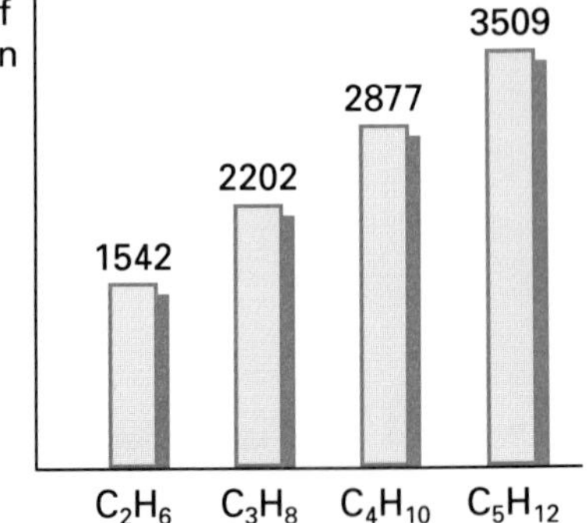

Figure 6.25 Enthalpies of combustion of alkanes

The roughly constant increase in enthalpy (over 600 kJ mol^{-1}) can be attributed to the addition of an extra $-CH_2-$ unit to each molecule. During combustion, each CH_2 is converted to CO_2 and H_2O and this appears to involve the release of a more or less fixed amount of energy. The same trend can be observed in the enthalpies of combustion of other series of compounds, e.g. alkanols.

Sample exercise

The enthalpies of combustion of ethene and but-1-ene are -1387 and -2711 kJ mol^{-1} respectively.
Predict the approximate enthalpy of combustion of propene.

Method

We must assume that the enthalpies of combustion of alkenes increase regularly with increasing chain length.
The ΔH_c for propene, C_3H_6, will be roughly half-way between the values for C_2H_4 and C_4H_8.
ΔH_c(propene) = ΔH_c(ethene) + difference between ΔH_c values for ethene and butene
$= -1387 + (-2711 + 1387)$ kJ mol^{-1}
$= -1387 + (-1324)$ kJ mol^{-1}
$= -(1387 - 662)$ kJ mol^{-1}
$= -2049$ kJ mol^{-1}
So predicted enthalpy of combustion of propene is -2049 kJ mol^{-1}.

6.9.1 Using bond enthalpies to predict ΔH

Bond enthalpies can be used to predict a value for the enthalpy change in a reaction. Take, for example, the combination of hydrogen and chlorine to form hydrogen chloride.

$$H_2(g) + Cl_2(g) \longrightarrow 2HCl(g)$$

The bond enthalpies are given in Table 6.5.

Bond	Bond enthalpy/kJ mol^{-1}
H—H	+436
Cl—Cl	+243
H—Cl	+431

Table 6.5 Bond enthalpies

The reaction itself can be split into two separate steps: bond breaking (endothermic) and bond forming (exothermic).

In the first step, the reactant molecules H_2 and Cl_2 dissociate to free atoms. In the second step, the freed atoms rearrange and combine to form HCl molecules.

$H_2(g)$ + $Cl_2(g)$ 2HCl(g)

→ 2H(g) + 2Cl(g) →

The enthalpy account may be presented as shown below.

Bonds broken: heat taken in		Bonds formed: heat given out	
H—H	+436	2(H—Cl)	2 × 431
Cl—Cl	+243		
Heat input:	+679	Heat output:	−862

$$\text{Balance} = (-862 + 679)\ \text{kJ mol}^{-1} = -183\ \text{kJ mol}^{-1}$$

$H_2(g) + Cl_2(g) \longrightarrow 2HCl(g) \quad \Delta H = -183\ \text{kJ mol}^{-1}$ of reaction as written

The enthalpy diagram for the conversion is outlined in Figure 6.26.

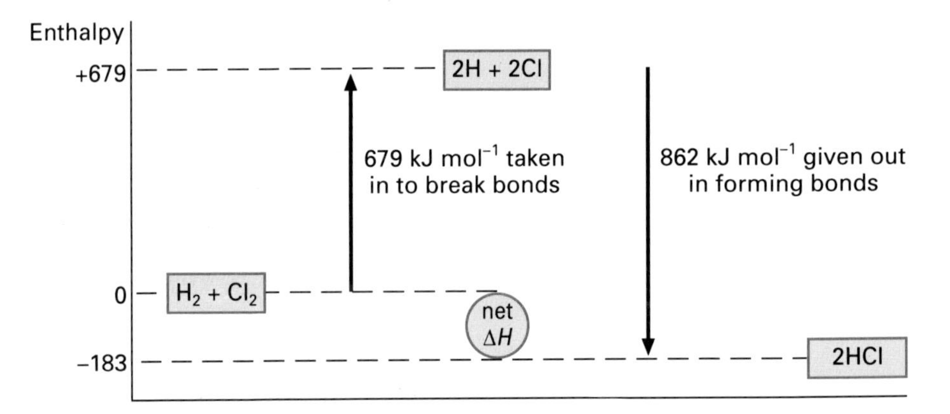

Figure 6.26 Enthalpy diagram for HCl formation (y axis not to scale)

The net amount of 183 kJ of heat energy is released by the formation of two moles of hydrogen chloride. The standard enthalpy of formation of hydrogen chloride would be the change associated with one mole of HCl and the thermochemical equation would be as follows:

$$\tfrac{1}{2}H_2(g) + \tfrac{1}{2}Cl_2(g) \longrightarrow HCl(g) \quad \Delta H\ \text{formation} = -91.5\ \text{kJ mol}^{-1}$$

6.9.2 Average bond enthalpies

Bond enthalpies apply to the dissociation and reforming of molecules in the gaseous state. Only if all reactants and products are gases can reaction enthalpies be predicted from bond enthalpy accounting. In practice, we seldom calculate ΔH(reaction) values from bond enthalpies if ΔH (formation) data is available instead. The reason for this reluctance is that, in most cases, it is not possible to assign an **exact** value to a bond enthalpy.

Take, for instance, the water molecule H_2O. The two O—H bonds are equivalent, but if the bonds are broken one at a time the enthalpy changes would not be the same.

	ΔH/kJ mol^{-1}
H—O—H(g) ⟶ HO(g) + H(g)	+501
HO(g) ⟶ H(g) + O(g)	+425

The second O—H bond is easier to break than the first bond since the fragment (OH) remaining after one H atom has been renewed is not as stable as the original water molecule.

In water's case the O—H bond enthalpy quoted would have to be an average value.

$$H_2O(g) \longrightarrow 2H(g) + O(g) \quad \Delta H = 501 + 425 = 926 \text{ kJ mol}^{-1}$$

Therefore average O—H bond enthalpy in water = 463 kJ mol^{-1}

The O—H bond is not of course confined to water but appears in molecules of a wide range of compounds, alcohols, carboxylic acids, etc. The O—H bond enthalpy listed in the databooks (458 kJ mol^{-1}) is an **average** or **mean** value derived from a large number of cases.

Only for diatomic molecules such as HBr, O_2, etc., can we assign exact bond enthalpies. For all others we must accept average values. Since such mean bond enthalpies are generalised figures, a ΔH value obtained by the use of these values must be regarded merely as an **estimate** of the enthalpy change for any particular reaction.

Sample exercise

Use bond enthalpies listed in a databook to estimate the enthalpy change for the following reaction:

$$C_2H_4(g) + H_2(g) \longrightarrow C_2H_6(g)$$

Method

In order to identify more easily the bonds which are broken and formed, it is worthwhile writing out structural formulae.

```
 H   H                       H   H
 |   |                       |   |
H—C=C—H  +  H—H  ⟶   H—C—C—H
                             |   |
                             H   H
```

The enthalpy account can now be drawn up.

Bonds broken: heat taken in		Bonds formed: heat given out	
C=C	+697	C—C	−337
H—H	+436	2(C—H)	−828
Heat input:	+1043	Heat output:	−1165

Balance=ΔH(reaction) = −122 kJ mol^{-1}

Summary of Unit 6

Having read and understood the information and ideas given in this unit, you should now be able to:

explain what is meant by enthalpy change and be able to calculate its value from an enthalpy diagram

distinguish endothermic and exothermic changes in terms of heat exchanged with surroundings and sign of enthalpy change

use the product, $cm\,\Delta T$, to calculate heat taken in or given out

describe how the enthalpy of combustion of a simple alkanol can be determined from a lab experiment

explain why most bond enthalpies are average values

use bond enthalpies to calculate the enthalpy change for a reaction

indicate how Hess's law is based on the principle of conservation of energy

- ▲ account for the similarities between enthalpies of neutralisation
- ▲ relate the enthalpy of solution of ionic compounds in water to enthalpies of lattice-breaking and hydration
- ▲ relate the enthalpy of formation of an ionic compound to the enthalpies of sublimation, ionisation, atomisation, electron gain and lattice-forming
- ▲ outline the connection between the stability of a compound and its enthalpy of formation
- ▲ apply Hess's law in the calculation of enthalpy changes (ΔH values)

PROBLEM SOLVING EXERCISES

1. Enthalpies of combustion of alkanes C_9–C_{11} are listed below.

Compound	Formula	ΔH(combustion)/kJ mol^{-1}
nonane	C_9H_{20}	−6118
decane	$C_{10}H_{22}$	−6770
undecane	$C_{11}H_{24}$	−7424

By using a graph or by calculation, predict a value for the enthalpy of combustion of dodecane, $C_{12}H_{26}$.

(PS skill 10)

2. Calcium oxide (quicklime) is produced industrially by decomposing calcium carbonate (limestone) in a kiln.

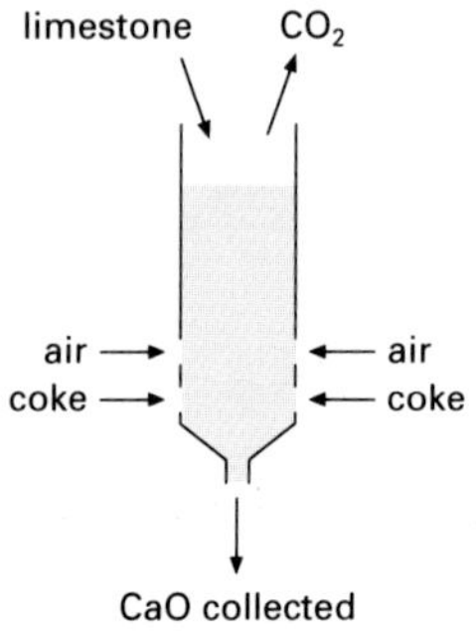

limestone CO$_2$
air → ← air
coke → ← coke
CaO collected

Figure 6.27

The reactions occurring in the kiln are as follows:

$CaCO_3(s) \rightarrow CaO(s) + CO_2(g)$ $\quad \Delta H = +178$ kJ mol^{-1}
$C(s) + O_2(g) \rightarrow CO_2(g)$ $\quad \Delta H = -393$ kJ mol^{-1}

Why add coke and air to the kiln rather than simply decomposing the limestone?

(PS skill 9)

3. The bond lengths and bond enthalpies of some 'interhalogen' compounds are given in the table.

Compound	ClF	BrF	BrCl	IF	ICl
Bond length/pm	163	176	216	191	232
Bond enthalpy/kJ mol^{-1}	253	237	218	278	208

A general trend can be observed: the longer the bond the lower the bond enthalpy. Which compound is **out of step** with the others?

(PS skill 1)

4. Why does ethanol, CH_3—CH_2—OH, have a different enthalpy of combustion from dimethyl ether, CH_3—O—CH_3?

(PS skill 7)

PROBLEM SOLVING EXERCISES

5. The thermochemical equation for the conversion of cyclopropane to 1-bromopropane is shown below.

$$C_3H_6 + HBr \longrightarrow C_3H_7Br \quad \Delta H = -3 \text{ kJ mol}^{-1}$$

By using the databook to find values for other bond enthalpies, calculate the mean bond enthalpy of the C—C bond in cyclopropane.

(PS skill 4)

6. The table below shows the approximate temperature rise ΔT when solutions of nitric acid and potassium hydroxide are mixed.

Acid	Alkali	ΔT/K
50 cm^3 1 mol l^{-1} HNO_3	50 cm^3 1 mol l^{-1} KOH	5
100 cm^3 1 mol l^{-1} HNO_3	100 cm^3 1 mol l^{-1} KOH	
50 cm^3 2 mol l^{-1} HNO_3	50 cm^3 2 mol l^{-1} KOH	

Complete the table.

(PS skill 6)

7. The enthalpies of lattice-breaking, ΔH_{lb}, for some ionic compounds are tabled below.

Compound	ΔH_{lb}/kJ mol^{-1}
NaCl	788
$MgCl_2$	2527
Na_2O	2570
MgO	3890

How would you explain this variation in enthalpy of lattice-breaking?

(PS skill 9)

8. The enthalpy changes associated with the dissociation of ammonia are listed below.

	ΔH/kJ mol^{-1}
$NH_3 \longrightarrow NH_2 + H$	427
$NH_2 \longrightarrow NH + H$	375
$NH \longrightarrow N + H$	356

Calculate the **average** bond dissociation enthalpy of the N—H bond in ammonia.

(PS skill 4)

9. The table below lists (a) enthalpies of formation for a range of metal oxides and (b) the times, measured or estimated, for these metals to form an oxide layer to a depth of 1 mm.

Metal	ΔH(formation)/ kJ per mol O_2	Time taken to form oxide layer/hours
Mg	−1162	more than 10^5
Ti	−848	less than 6
Fe	−508	24
Ni	−439	600
Co	−422	7
Ag	−5	very long
Au	+80	infinite

What correlation is there between the **energetic stability** of an oxide [as indicated by its ΔH(formation)] and the **rate** of its formation? Justify your answer.

(PS skill 8)

10. The enthalpy change when one mole of a metal carbonate is precipitated from aqueous solution is described as its enthalpy of precipitation. Three such ΔH precipitation values are given below. What value would you predict for the ΔH(precipitation) of $CaCO_3$?

Metal carbonate	ΔH (precipitation)/kJ mol^{-1}
$SrCO_3$	+2.9
$BaCO_3$	−5.0
$MgCO_3$	+14.2

(PS skill 10)

11. In a petrol engine, air and petrol are mixed before undergoing combustion. The ratio (proportion) of air to petrol in the fuel mixture affects the amounts of CO and NO_x gases produced by combustion and emitted in the exhaust. The relationships are shown by the graphs.

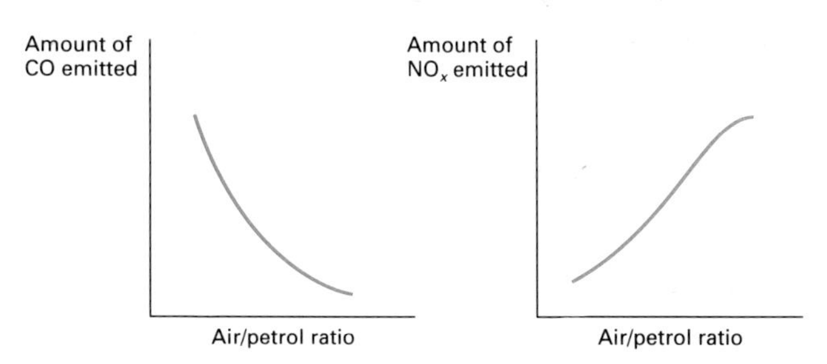

Can you explain the shape of each graph?

(PS skill 9)

Unit 7
Chemical Equilibrium

ASSUMED KNOWLEDGE AND UNDERSTANDING

Before starting on Unit 7 you should know and understand:

the nature and characteristics of acids and bases (alkalis)

the concept of neutralisation

the pH scale

the industrial synthesis of ammonia (Haber process)

the factors which affect reaction rates

the function and effect of catalysts

the structures of carboxylic acids and simple amines

how to carry out calculations involving moles and concentrations of solutions.

OUTLINE OF THE CORE AND EXTENSION MATERIAL

As you progress through Unit 7 you should, at least, try to grasp the fundamental content listed under **core** and, if possible, pick up the extra and/or more difficult content listed under **extension**.

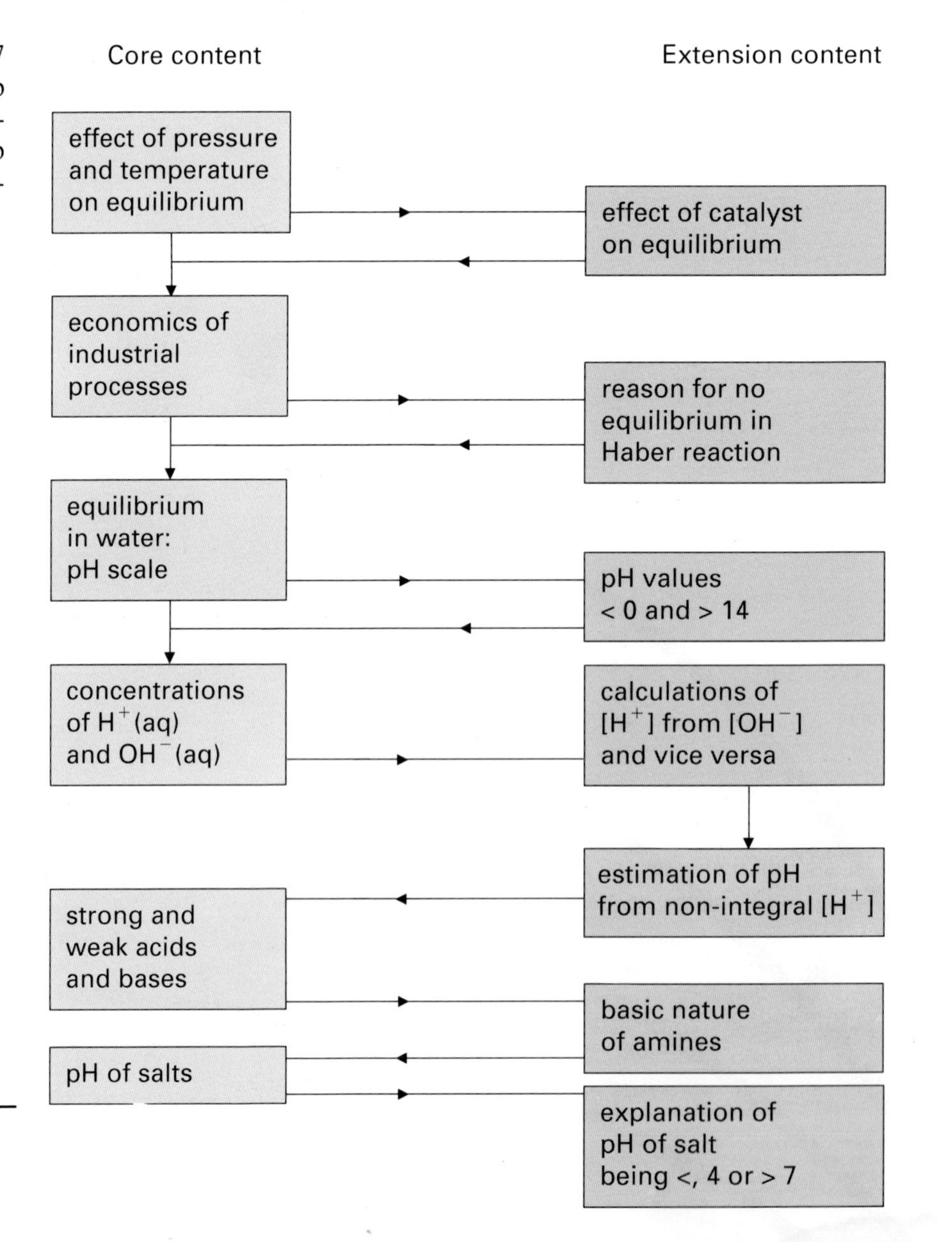

7 CHEMICAL EQUILIBRIUM

7.1 Reversible changes

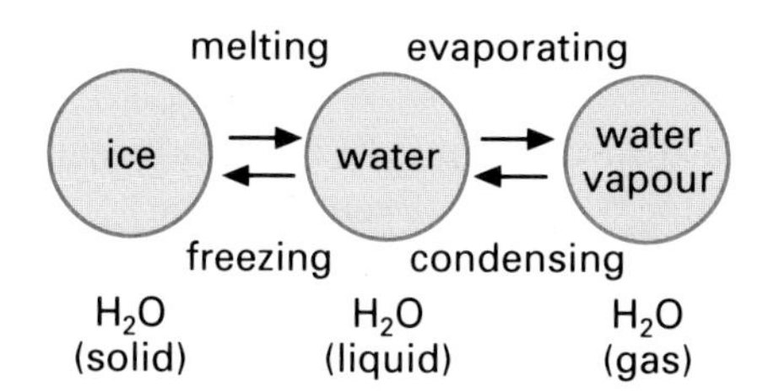

Figure 7.1 Reversible physical changes

In chemistry we are interested in the changes of form and composition of substances. Physical processes such as changes of state are examples of changes which can be easily reversed: in many cases the change can be turned around simply by altering the temperature of the substance. Water is a case in point.

Some chemical reactions are also easily reversed. When steam is passed over heated iron, iron oxide and hydrogen gas are produced. On the other hand, if hydrogen gas is passed over heated iron oxide, iron and steam are produced. The two changes can be collected into one equation with two arrows.

$$3Fe(s) + 4H_2O(g) \rightleftharpoons Fe_3O_4(s) + 4H_2(g)$$

Some reversible reactions which feature in biological systems or in chemical manufacturing are listed in Table 7.1.

Reversible reaction	Equation
photosynthesis and respiration	$6CO_2 + 6H_2O \rightleftharpoons C_6H_{12}O_6 + 6O_2$
formation and decomposition of sulphur trioxide	$SO_2 + \frac{1}{2}O_2 \rightleftharpoons SO_3$
decomposition and formation of chalk	$CaCO_3 \rightleftharpoons CaO + CO_2$
electrolytic refining of copper	$Cu \rightleftharpoons Cu^{2+} + 2e^-$

Table 7.1 Reversible reactions

Other chemical reactions may be much less reversible. In some cases the conditions which would bring about an observable reverse change may not be known. Such reactions are described as **irreversible**.

7.2 Equilibrium

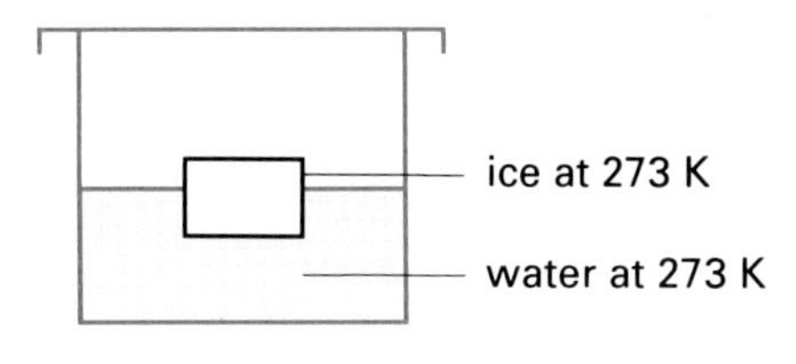

Figure 7.2 Ice–water equilibrium

The fact that changes are reversible leads to a situation called **equilibrium**. Consider, for example, the following situation. An ice cube is placed into water, contained inside a fully insulated flask.

At 273 K, ice melts to form water and water freezes to form ice. At the ice–water interface (boundary) the H_2O molecules are involved in the reversible change.

$$H_2O(s) \underset{\text{freezing}}{\overset{\text{melting}}{\rightleftharpoons}} H_2O(l)$$

After a time the rates of melting and freezing equalise. This can be inferred from observing that the amount of ice remains constant: it neither enlarges nor diminishes in size, despite the constant activity at its surface with water.

The ice and water at the interface are said to be in a **state of equilibrium**. This state is conveyed by an equation using **half arrows** as shown below.

$$H_2O(s) \rightleftharpoons H_2O(l)$$

A similar solid–liquid equilibrium is set up when a solid is surrounded by a saturated solution of itself in some solvent.

If grey crystals of iodine are placed in a dish of ethanol, a reddish-brown colour will appear in the liquid as solid iodine dissolves. Eventually the solid iodine stops diminishing in size and the liquid stops increasing in reddish-brown colour. No further change can be observed. At a molecular level, a state of equilibrium has been set up. Iodine is crystallising out of solution as fast as it is dissolving.

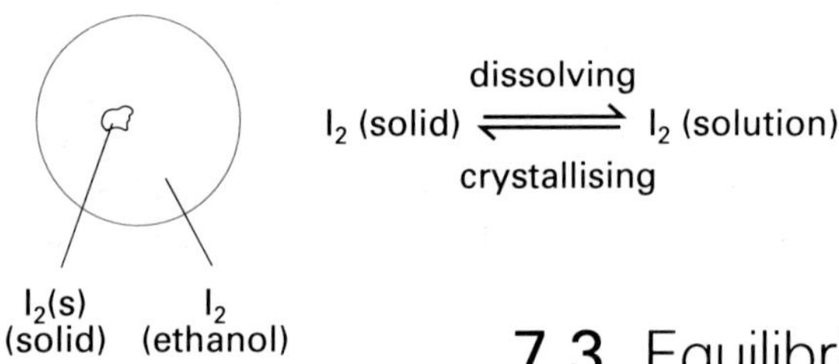

Figure 7.3 Iodine and equilibrium

7.3 Equilibrium in a chemical reaction

Chemical reactions involve changing one set of substances, the reactants, into a different set of substances, the products. If most reactions are reversible, at least to some slight extent, the labels 'reactants' and 'products' need to be carefully defined; the 'products' also may be reacting. We shall use the term 'reactants' to describe the substances which appear on the left-hand side of the reaction equation, and 'products' to describe the substances which appear on the right-hand side.

In a reversible reaction the rate of the forward reaction is, at first, greater than the rate of the backward reaction: this is simply because there is a much greater concentration of reactants than of products in the mixture. As time goes on, the concentration of reactants decreases while the concentration of products increases. This causes the forward reaction rate to fall and the rate of the backward reaction to rise. This continues until **the two rates become equal**. At this point the system is said to be in chemical equilibrium. At equilibrium. **as long as the same reaction conditions prevail**, the concentrations of reactants and products undergo no further change.

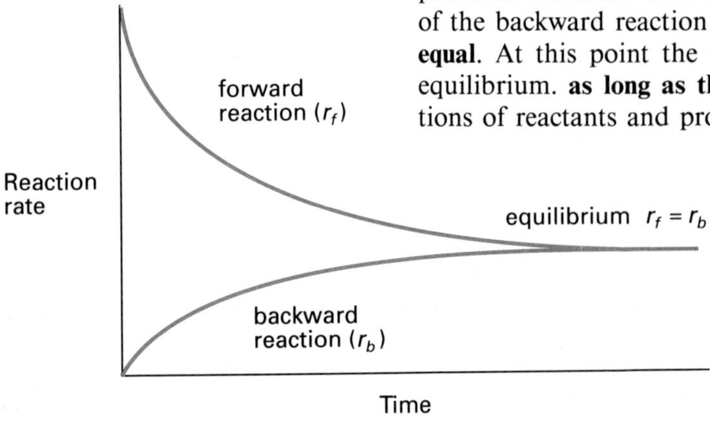

Figure 7.4 Rates and equilibrium

In a reversible reaction, equilibrium is reached when the rate of the backward reaction is equal to the rate of the forward reaction.

7.3.1 The Fe^{2+}/Ag^{+} : Fe^{3+}/Ag equilibrium

When silver(I) sulphate solution is added to iron(II) sulphate solution, silver is precipitated and iron(III) sulphate forms. This redox reaction does not proceed to 100 per cent completion. Instead it reaches an equilibrium as shown below.

$$Fe^{2+}(aq) + Ag^{+}(aq) \rightleftharpoons Fe^{3+}(aq) + Ag(s)$$

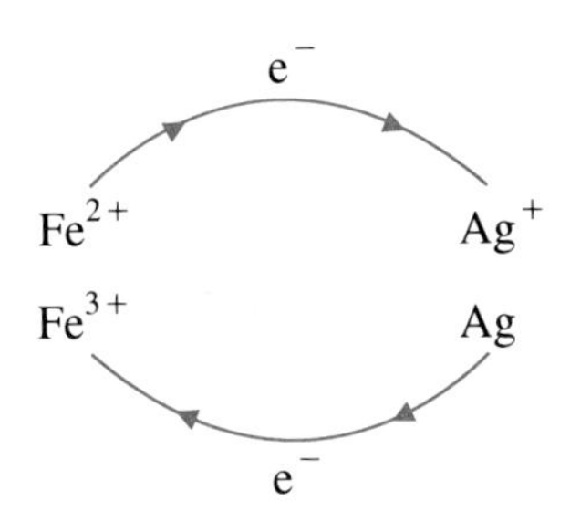

The two opposing reactions continue to take place after the constant concentrations have been attained. As fast as electrons are transferred from $Fe^{2+}(aq)$ to $Ag^{+}(aq)$, electrons are transferred from Ag(s) to $Fe^{3+}(aq)$.

To an external observer the reaction appears to have stopped. No further precipitation of silver takes place and the yellowish-brown colour of the solution, due to $Fe^{3+}(aq)$, does not change in intensity.

Chemical equilibrium is, however, not static but **dynamic**: the word implies continuing change and motion. The dynamic nature of this equilibrium can be proved with the help of a radioactive isotope. A tiny quota of **radioactive silver** solid, $^{*}Ag(s)$, is added to the equilibrium mixture. After some time radioactivity is detected in the solution. This is attributed to radioactive silver ions, $^{*}Ag^{+}(aq)$. The conversion from silver atoms to silver ions must therefore be taking place continually during the state of equilibrium,

together with continuous interchange of iron ions, $Fe^{3+} \rightleftharpoons Fe^{2+}$.

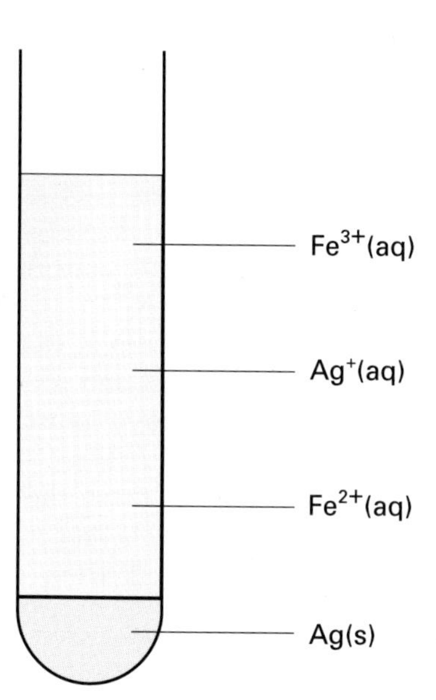

Figure 7.5 Equilibrium mixture

7.4 Position of equilibrium

In an **irreversible** reaction, which runs to 100 per cent completion, the concentration of reactants starts at a maximum and decreases eventually to zero. At the same time, the concentration of products starts from zero and rises to a maximum. Figure 7.6 shows diagrammatically the simultaneous run-down of reactants and build-up of products in an irreversible reaction.

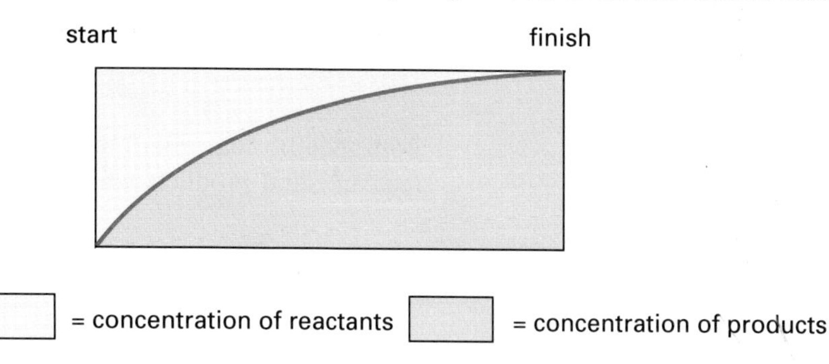

Figure 7.6 Outline of fall in reactants and rise in products

In a **reversible** reaction, reactants $\rightleftharpoons$ products, a state of equilibrium is set up before the reaction has been able to run its full course. In some cases the forward reaction is nearly complete before the rate of the reverse reaction is high enough to establish equilibrium. We say the reaction is very much **to the right**.

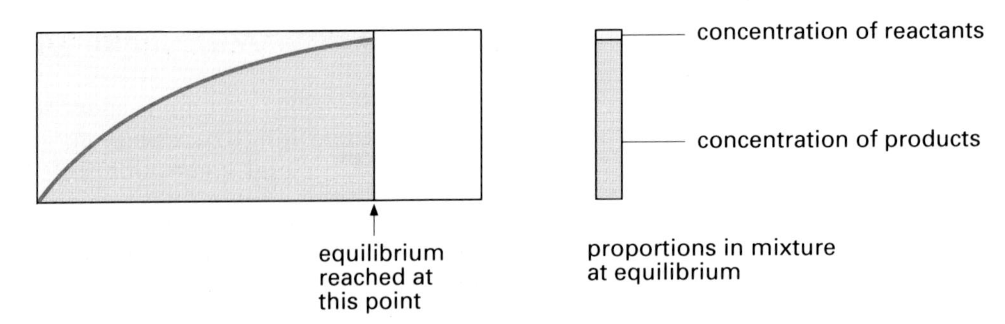

Figure 7.7 'To the right' reaction

In other cases the forward reaction is barely under way when equilibrium is established. This type of reaction would be described as being very much **to the left**.

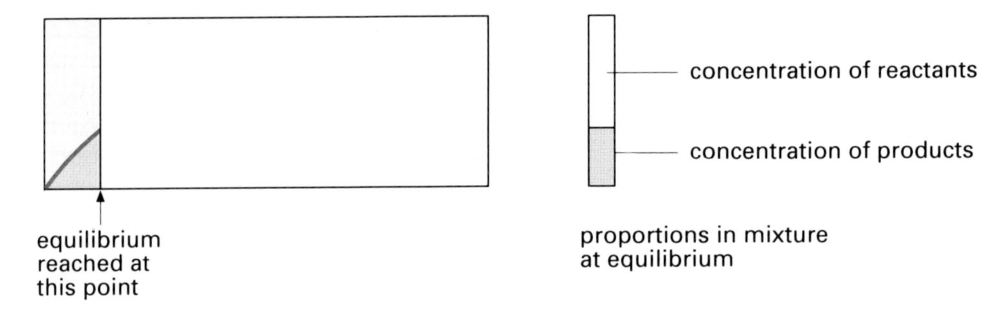

Figure 7.8 'To the left' reaction

In yet other reactions, both the forward and reverse reactions are given a decent run before equilibrium is set up. Neither reaction is particularly favoured or dominant and the equilibrium mixture will contain considerable concentrations of both reactants and products.

It must also be said that very few reactions reach equilibrium at exactly a half-way point with the mixture divided equally into reactants and products. At equilibrium, forward and back reaction **rates** are equal: the proportions of reactants and products are not.

Provided other conditions remain the same, a system will reach the same position of equilibrium whether it starts from 'reactants' or whether it starts from 'products'. An equilibrium state can be approached from either direction, as in the iron–silver system.

$$Fe^{2+} + Ag^{+} \longrightarrow \boxed{Fe^{2+} + Ag^{+} \rightleftharpoons Fe^{3+} + Ag} \longleftarrow Fe^{3+} + Ag$$

Both equilibrium mixtures will have exactly the same composition of reactants and products whichever route is taken.

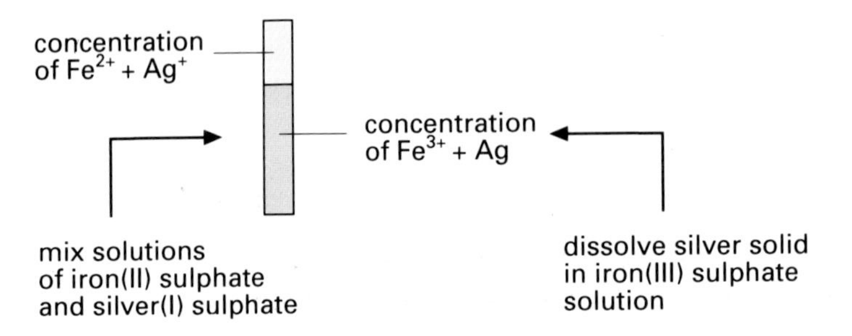

Figure 7.9 Two ways of producing an equilibrium mixture

7.5 Factors that disturb the position of equilibrium

A chemical system in a state of equilibrium consists of two reactions, a forward and an opposing backward reaction. It seems reasonable to suggest that the factors which affect rates of reaction **might** alter the delicate balance between proportions of reactants and products, and thus disturb the equilibrium. These factors are:

(a) catalysts
(b) concentration (solutions)
(c) pressure (gases)
(d) temperature.

7.5.1 Effect of a catalyst on the position of equilibrium

A catalyst reduces the activation energies of forward and back reactions to the same extent.

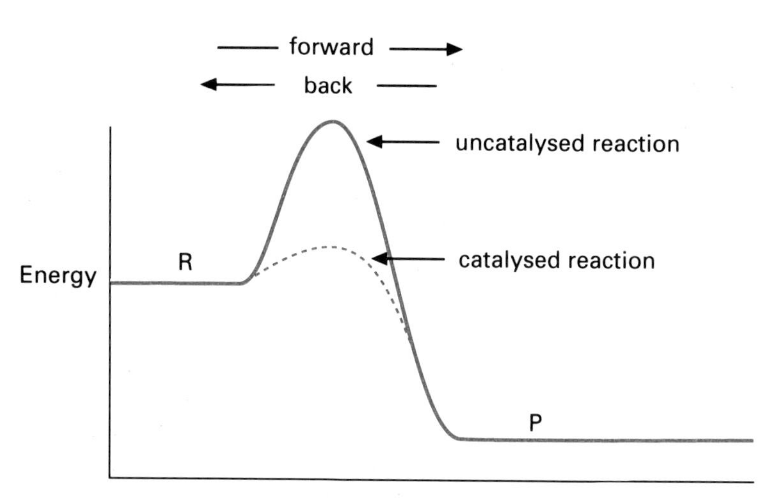

Figure 7.10 Catalysed and uncatalysed reactions

The **rates** of the forward and back reactions in the catalysed system are therefore increased equally. The time taken for the system to reach equilibrium will be shortened by using a catalyst (Figure 7.11).

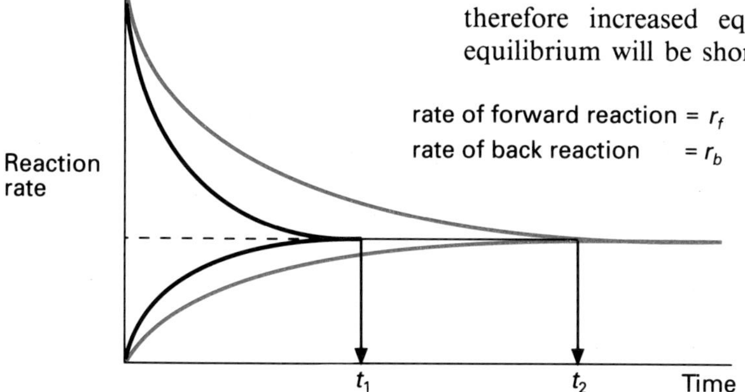

t_1 = time taken to reach equilibrium in the catalysed reaction
t_2 = time taken to reach equilibrium in the uncatalysed reaction

Figure 7.11 Time taken to reach equilibrium

Although the state of equilibrium will be established **more quickly** in a catalysed reaction, the relative concentrations of reactants and products in the mixture will be **exactly the same** as in the uncatalysed reaction.

> ***A catalyst therefore shortens the time taken to reach equilibrium but has no effect on the position of the equilibrium.***

7.5.2 Effect of a change of concentration on the position of equilibrium

Ions in aqueous solution cannot enter or leave the solution of their own accord. While gaseous products may bubble out of solution and escape from a state of equilibrium, the concentration of ions cannot spontaneously alter in this way. It is necessary to add substances to the solution. These may have the effect of increasing the concentration of some type of ion or other. On the other hand, the added substance may have the effect of removing ions from solution. Either way, this will disturb the existing equilibrium.

Suppose the composition of an equilibrium mixture in a solution is as shown in Figure 7.12.

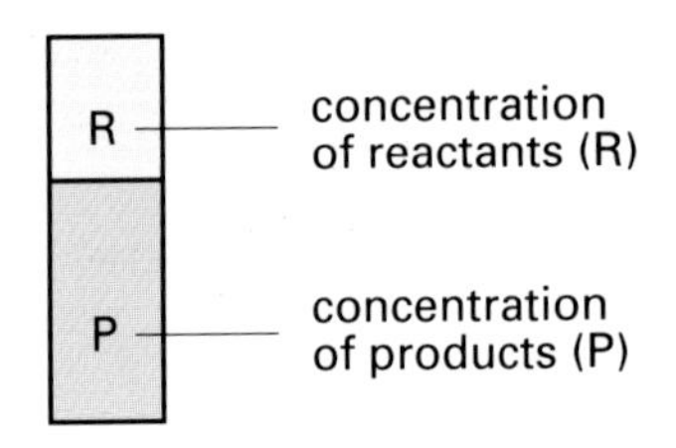

Figure 7.12 Equilibrium mixture

If **more reactant** is added to the solution, this will immediately raise the concentration of R. This, in turn, will increase the rate of the forward reaction but not that of the reverse reaction. Equilibrium will be lost but will gradually re-establish. However, the position of equilibrium will have shifted to a new position somewhat further to the right. That is to say the new equilibrium mixture will contain a **higher proportion of product** than it did before.

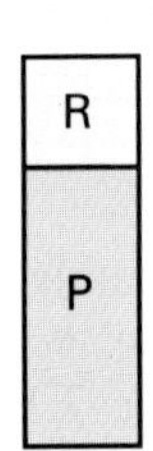

Figure 7.13 New equilibrium mixture after adding more R

The same effect, i.e. a shift to an equilibrium mixture containing relatively more product, can be achieved by another means. If **some product is removed** from the equilibrium mixture this will create an imbalance. The concentration of P will have been decreased and this will lower the rate of the back reaction. The R $\longrightarrow$ P reaction will be faster in comparison. Again equilibrium will be lost but eventually re-established, giving a mixture containing relatively **more** P and **less** R.

To increase the proportion of product in an equilibrium mixture the chemist has two options.

(a) To raise, by addition, the concentration of reactant(s).
(b) To decrease, by removal, the concentration of product(s).

Sample exercise

If alkali is added to aqueous potassium dichromate, the following equilibrium is set up.

$$\mathbf{Cr_2O_7^{2-}(aq)} + OH^-(aq) \rightleftharpoons \mathbf{2CrO_4^{2-}(aq)} + H^+(aq)$$

dichromate ion — orange; chromate ion — yellow

What changes in colour would you observe if you added to the solution
(a) some NaOH (aq)?
(b) some HCl(aq)?

Method

(a) If NaOH(aq) is added to the mixture, the concentration of $OH^-(aq)$ will be raised. This will have the effect of shifting the position of equilibrium to the right. In the new equilibrium mixture there will be a greater proportion of chromate ion, $CrO_4^{2-}(aq)$.
The solution will thus appear less orange and **more yellow** than before.

(b) If HCl(aq) is added, the concentration of $H^+(aq)$ will rise and this will shift the position of equilibrium to the left. A new mixture will be created, with relatively more dichromate ion, $Cr_2O_7^{2-}(aq)$.
The solution will therefore appear **more orange** and less yellow than before.

7.5.3 Effect of a change of pressure on the position of equilibrium

A change in **pressure** only affects equilibrium mixtures in which **gases** are involved. Altering the pressure may have the effect of favouring one of the two reactions (forward and backward) at the expense of the other.

$$\text{reactant gases} \underset{\text{backward}}{\overset{\text{forward}}{\rightleftharpoons}} \text{product gases}$$

Increasing the pressure favours whichever reaction brings out a **reduction** in the number of gas molecules. The equilibrium is lost but then re-established with the mixture now containing, in total, fewer gas molecules.

Increasing the pressure causes a shift to an equilibrium mixture with a smaller number of moles of gas molecules.

Conversely, **decreasing the pressure** causes the equilibrium to shift to a position where there is a **larger** overall number of moles of molecules in the mixture.

What happens if the equilibrium system contains equal amounts of reactant and product molecules? In this case a change in pressure will have **no effect** on the position of equilibrium.

Methanol is one of the major chemicals manufactured today. It is produced by combining carbon monoxide and hydrogen, the components of syngas. The equilibrium is represented as follows.

$$CO(g) + 2H_2(g) \rightleftharpoons CH_3OH(g)$$

3 moles of molecules — 1 mole of molecules

High pressure will favour the forward reaction as it creates a reduction in the amount of molecules. High pressure should therefore increase the yield of methanol.

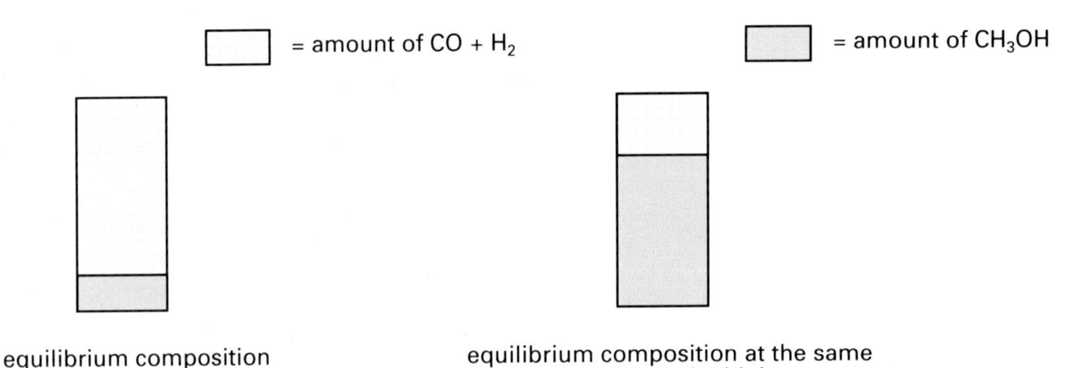

Figure 7.14 Effect of high pressure in methanol synthesis

The original high pressure process for making methanol was developed in Germany in 1923. The syngas mixture was compressed to 300 atmospheres before entering the catalytic converter.

In 1966 ICI introduced a more efficient catalyst which allowed the process to be run at lower pressures, between 50 and 100 atmospheres. At lower pressures the equilibrium yield is lower but the plant is simpler to build, cheaper to run and safer.

Syngas is itself generated by the reaction of steam with hydrocarbons such as methane.

$$CH_4(g) + H_2O(g) \rightleftharpoons CO(g) + 3H_2(g)$$

2 moles of molecules — 4 moles of molecules

In this reaction, raising the pressure would favour the back reaction and thus decrease the yield of syngas. Consequently the conversion is carried out at **normal** atmospheric pressure.

7.5.4 Effect of a change of temperature on the position of equilibrium

Chemical reactions are either exothermic or endothermic. Reversible reactions are exothermic in one direction and endothermic in the other direction. In a state of equilibrium both types of reaction must be occurring simultaneously.

Either (a) $R \underset{\text{exo}}{\overset{\text{endo}}{\rightleftharpoons}} P$ or (b) $R \underset{\text{endo}}{\overset{\text{exo}}{\rightleftharpoons}} P$

What effect will an increase in the temperature of an equilibrium system have on the rates of forward and back reactions? Both rates will be increased but **not equally**. A rise in temperature will increase the rate of an **endothermic** reaction **more** than that of the reverse exothermic reaction.

Raising the temperature of an equilibrium mixture therefore increases the rate of the endothermic reaction more than it increases the rate of the exothermic reaction opposing it. This will disturb the existing equilibrium and, although it is subsequently regained, a mixture of different composition will be formed. The direction of the shift in equilibrium is given as follows.

An increase in temperature shifts the equilibrium position to a mixture which is formed by absorbing heat.

Decreasing the temperature will have the opposite effect. A fall in temperature will favour the exothermic reaction. The equilibrium position will shift in the direction which brings about evolution of heat.

The synthesis of methanol is an **exothermic** reaction.

$$CO(g) + 2H_2(g) \underset{\text{endo}}{\overset{\text{exo}}{\rightleftharpoons}} CH_3OH \qquad \Delta H = -91 \text{ kJ mol}^{-1}$$

An increase in temperature will favour the back reaction and produce a mixture with less methanol than before. To obtain a high equilibrium yield of methanol the reaction ought, in theory, to be carried out at **low temperature**. However, low temperature means low rates of reaction and slow approach to equilibrium. To achieve a worthwhile **rate**, a moderately high temperature must be used even if this sacrifices some yield of methanol. The introduction, in 1966, of a much improved catalyst (copper zinc oxide) enabled the formation of methanol to be carried at lower temperatures. The ICI process is carried out at a temperature of 475–575 K compared to the 625–675 K required by the original German process.

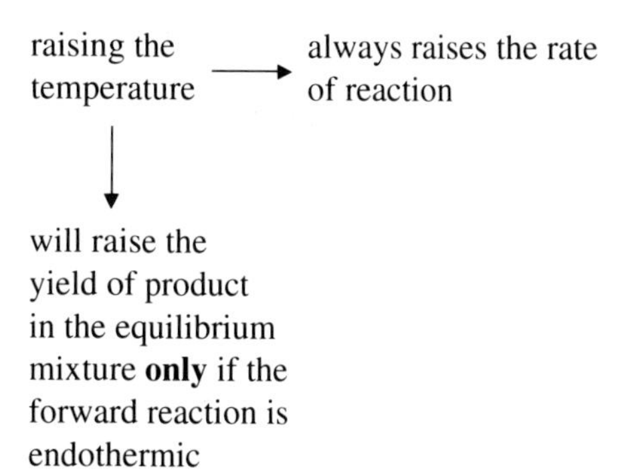

When the synthesis of a product occurs **endothermically**, then moving to **high temperatures** will increase yield of product, as well as ensuring a high reaction rate. One such example is the steam reforming of methane, to form syngas.

$$CH_4(g) + H_2O \underset{\text{endo}}{\overset{\text{exo}}{\rightleftharpoons}} CO(g) + 3H_2(g) \qquad \Delta H = +206 \text{ kJ mol}^{-1}$$

In practice, the reacting gases are passed over a nickel catalyst, in a furnace heated to temperatures as high as 1100 K.

In industry, to achieve good yield (high conversion) in an exothermic reaction, heat must be continually removed from the reactor. This prevents the temperature from rising as heat is given out during the reaction. In an endothermic reaction, heat must be continually supplied to the reactor in order to prevent the temperature from falling as heat is absorbed during the course of the reaction.

Sample exercise

The hydration of ethene to form ethanol is a gas phase catalysed reaction.

$$C_2H_4(g) + H_2O(g) \rightleftharpoons C_2H_5OH(g) \qquad \Delta H = -44 \text{ kJ mol}^{-1}$$

What conditions will improve the equilibrium yield of ethanol?

Method

(a) The formation of ethanol is exothermic (negative ΔH).
High temperature would favour the reverse (endothermic) reaction which is the dehydration of ethanol.
Low temperatures are therefore advised although it is accepted that temperature may have to be raised a little simply to achieve a decent rate of reaction.

(b) The formation of ethanol occurs with a decrease in the number of moles of gas (2 moles of molecules → 1 mole of molecules).
Raising the pressure on the system will therefore favour the production of ethanol.

7.6 Equilibrium and industrial processes

In designing and controlling large scale industrial reactions, it is important to determine the conditions which lead to a maximum yield of the product. The profitability of a process depends on forming as much of the product as quickly and as cheaply as possible.

The **Haber** process of synthesising ammonia is a good example of the application of chemical principles to an industrial situation. Fritz Haber, a German chemist, saw the formation of ammonia from its elements as an intellectual challenge. The reaction is reversible.

$$N_2(g) + 3H_2(g) \rightleftharpoons 2NH_3(g) \qquad \Delta H = -92 \text{ kJ mol}^{-1}$$

Figure 7.15 ICI's ammonia plant at Severnside, near Bristol; the tall pressure vessels in which the nitrogen and hydrogen react are conspicuous

In the absence of a catalyst nitrogen and hydrogen barely combine, even at high temperatures. High temperature would, in any case, force the equilibrium to the left and little ammonia would be formed. An **iron catalyst** was developed to produce not only a reasonably rapid reaction, but at a temperature sufficiently low to ensure a reasonably high yield of ammonia.

The formation of ammonia brings about a decrease in the total amount of gas. Haber thus constructed a **high pressure** apparatus in his lab to force equilibrium to the right. In 1909 he succeeded in producing a small amount of ammonia. The chemical company BASF bought rights to his process and their chemical engineer Carl Bosch organised the scaling up of Haber's lab process to industrial plant size. In 1913 the first full-scale ammonia plant went into operation at Ludwigshafen-Oppau, near Mannhein.

Both Haber and Bosch were awarded Nobel prizes for chemistry, Haber for his theoretical contribution and Bosch for his practical contribution of high pressure technology.

7.6.1 Practical aspects of ammonia production

Conditions of high pressure and low temperature are, in theory, required to guarantee a good concentration of ammonia in an equilibrium mixture with nitrogen and hydrogen. Industrialists, however, have to take other economic factors into consideration when deciding what the operating conditions should be. High pressures may well improve equilibrium yield of ammonia but we must remember that considerable power is required to drive compressors and that the chemical plant must be pressure tight.

Low temperatures are effective in reducing the amount of ammonia which decomposes, but they also reduce the rate at which reaction takes place. It takes time for gases to come to a state of equilibrium and prolonging this time does not make economic sense. A faster reaction rate can be achieved using higher temperatures, but higher temperatures can reduce the life and activity of the iron catalyst as well as decreasing the yield of ammonia.

In chemical manufacturing, compromises and balances have to be struck between all the competing factors. The pros and cons (advantages and disadvantages) are summarised in Table 7.2.

Condition	Pro	Con
high pressure	good equilibrium yield of NH_3	costly to build and operate
low temperature	good equilibrium yield of NH_3 and easy on catalyst	reaction slow to reach equilibrium

Table 7.2 Advantages and disadvantages of specific conditions

7.6.2 Equilibrium versus recycling

The percentage yield of ammonia in an equilibrium mixture at various pressures and temperatures is summarised in Figure 7.16.

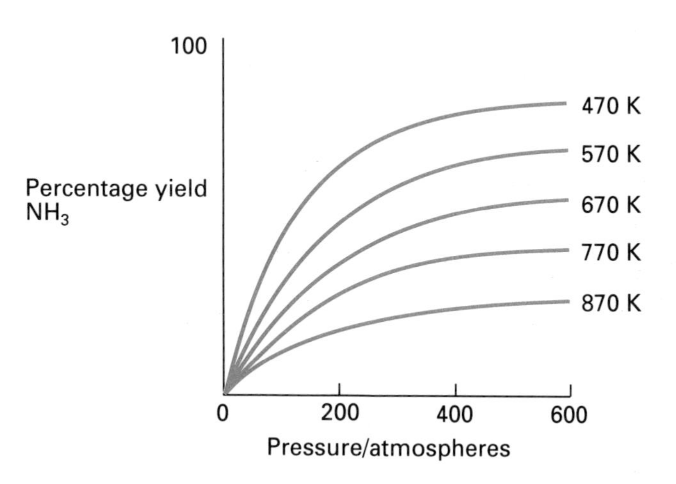

Figure 7.16 Equilibrium yields of ammonia at five different temperatures and varying pressures

A modern ammonia plant, using the ICI process, is likely to be operating at a pressure just over 80 atmospheres and temperatures around 500 K. Under these conditions equilibrium yield of ammonia would appear to be roughly 25 per cent. However, the actual fraction of nitrogen (and hydrogen) converted to ammonia in a typical plant is about 14 per cent. How does one explain this difference? The time spent by the gases in the catalytic converter is **too short** for an equilibrium state to be attained. Even with good catalysts the rate of reaction is too slow to justify waiting for equilibrium to be established.

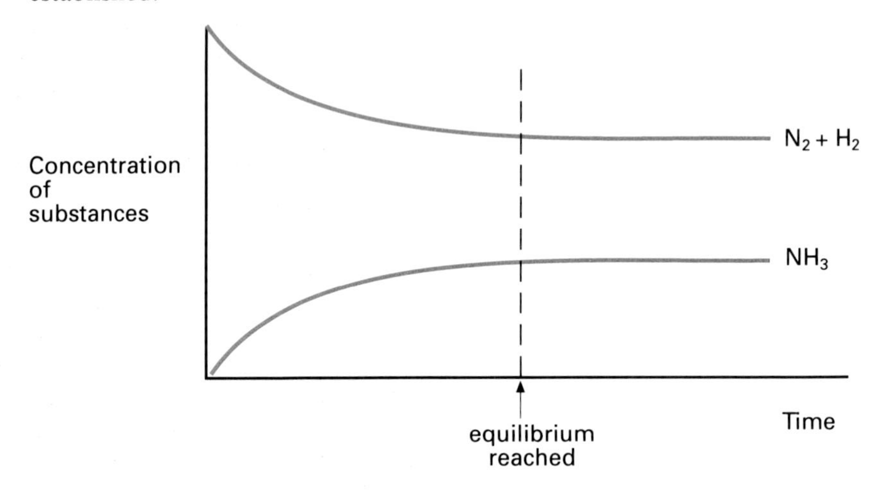

Figure 7.17 Variation in concentration of N_2+H_2 and NH_3 with time

It is more economical to use a **high rate of flow of reactants** over the catalyst, remove the ammonia formed and recycle the unreacted gases than to use a slower rate of flow, allow time for equilibrium to establish and obtain a higher yield of ammonia per 'pass'. The critical figure for chemical engineers is output per hour.

$$\text{output per hour} = \text{rate of flow} \times \text{conversion fraction}$$

A better output per hour is achieved by having a lower conversion fraction (14 per cent) but a much higher flow rate than would be the case if one waited for equilibrium to be attained every time.

A flow diagram of the latter stage of ammonia manufacture is given in Figure 7.18.

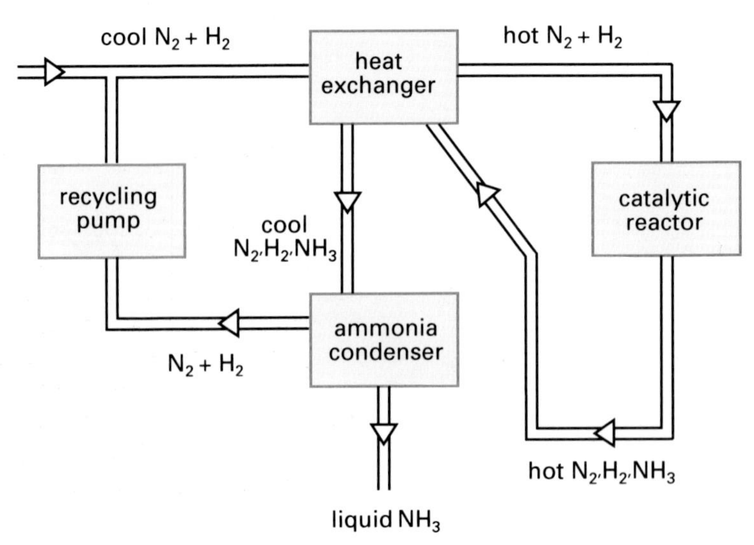

Figure 7.18 Later stages of the Haber process

The nitrogen–hydrogen mixture (1:3) is passed through the catalytic reactor, under high pressure and moderately high temperatures. The hot gases leaving the reactor contain about 14 per cent ammonia. Their heat is transferred to the cool incoming gases inside the heat exchanger before the gases are passed to a refrigeration unit. Here the ammonia is condensed and removed as a liquid.

The unreacted nitrogen and hydrogen gases are pumped back into the incoming gas stream and **recycled**. Repeated recycling enables plant operators to achieve a reaction efficiency (moles of NH_3 molecules finally formed per 100 moles of N atoms reacted) which is ultimately very high – up to 98 per cent. Continuous recycling can therefore make up for low conversion to ammonia on each single passage of reactants.

7.6.3 Availability of raw materials for ammonia manufacture

The **nitrogen** required for ammonia production is obtained from air. The other reactant, **hydrogen**, may be derived from a variety of sources. The first ammonia plants obtained their hydrogen from electrolysis of water, a procedure which used up considerable energy and was therefore very costly.

Nowadays, hydrogen gas is obtained from **syngas**, a mixture of carbon monoxide and hydrogen. The syngas is itself produced from natural gas, petroleum oil fractions or coal; it all depends on which natural resources are available to the country in which the ammonia plant is sited. Nearly all British production of syngas is currently from natural gas (North Sea) but India, for example, depends on naphtha as a source of syngas and South Africa is obliged to use coal.

The reactant gases must then be carefully purified in order to remove the carbon monoxide and substances which might poison the iron catalyst at the reactor stage. Compressors raise the gases to whatever high pressure is desired.

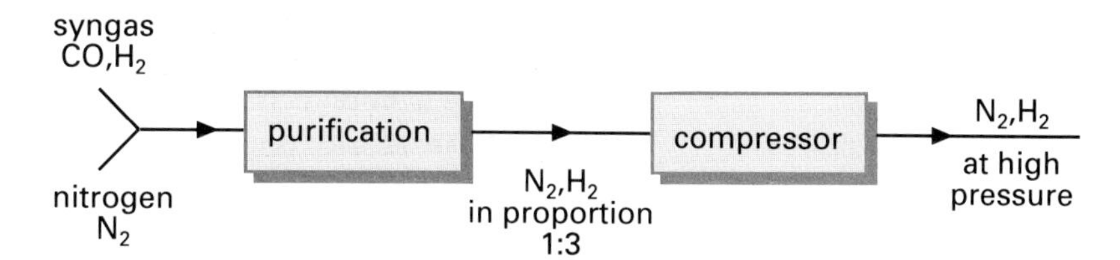

Figure 7.19 Production of reactants mixture

7.6.4 Marketability of ammonia

Ammonia is one of the most important manufactured chemicals as it is the starter chemical for the production of nitrogen-containing fertilisers such as liquid ammonia, ammonium sulphate, urea and nitrates.

The conversion of atmospheric nitrogen, N_2, to ammonia, NH_3, is called **nitrogen fixation**. In the Haber process, the tightly held nitrogen atoms of N_2 are torn apart and then attached to three less tightly held hydrogen atoms. The nitrogen is now more likely to undergo chemical change. These chemical changes result in its incorporation into nitrate ions, amino acid molecules and finally the protein chains of plants and other living organisms.

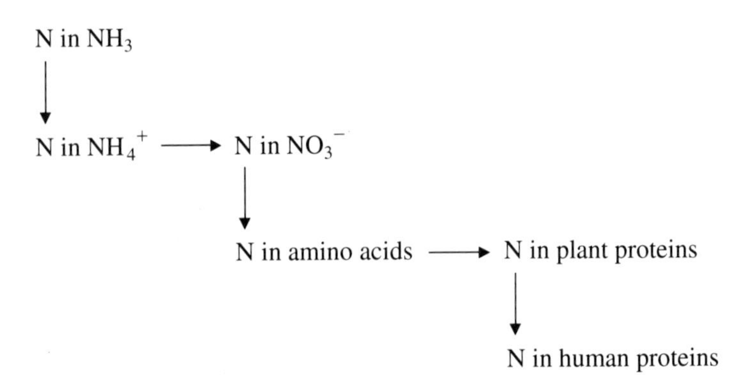

Since ammonia is a source of usable nitrogen for crops, it has to be manufactured on an enormous scale. It provides the fertilisers which help to grow the crops which feed the ever-increasing population of the world. It is also a starter molecule for the production of nitrogen-containing organic molecules in plastics, dyestuffs, explosives and pharmaceuticals.

7.7 Equilibrium in aqueous solutions

Careful measurements show that pure water conducts electric current to a very slight extent. This tiny conductivity is due to a **very slight dissociation** of water molecules into ions.

$$H_2O(l) \rightleftharpoons H^+(aq) + OH^-(aq)$$

$H^+(g)$ is a bare proton. Bare protons are not found in water. Because of their small size and high charge/mass ratio they are always bonded to at least one water molecule. The equation given above for the ionisation of water is a simple version. In water, protons can be **transferred from one water molecule to another** as shown by the more accurate equation below.

$$H_2O(l) + H_2O(l) \rightleftharpoons H_3O^+(aq) + OH^-(aq)$$

As a result, a minute fraction of molecules (one in 550 million) are in the ionised form at any one time. No molecule remains in that condition for long: ions form and disappear very rapidly, with H^+ transfer occurring at the rate of about 1000 times a second. For simplicity's sake, however, we shall use the symbol $H^+(aq)$ to represent hydrated protons.

7.7.1 Reaction of metals in water

When reactive metals, such as lithium, are placed in water a reaction occurs, an alkaline solution forms and hydrogen gas is evolved (Figure 7.20).

$$Li(s) + H_2O(l) \longrightarrow LiOH(aq) + \tfrac{1}{2}H_2(g)$$

This is a redox reaction which involves oxidation of the metal and reduction of hydrogen ions.

$$Li(s) \longrightarrow Li^+(aq) + e^-$$
$$e^- + H^+(aq) \longrightarrow \tfrac{1}{2}H_2(g)$$

This begs the question: if there are so few hydrogen ions in water, how can

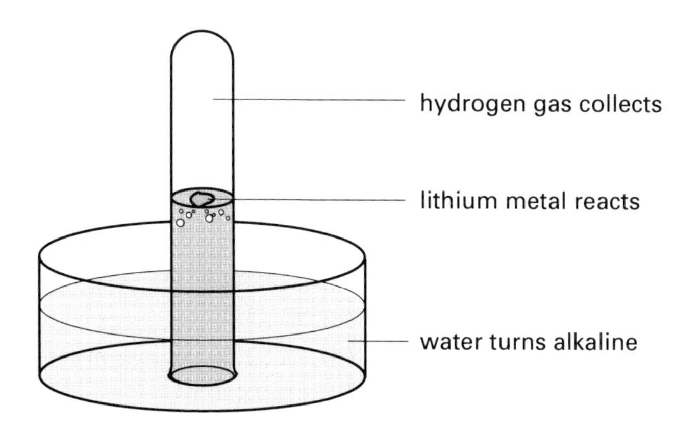

Figure 7.20 Reaction of lithium and water

a metal **continually** find H^+ ions to reduce and displace as molecules of gas? The answer lies with the equilibrium between the molecules of water and its ions.

$$H_2O(l) \rightleftharpoons H^+(aq) + OH^-(aq)$$

As each H^+ ion is reduced and removed from solution as part of a molecule of gas (H_2) the existing equilibrium in water is momentarily lost but immediately re-established by ionisation of another water molecule. This process continues as long as H^+ ions are removed, and leads to an accumulation of **surplus OH^- ions** in solution. The full **ionic** equation for the reaction is therefore:

$$Li(s) + H_2O(l) \longrightarrow Li^+(aq) + OH^-(aq) + \tfrac{1}{2}H_2(g)$$

7.8 The pH scale

When pure water dissociates, one H^+ ion is produced for every OH^- ion. In water, therefore, the concentrations of $H^+(aq)$ and $OH^-(aq)$ are equal. If we use the square brackets symbol $[x]$ to denote 'concentration of x', then

$$[H^+] = [OH^-]$$

Pure water is said to be **neutral**. At 298 K the concentration of $H^+(aq)$ in pure water is 1×10^{-7} mol l^{-1}. In **aqueous solutions** the concentration of hydrogen ions may vary over a very wide range. To avoid the inconvenience of working with very small numbers and negative indices such as 10^{-7}, the Danish scientist S.P.L. Sorenson devised, in 1909, the logarithmic **pH scale**. pH is defined as the negative logarithm, to the base 10, of the concentration of hydrogen ion, $H^+(aq)$, in moles per litre.

$$pH = -\log [H^+]$$

The square brackets, [ ], stand for concentration in mol l^{-1}.

For water or any neutral aqueous solution where $[H^+] = 1 \times 10^{-7}$,

$$pH = -\log (1 \times 10^{-7}) = 7$$

For a solution that contains 0.01 mol $H^+(aq)$ per litre,

$$[H^+] = 0.01 = 1 \times 10^{-2}$$

Therefore $pH = -\log (1 \times 10^{-2}) = 2$

The pH values for a range of H^+(aq) concentrations are given in Figure 7.21.

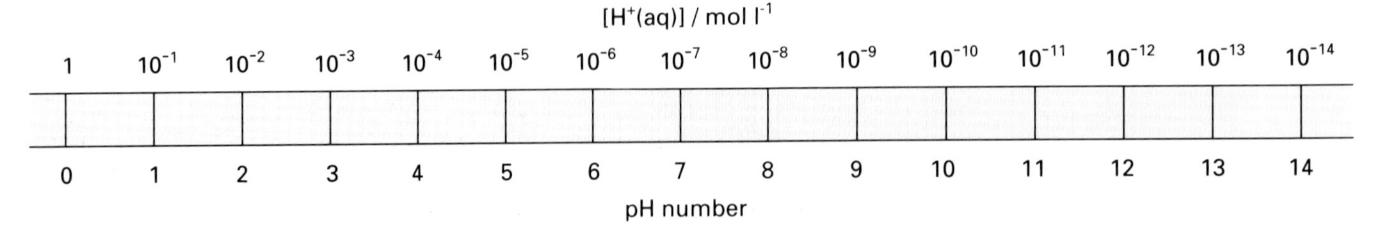

Figure 7.21 Hydrogen ion concentration and the pH scale

It is clear that a **tenfold** change in $[H^+]$ produces a change in pH number of **one** unit. $[H^+]$ can be systematically reduced by progressively diluting an acid solution with water.

The change in $[H^+]$ can be followed by measuring the electrical potential difference in the solution (pH meter) or the change in colour of certain dyes (indicator solution or paper). These measuring devices can be calibrated to give direct readings of pH.

An aqueous solution of HCl with a concentration of 0.001 mol l^{-1} would have a H^+(aq) concentration of 1×10^{-3} mol l^{-1} and consequently a pH of 3. To produce a change in pH of **one** unit the concentration of H^+(aq) must be altered by a factor of **ten**. The aqueous HCl of pH 3 would need to be diluted to one tenth of its previous concentration to produce a solution of pH 4, and so on.

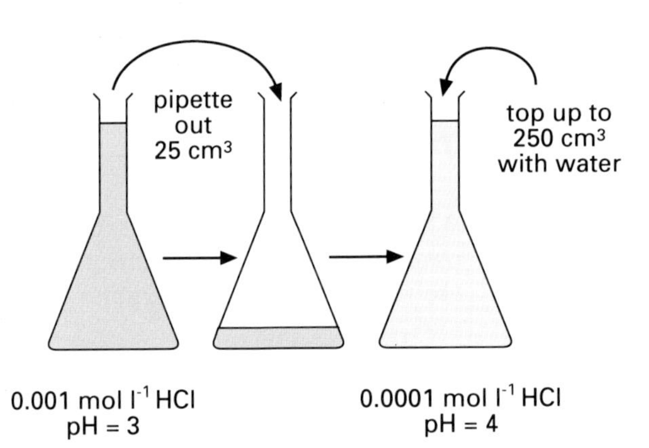

Figure 7.22 Change of pH with tenfold dilution

7.8.1 The ionic product of water

The concentrations of H^+(aq) and OH^-(aq) in aqueous solutions vary widely, but these ions are always involved in an equilibrium with water molecules. One fundamental constant in this equilibrium is the product of the concentrations of H^+(aq) and OH^-(aq).

$$[H^+] \times [OH^-] = \text{a constant}$$

This constant is called the **ionic product of water** and varies slightly with temperature.

In pure water all H^+ and OH^- ions arise from the dissociation of water molecules and their concentrations are therefore equal.

At 298 K $[H^+] = [OH^-] = 1 \times 10^{-7}$ mol l^{-1}
Ionic product of water $= [H^+] \times [OH^-]$
$= 1 \times 10^{-7} \times 1 \times 10^{-7}$ mol^2 l^{-2}
$= 1 \times 10^{-14}$ mol^2 l^{-2}

If we define an acid as a substance which **increases** the $[H^+]$ in aqueous solution, then at the same time there must be a **decrease** in $[OH^-]$. Similarly when $[OH^-]$ is increased, by the addition of a base to water, the $[H^+]$ must decrease proportionally. The crucial point is that the relationship

$$[H^+] \times [OH^-] = 1 \times 10^{-14} \text{ mol}^2 \text{ l}^{-2}$$

must hold true at all times in all aqueous solutions.

Table 7.3 illustrates this reciprocating relationship of $[H^+]$ and $[OH^-]$ which underpins the whole pH scale.

Concentration of H^+(aq)/mol l^{-1}	$[H^+]$	pH	$[OH^-]$	
10	1×10^1	−1	1×10^{-15}	
1	1×10^0	0	1×10^{-14}	
0.1	1×10^{-1}	1	1×10^{-13}	
0.01	1×10^{-2}	2	1×10^{-12}	
0.001	1×10^{-3}	3	1×10^{-11}	
0.000 1	1×10^{-4}	4	1×10^{-10}	
0.000 01	1×10^{-5}	5	1×10^{-9}	
0.000 001	1×10^{-6}	6	1×10^{-8}	
0.000 000 1	$\mathbf{1 \times 10^{-7}}$	7	$\mathbf{1 \times 10^{-7}}$	**0.000 000 1**
	1×10^{-8}	8	1×10^{-6}	0.000 001
	1×10^{-9}	9	1×10^{-5}	0.000 01
	1×10^{-10}	10	1×10^{-4}	0.000 1
	1×10^{-11}	11	1×10^{-3}	0.001
	1×10^{-12}	12	1×10^{-2}	0.01
	1×10^{-13}	13	1×10^{-1}	0.1
	1×10^{-14}	14	1×10^0	1
	1×10^{-15}	15	1×10^1	10
	$[H^+]$	pH	$[OH^-]$	concentration of OH^-(aq)/mol l^{-1}

Table 7.3 Relationship between $[H^+]$, pH and $[OH^-]$

7.8.2 Calculating pH of solutions

The ion product constant enables us to calculate the pH of a solution whose $[OH^-]$ is known. In this and other calculations we can neglect the very small contribution of H^+ and OH^- ions which arise from the ionisation of water.

Sample exercise

What is the pH of an aqueous solution of 0.005 mol l^{-1} calcium hydroxide?

Method

If we assume that calcium hydroxide is fully dissociated into ions,

$$Ca(OH)_2(aq) \longrightarrow Ca^{2+}(aq) + 2OH^-(aq)$$

then we can determine the concentration of $OH^-(aq)$,

$$\begin{aligned} [OH^-] &= (2 \times 0.005)\ mol\ l^{-1} \\ &= 0.01\ mol\ l^{-1} \\ &= 1 \times 10^{-2}\ mol\ l^{-1} \end{aligned}$$

Ion product $[H^+]\,[OH^-] = 1 \times 10^{-14}\ mol^2\ l^{-2}$

Therefore $[H^+] = \dfrac{1 \times 10^{-14}}{1 \times 10^{-2}}\ mol\ l^{-1}$

$= 1 \times 10^{-12}\ mol\ l^{-}$

Consequently, $pH = -\log(1 \times 10^{-12})$

$= 12$

The pH of a solution may be limited by the solubility of the substance in water. Calcium hydroxide is about 100 times more soluble than magnesium hydroxide. This difference in solubility is reflected in the different pH values of their **saturated** solutions.

Saturated solutions	pH
$Mg(OH)_2$	10.5
$Ca(OH)_2$	12.4

The low maximum concentration of OH^- ions means that a solution of magnesium hydroxide cannot do severe damage to the body. A slurry of magnesium hydroxide in water (milk of magnesia) is often taken internally as a mild laxative or antacid.

7.8.3 Estimating pH of solutions

Concentrations of aqueous solutions do not always conveniently appear as neat and tidy tenfold fractions of 1 mol l^{-1} such as 0.01 mol l^{-1}. Take, for instance, a 0.025 mol l^{-1} solution of $H_2SO_4(aq)$. If we assume dilute sulphuric acid is fully dissociated into ions, then we can calculate the concentration of $H^+(aq)$ in the solution.

$$H_2SO_4(aq) \longrightarrow 2H^+(aq) + SO_4^{2-}(aq)$$

$$\begin{aligned} [H^+] &= (2 \times 0.025)\ mol\ l^{-1} \\ &= 0.05\ mol\ l^{-1} \end{aligned}$$

How can we estimate the pH of this acid solution? The concentration, 0.05 mol l^{-1} $H^+(aq)$, is between the upper concentration of 0.10 mol l^{-1} $H^+(aq)$, [pH = 1], and the lower concentration of 0.01 mol l^{-1} $H^+(aq)$, [pH = 2].

[H(aq)]	pH
0.10	1
0.05	
0.01	2

The pH of 0.05 mol l^{-1} H^+(aq) is therefore between pH 1 and pH 2. The actual pH can be calculated and compared to our estimate.

$$\begin{aligned} pH &= -\log [H^+] \\ &= -\log (0.05) \\ &= -(-1.3) \quad \text{(using the log function on a calculator)} \\ &= 1.3 \end{aligned}$$

As predicted, the pH is between 1 and 2.

7.8.4 pH values beyond 0 and 14

Where the concentration of H^+(aq) is larger than 1 mol l^{-1}, the pH of the acid solution will have a **negative** value (less than zero). Similarly, if the concentration of OH^-(aq) is larger than 1 mol l^{-1}, the pH of the solution will be **greater than 14**. The pH scale therefore extends (slightly) beyond 0−14 (see Table 7.3).

7.9 Acids

Compounds whose aqueous solutions contain ions are traditionally classed as acids, bases or salts.

Hydrogen chloride is a colourless gas which dissolves easily in water forming a conducting solution known as hydrochloric acid. The same hydrogen chloride also dissolves in the non-polar solvent toluene (methylbenzene), but here the solution is non-conducting. Clearly water has the ability to bring about ionisation of the HCl molecules while toluene has not.

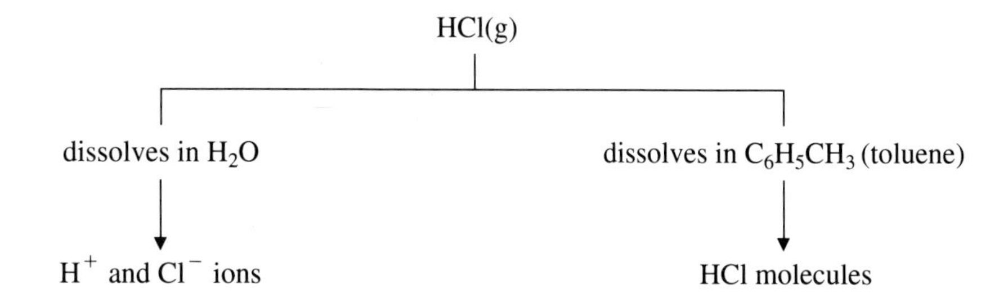

The reaction between hydrogen chloride and water can be described as dissociating in water to produce hydrogen ions.

$$HCl(g) \longrightarrow H^+(aq) + Cl^-(aq)$$

Acids are substances which dissociate in water to form hydrogen ions, H^+(aq).

7.9.1 Strong acids

The dissociation of an acid in water can be represented generally as follows:

$$HA(aq) \rightleftharpoons H^+(aq) + A^-(aq)$$

The further the equilibrium position lies to the right, the greater the concentration of H^+ ions present in the solution, and the stronger the acid is considered to be. Acids like hydrochloric acid are classified as **strong acids** because they are almost **completely dissociated into ions**. The reverse arrow in the equation is replaced by a one-way arrow to indicate that the ionisation process goes virtually all the way to completion. The ions show little tendency to recombine into molecules.

There are relatively few strong acids. Table 7.4 lists the most important ones.

Strong acid	Equation
hydrochloric acid	$HCl \longrightarrow H^+ + Cl^-$
hydrobromic acid	$HBr \longrightarrow H^+ + Br^-$
hydriodic acid	$HI \longrightarrow H^+ + I^-$
perchloric acid	$HClO_4 \longrightarrow H^+ + ClO_4^-$
nitric acid	$HNO_3 \longrightarrow H^+ + NO_3^-$
sulphuric acid	$H_2SO_4 \longrightarrow 2H^+ + SO_4^{2-}$
chromic acid	$H_2CrO_4 \longrightarrow 2H^+ + CrO_4^{2-}$

Table 7.4 Strong acids

7.9.2 Weak acids

The vast majority of acids, when dissolved in water, only **slightly dissociate** into ions. Such acids are said to be **weak acids**. An equilibrium is set up between un-ionised molecules and ions as shown by the general equation below.

$$HA(aq) \rightleftharpoons H^+(aq) + A^-(aq)$$

Table 7.5 (opposite) lists a selection of weak acids.

Many weak acids are compounds composed largely of carbon and hydrogen. Generally speaking, the H atoms attached to C atoms are not ionisable in water. Instead, the ionisable hydrogens are usually found bound to oxygen atoms. In **carboxylic acids**, for example, there is always partial ionisation of the O–H bond next to the carbonyl group, $>C{=}O$, when the compound is dissolved in water.

$$\bigcirc\!-\!\overset{\overset{\displaystyle O}{\|}}{C}\!-\!O\!-\!H \rightleftharpoons \bigcirc\!-\!\overset{\overset{\displaystyle O}{\|}}{C}\!-\!O^- + H^+$$

The carbonyl group has the ability to attract electrons away from the hydrogen atom and thus assist its release as a H^+ ion.

Table 7.5 Weak acids

Weak acid	Equation
methanoic acid	$HCOOH \rightleftharpoons HCOO^- + H^+$
ethanoic acid	$CH_3COOH \rightleftharpoons CH_3COO^- + H^+$
propanoic acid	$C_2H_5COOH \rightleftharpoons C_2H_5COO^- + H^+$
butanoic acid	$C_3H_7COOH \rightleftharpoons C_3H_7COO^- + H^+$
phenol	$C_6H_5OH \rightleftharpoons C_6H_5O^- + H^+$
benzoic acid	$C_6H_5COOH \rightleftharpoons C_6H_5COO^- + H^+$
nitrous acid	$HNO_2 \rightleftharpoons H^+ + NO_2^-$
hydrocyanic acid	$HCN \rightleftharpoons H^+ + CN^-$
hydrofluoric acid	$HF \rightleftharpoons H^+ + F^-$
carbonic acid	$H_2O + CO_2 \rightleftharpoons 2H^+ + CO_3^{2-}$
sulphurous acid	$H_2O + SO_2 \rightleftharpoons 2H^+ + SO_3^{2-}$
phosphoric acid	$H_3PO_4 \rightleftharpoons 3H^+ + PO_4^{3-}$

The term 'weak' refers to the degree of ionisation of an acid. It does not imply that a weak acid is incapable of doing damage. For example, methanoic acid, HCOOH, is a weak acid, yet it provides much of the venom which is injected by stinging ants and caterpillars.

Figure 7.23 The extraordinary caterpillar of the puss moth; it protects itself from predators by emitting a spray of methanoic acid

7.9.3 Carbonic and sulphurous acids

Figure 7.24 Celebration! the effect of 'undissolving' the carbon dioxide in champagne

Carbon dioxide dissolves slightly in water. Some of the CO_2 molecules which manage to dissolve react with water to produce hydrogen and carbonate ions (see Table 7.5). The amount of CO_2 which you can dissolve depends on the positions of two linked equilibria.

$$CO_2(g) \rightleftharpoons CO_2(aq) \rightleftharpoons H^+(aq) + CO_3^{2-}(aq)$$

undissolved molecules	dissolved molecules	dissolved ions

To produce fizzy drinks, CO_2(g) is passed under high pressure into the aqueous solution of lemonade, etc. This shifts the equilibria **to the right** and more CO_2 dissolves. When the cap or cork is removed from a bottle of fizzy drink the pressure is reduced and the equilibria shift back **to the left**. Some CO_2 'undissolves', and gas bubbles out of solution.

The solubility of CO_2 is increased in alkaline solution where the concentration of H^+(aq) is lower. This decrease in $[H^+]$ shifts the equilibria **to the right**. Carbon dioxide is much more soluble in seawater (pH 8) or blood (pH 7.4) than it is in pure water.

The formulae H_2CO_3(aq) and $H_2SO(_3$(aq) are often applied to carbonic and sulphurous acids. This is useful chemical shorthand as long as you appreciate that hexatomic molecules of structure H_2CO_3 or H_2SO_3 have never been isolated.

Table 7.6 Dissociation of carbonic and sulphurous acids

0.1 mol l^{-1} acid	Per cent dissociation
carbonic	0.2
sulphurous	20

Sulphur dioxide is very soluble in water and, like carbon dioxide, forms a weakly acidic solution (mixture of ions and molecules). SO_2 is released into the air from combustion of fossil fuels in power stations and from natural sources such as volcanoes.

Rain is already slightly acidic due to dissolved CO_2. The reactions of SO_2 and oxides of nitrogen with ozone and water vapour in the atmosphere lead to formation of **acid rain**: a dilute but now more corrosive solution of sulphurous, sulphuric, nitrous and nitric acids.

Figure 7.25 Acid rain is thought to be responsible for the crumbling away of this ancient gargoyle

7.10 Experiments to compare strengths of acids

Strong and weak acids may be compared in terms of:

(a) pH
(b) electrical conductivity
(c) rate of reaction
(d) amount of base which can be neutralised.

For comparison to be fair, it is necessary to dissolve similar (equimolar) amounts of each acid in water. The acids will dissociate into ions to differing extents. These differences may then be demonstrated by comparing measurements of pH, etc.

7.10.1 Comparing pH values

The pH values of various acid solutions can be compared by testing samples with a pH meter. To ensure all solutions are equimolar with respect to **potential** H^+ ions, we must use concentrations for diprotic acids [2 mol H^+/mol acid] which are **half** the concentrations used for monoprotic acids [1 mol H^+/mol acid].

Assuming full ionisation,

1 mol HCl $\rightarrow$ 1 mol ($H^+ + Cl^-$) $\rightarrow$ **1 mol H^+** + 1 mol Cl^-
0.5 mol H_2SO_4 $\rightarrow$ 0.5 mol ($2H^+ + SO_4^{2-}$) $\rightarrow$ **1 mol H^+** + 0.5 mol SO_4^{2-}

Results for some acids (all 0.1 mol l^{-1} with respect to **potential** H^+ ions) are given in Table 7.7.

Acid	Reading on pH meter
HNO_3	1.1
HCl	1.1
H_2SO_4	1.2
H_2O/SO_2 (H_2SO_3)	1.5
HCOOH	2.3
CH_3COOH	2.9
H_2O/CO_2 (H_2CO_3)	3.8
HCN	5.1

Table 7.7 Comparing pH values

The presence of 0.1 mol H^+ (aq) per litre of solution would be shown by a pH of 1.0. Clearly the first three acids in the table are almost fully ionised. Nitric, hydrochloric and sulphuric acids are therefore confirmed as **strong** acids. The other acids have higher pH values which indicate lower concentrations of H^+(aq). Thus sulphurous, methanoic, ethanoic, carbonic and hydrocyanic acids are confirmed as **weak** acids.

The higher the pH value of the solution, the weaker the acid.

The above generalisation is valid provided we compare acid solutions which are equimolar with respect to potential H^+(aq).

7.10.2 Comparing electrical conductivities

The electrical conductivity of a solution depends upon the concentration of ions present. A simple circuit for comparing conductivities of solutions is shown in Figure 7.26.

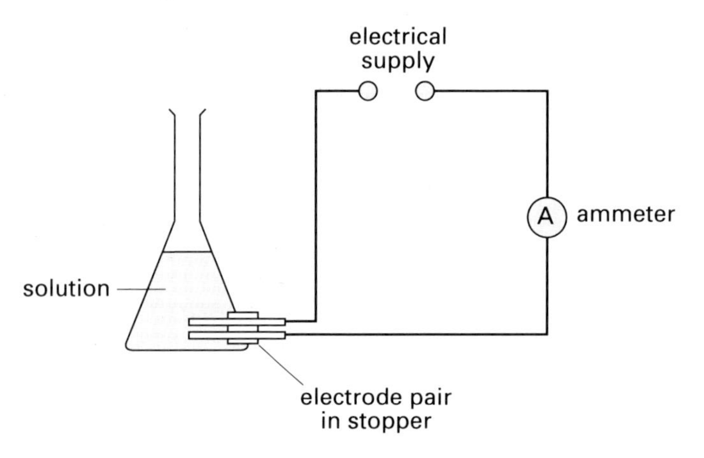

Figure 7.26 Circuit for comparing conductivities

Solutions of HCl(aq) and CH_3COOH(aq) of equal concentration, and under the same electrical potential difference, pass different amounts of current, as measured on the ammeter.

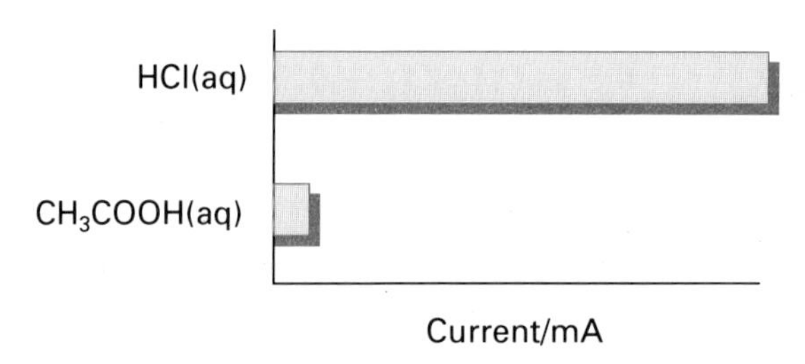

Figure 7.27 Relative conductivities of strong and weak acids

7.10.3 Comparing reaction rates

The reactions between dilute acid and a reactive metal, such as magnesium, or a carbonate, such as magnesium carbonate, can be summarised as follows.

$$2H^+(aq) + Mg(s) \longrightarrow H_2(g) + Mg^{2+}(aq)$$
$$2H^+(aq) + CO_3^{2-}(s) \longrightarrow H_2O(l) + CO_2(g)$$

The **rates** of these reactions will depend on the concentration of H^+(aq) in the acid solution. The rates of reaction of HCl(aq) and CH_3COOH(aq) with either metal or carbonate can be compared. This may be done by measuring the times taken to complete reaction and collect all the gas released.

$$\frac{\text{rate of HCl reaction}}{\text{rate of } CH_3COOH \text{ reaction}} = \frac{\text{time taken to complete } CH_3COOH \text{ reaction}}{\text{time taken to complete HCl reaction}}$$

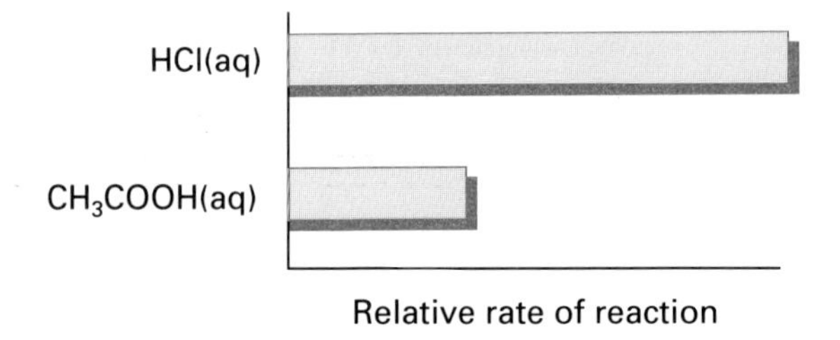

Figure 7.28 Relative reaction rates of strong and weak acids

The faster rate of reaction with hydrochloric acid is a result of the higher concentration of $H^+(aq)$ present. Ethanoic acid contains only a small proportion of $H^+(aq)$ ions at any given time, and its reactions with metal or carbonate are therefore slower.

7.10.4 Comparing amounts of alkali neutralised

The neutralisation of an alkali by an acid is conveyed by the simple equation:

$$H^+(aq) + OH^-(aq) \longrightarrow H_2O(l)$$

Strong acids are more or less fully dissociated into ions. Weak acids, on the contrary, are only partly ionised and thus are mainly molecular. We might reasonably predict that, all else being equal, a weak acid would neutralise a much **smaller** amount of alkali than a strong acid.

100 cm^3 volumes of 0.01 mol l^{-1} solutions of HCl, NaOH and CH_3COOH have roughly the composition (in millimoles) described in Figure 7.29.

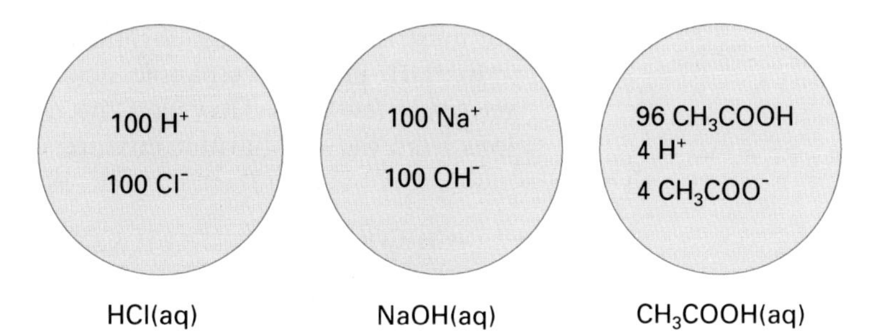

Figure 7.29 Composition of aqueous solutions

HCl is 100 per cent ionised. CH_3COOH is only 4 per cent ionised. Will HCl neutralise 25 times more NaOH than will CH_3COOH? The prediction can be experimentally tested using titration apparatus as shown in Figure 7.30.

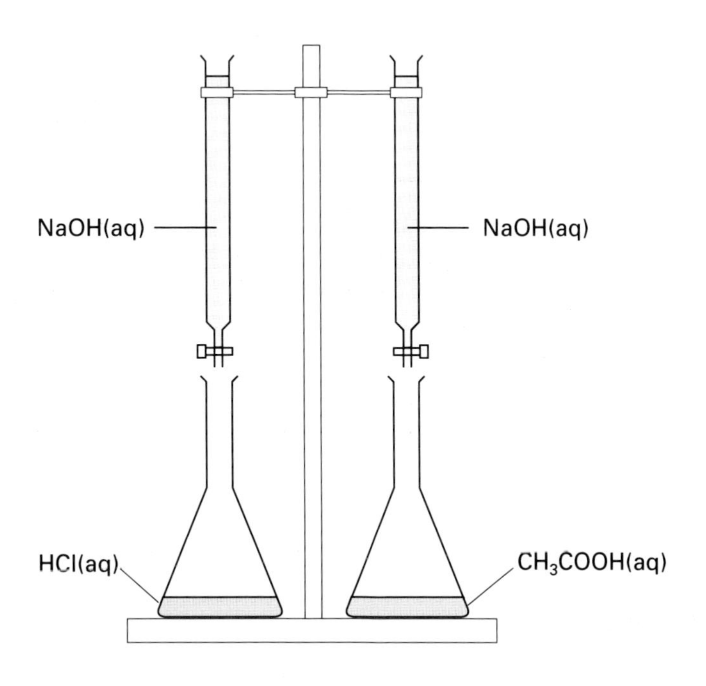

Figure 7.30 Titration apparatus

Equal volumes of equimolar (0.01 mol l^{-1}) HCl and CH_3COOH are titrated with NaOH solutions of similar concentration. The end-point of the neutralisation is observed by adding bromothymol blue indicator. It turns colour from yellow to blue when all the acid in the flask has been neutralised by the alkali added. Results are given in Table 7.8.

	Volume of acid in flask	Volume of alkali neutralised
HCl	25 cm^3	25 cm^3
CH_3COOH	25 cm^3	25 cm^3

Table 7.8 Titration results

The results are **not** as predicted. How can a slightly ionised weak acid neutralise exactly as much alkali as a fully ionised strong acid? For the answer we must look at the equilibrium within the weak acid solution.

In 0.01 mol l^{-1} ethanoic acid solution there are, **at any given time**, relatively few ions. However, the dropwise addition of alkali removes H^+ ions from the equilibrium mixture.

$$CH_3COOH(aq) \rightleftharpoons CH_3COO^-(aq) + H^+(aq)$$

($H^+(aq)$ removed by $OH^-(aq)$ to form $H_2O(l)$)

This removal of a product shifts the position of equilibrium **to the right**. More CH_3COOH molecules undergo ionisation to replace the H^+ ions which have been taken out by neutralisation. **Eventually the weak acid ends up with all its molecules dissociated** and able to neutralise the same amount of alkali as any strong acid.

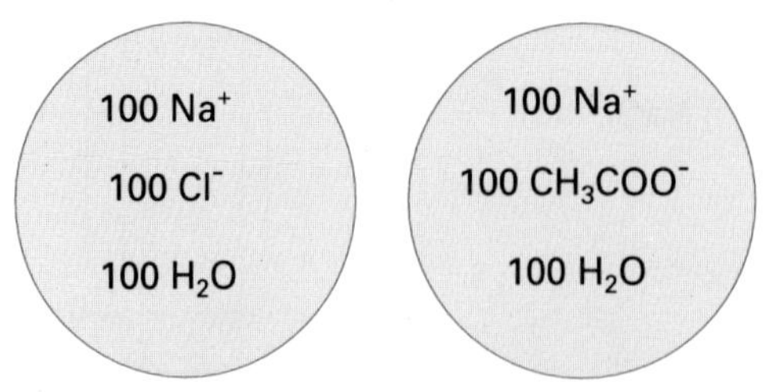

Figure 7.31 Composition of products at equivalence-points

A **weak** acid can, therefore, be distinguished from a strong acid in having:

(a) **higher** pH
(b) **lower** conductivity
(c) **slower** reaction rate.

A weak acid, however, neutralises the **same** volume of alkali as a strong acid.

7.11 Bases

Bases can also be defined by reference to hydrogen ions.

> ***Bases are substances which can react with hydrogen ions, $H^+(aq)$, to form water.***

Metal oxides, hydroxides and carbonates can all accept H^+ ions from acidic solutions, forming water, and must therefore be categorised as bases.

$$O^{2-}(s) + 2H^+(aq) \longrightarrow H_2O(l)$$
$$OH^-(s) + H^+(aq) \longrightarrow H_2O(l)$$
$$CO_3^{2-}(s) + 2H^+(aq) \longrightarrow H_2O(l) + CO_2(g)$$

Bases which are **soluble in water** and dissociate to produce OH^- ions in aqueous solution are classed as **alkalis**. Both copper(II) hydroxide and magnesium hydroxide are bases since they react with acids to produce water.

$$Cu(OH)_2(s) + 2HCl(aq) \longrightarrow CuCl_2(aq) + 2H_2O(l)$$
$$Mg(OH)_2(s) + 2HCl(aq) \longrightarrow MgCl_2(aq) + 2H_2O(l)$$

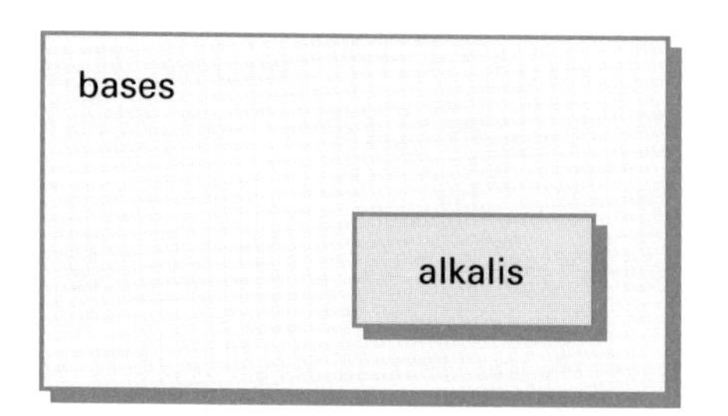

Figure 7.32 Alkalis are water-soluble bases

Copper hydroxide, however, is not soluble in water and cannot dissociate to form hydroxide ions. On the other hand magnesium hydroxide is soluble in water and is able to dissociate to form $OH^-(aq)$.

$$Mg(OH)_2(s) \xrightarrow{H_2O} Mg^{2+}(aq) + 2OH^-(aq)$$

$Cu(OH)_2$ is a base but not an alkali.
$Mg(OH)_2$ is a base and is also an alkali.
Alkalis form a subset within bases.

7.11.1 Strong bases

The characteristic property of soluble bases is their ability to produce more hydroxide ions in water, i.e. form alkaline solutions. Strong bases, like strong acids, are more or less fully dissociated into ions when dissolved in water. Unlike strong acids, strong bases are ionic rather than molecular compounds. In dissolving, the ions which have been fixed in a lattice dissociate to form independent hydrated ions.

Table 7.9 lists the major strong bases.

Strong base	Equation
lithium hydroxide	$Li^+OH^-(s) \longrightarrow Li^+(aq) + OH^-(aq)$
sodium hydroxide	$Na^+OH^-(s) \longrightarrow Na^+(aq) + OH^-(aq)$
potassium hydroxide	$K^+OH^-(s) \longrightarrow K^+(aq) + OH^-(aq)$
magnesium hydroxide	$Mg^{2+}(OH^-)_2(s) \longrightarrow Mg^{2+}(aq) + 2OH^-(aq)$
calcium hydroxide	$Ca^{2+}(OH^-)_2(s) \longrightarrow Ca^{2+}(aq) + 2OH^-(aq)$
barium hydroxide	$Ba^{2+}(OH^-)_2(s) \longrightarrow Ba^{2+}(aq) + 2OH^-(aq)$

Table 7.9 Strong bases

Figure 7.33 The Steetley magnesia plant, Hartlepool

Of the bases listed above only NaOH and KOH are sufficiently soluble in water to form highly concentrated aqueous solutions. These solutions can be very corrosive to skin and have the traditional names of 'caustic soda' and 'caustic potash'.

Magnesium hydroxide (magnesia) is used as a lining for furnaces and in various chemical processes. It is obtained, from seawater, by Steetley Ltd in a huge plant at Hartlepool. The process consists of heating dolomite ore ($CaCO_3.MgCO_3$) to produce dolime (CaO.MgO) and mixing the dolime with seawater. Magnesium hydroxide, because of its low solubility, precipitates out as a white creamy slurry.

$$CaO.MgO + \left\{\begin{matrix} MgCl_2(aq) \\ MgSO_4(aq) \end{matrix}\right. \longrightarrow \left\{\begin{matrix} CaCl_2(aq) \\ CaSO_4(aq) \end{matrix}\right. + Mg(OH)_2(s)$$

Mg salts in seawater; precipitate of magnesia

7.11.2 Weak bases

Ammonia gas is a base. It dissolves in water, extracting H^+ ions from H_2O and leaving OH^- ions. The reaction, however, reaches equilibrium rather than going to completion.

$$NH_3(aq) + H_2O(l) \rightleftharpoons NH_4^+(aq) + OH^-(aq)$$

Aqueous ammonia, being only **partly ionised**, is therefore a **weak base**. The aqueous solution, a mixture of molecules and ions, is often referred to as ammonium hydroxide and given the formula $NH_4OH(aq)$. As with H_2CO_3 and H_2SO_3, this is a convenient label as long as one remembers that no such heptatomic molecule actually exists.

The ability of NH_3 to extract protons (H^+) from other substances is due to a **pair of non-bonded electrons** on the nitrogen atom. These electrons allow the N atom to form a fourth bond, to the H^+ ion which has been extracted.

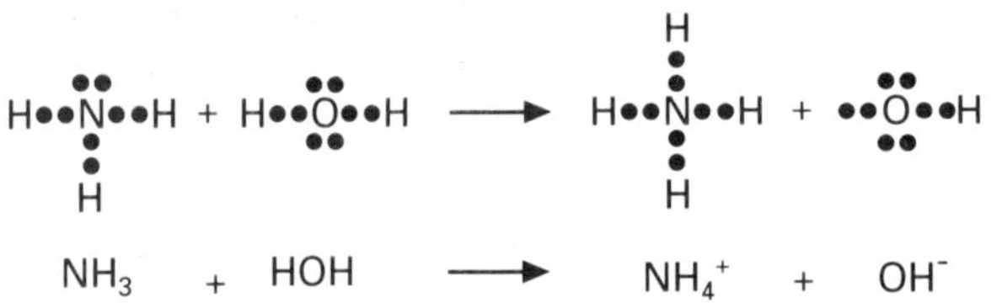

$$NH_3 + HOH \longrightarrow NH_4^+ + OH^-$$

Figure 7.34 Reaction of ammonia in water

If one or more of the N—H bonds in ammonia is replaced with N—C bonds, we obtain a class of compounds called **amines**. Where only one bond is replaced the amine is described as a **primary** amine.

Many drugs are primary amines. Three examples are shown in Figure 7.35.

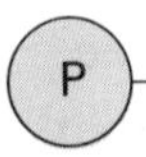

(P) = penicillin base structure

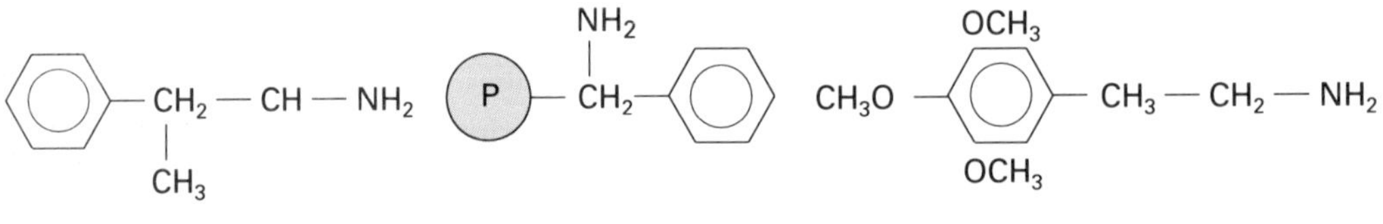

(a) Amphetamine, a stimulant: C_6H_5—CH_2—CH(CH_3)—NH_2

(b) Ampicillin, an antibiotic: P—CH(NH_2)—C_6H_5

(c) Mescaline, a hallucinatory drug: CH_3O—$C_6H_2(OCH_3)_2$—CH_3—CH_2—NH_2

Figure 7.35

Amines also contain N atoms with unshared pairs of electrons. They are, therefore, potential H^+ ion acceptors, i.e. bases. Like ammonia, ionisation is only partial and amines belong to the category of **weak** base. The dissociation equilibria of the three simplest primary amines are given below.

methylamine: $CH_3NH_2 + H_2O \rightleftharpoons CH_3NH_3^+ + OH^-$

ethylamine: $C_2H_5NH_2 + H_2O \rightleftharpoons C_2H_5NH_3^+ + OH^-$

propylamine: $C_3H_7NH_2 + H_2O \rightleftharpoons C_3H_7NH_3^+ + OH^-$

7.11.3 Experiments to compare the strengths of bases

As with acids, strong and weak bases differ in pH and electrical conductivity. The relative figures for 0.1 mol l^{-1} sodium hydroxide and ammonia solutions are shown in Figure 7.36.

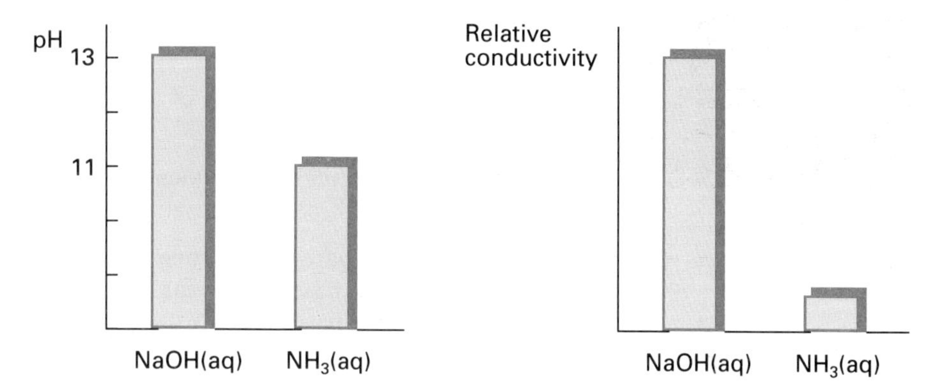

Figure 7.36 Comparison of strong and weak bases

Weak bases have lower pH values (nearer 7) than equivalent strong bases as they have lower concentrations of OH^-(aq). The lower concentration of ions in weak bases also accounts for their lower electrical conductivity.

We **cannot** distinguish strong and weak bases by comparing the amounts of acid they are able to neutralise. For example, 25 mol 0.1 mol l^{-1} HCl(aq) is neutralised completely

either by 25 ml 0.1 mol l^{-1} NaOH(aq)
or by 25 ml 0.1 mol l^{-1} NH_3(aq).

A strong base such as NaOH(aq) is fully dissociated into ions at the outset of the titration with acid. A weak base such as NH_3(aq), on the other hand, is only slightly ionised to start with, but responds to the addition of acid by ionising further.

$$NH_3(aq) + H_2O(l) \rightarrow NH_4^+(aq) + OH^-(aq)$$

↓ removed by H^+(aq) to form $H_2O(l)$

Whether the base is strong or weak, the net result is the same: equal amounts of H^+(aq) are neutralised.

7.12 Salts

A base reacts with an acid to form water and a salt solution.

$$\text{base} + \text{acid} \longrightarrow \text{water} + \text{salt}$$

The reaction is described as neutralisation because the combination of OH^-(aq) and H^+(aq) produces neutral water. The salt solution, however, may or may not be neutral. When tested with a pH meter or indicator, salt solutions are found to be acidic (pH less than 7), neutral (pH 7) and basic (pH greater than 7).

Acidic salts	Neutral salts	Basic salts
NH_4Cl	$LiNO_3$	KF
$(NH_4)_2SO_4$	$MgCl_2$	Na_2CO_3
NH_4NO_3	K_2SO_4	CH_3COONa
	$CaCl_2$	NaCN
	Na_2SO_4	Na_2SO_3
	$BaCl_2$	HCOOK

Table 7.10 Examples of acidic, neutral and basic salts

If we examine the chemical composition of these salts, we can establish the following three generalisations.

(a) Weak bases and strong acids form acidic salts.
(b) Strong bases and weak acids form basic salts.
(c) Strong bases and strong acids form neutral salts.

Why should this pattern emerge? The answer lies in the reaction between the salt and water.

7.12.1 Hydrolysis of salts

When salts dissolve in water they are completely dissociated into ions. Those ions may or may not interact with water and disturb its equilibrium.

$$H_2O(l) \rightleftharpoons H^+(aq) + OH^-(aq)$$

If interaction with water, known as **hydrolysis**, takes place, this creates surplus production of either H^+ ions or OH^- ions in the solution. The pH is therefore altered from that of pure water (pH 7). Let us take an example of each type of salt solution.

Acidic salt: NH_4Cl(aq)

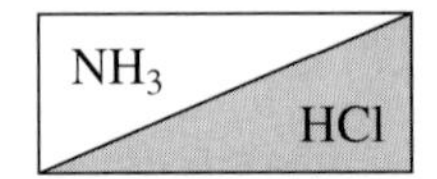

NH_3(aq) is a weak base: its ions have some tendency to recombine into molecules, i.e. associate. HCl(aq) is a strong acid so its ions have no tendency to associate. When ammonium chloride is dissolved in water, the NH_4^+ ions are fully dissociated from Cl^- ions but **interact with OH^- ions** from the water. Some association (*) occurs and this disturbs the water equilibrium.

$$NH_4^+ + Cl^-$$

$$H_2O \rightleftharpoons H^+ + OH^-$$

$$^*NH_4^+ + OH^- \longrightarrow NH_3 + H_2O$$

More water molecules ionise to replace the OH^- ions removed by association and this produces an **excess of H^+ ions**. The solution is therefore acidic and has a pH less than 7.

Neutral salt: $KNO_3(aq)$

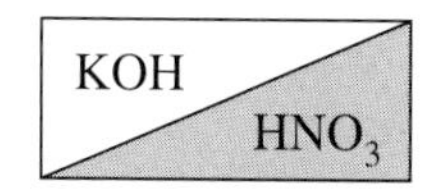

KOH(aq) and $HNO_3(aq)$ are a strong base and strong acid respectively. Their ions have no tendency to recombine. When potassium nitrate is dissolved in water, its ions are fully dissociated and have no inclination to associate with either $H^+(aq)$ or $OH^-(aq)$.

$$K^+ + NO_3^-$$

$$H_2O \rightleftharpoons H^+ + OH^-$$

The water equilibrium remains undisturbed and the solution is neutral with pH equal to 7.

Basic salt: $CH_3COONa(aq)$

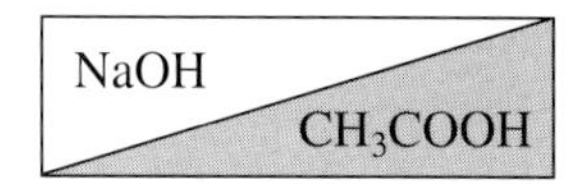

NaOH(aq) is a strong base which completely dissociates into ions which have no inclination to associate. $CH_3COOH(aq)$, on the other hand, is a weak acid; its ions are liable to recombine to form molecules. When sodium ethanoate dissolves in water, the CH_3COO^- ions associate (*) with H^+ ions from the water and this shifts the equilibrium towards a **surplus of OH^- ions.**

$$Na^+ + CH_3COO^-$$

$$H_2O \rightleftharpoons H^+ + OH^-$$

$$^*\ CH_3COO^- + H^+ \longrightarrow CH_3COOH$$

The solution is therefore basic and shows a pH value above 7.
Soaps are sodium or potassium salts of a family of weak acids called fatty acids. Soaps are therefore basic.

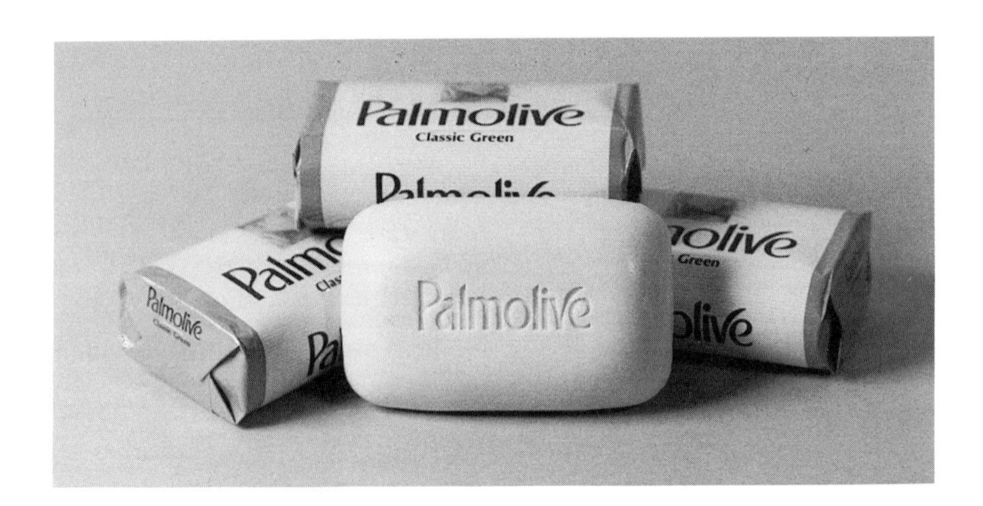

Figure 7.37 Soaps – sodium and potassium salts of fatty acids – have a pH value above 7

Summary of Unit 7

Having read and understood the information and ideas given in this unit, you should now be able to:

explain the concept of dynamic equilibrium in a reversible chemical reaction

account for the effect of changing pressure or temperature on the equilibrium yield of ammonia from the Haber process

given the state equation and ΔH value, predict the effect which a deliberate change in temperature, pressure or concentration will have on the composition of an equilibrium mixture

explain that the operating conditions in an industrial process are usually a compromise between what is theoretically advantageous and what is economically sensible

outline the factors which determine which one of a choice of routes will be selected for making some chemical product

explain metal–water reactions in terms of a shift in the position of the water equilibrium

relate integral (whole number) pH values to concentrations of $H^+(aq)$ in moles per litre

account for water having a pH of 7 and being classified as 'neutral'

define strong and weak acids/bases, in terms of dissociation into ions

contrast strong and weak acids/bases, in terms of pH, conductivity and reaction rate

state that strong and weak acids/bases do not differ in the amount of base/acid they can neutralise

identify the acidic hydrogen in carboxylic acids

explain both the weakly acidic nature of $CO_2(aq)$ or $SO_2(aq)$ and the weakly alkaline nature of $NH_3(aq)$

predict whether a salt solution will be acidic, neutral or alkaline (basic), given the strength of the acid and base used to make the salt

▲ describe the effect of a catalyst on equilibrium

▲ explain why, in the Haber process, gases are rapidly recycled without reaching a state of equilibrium in the reactor

▲ account for pH values below 0 and above 14

▲ use the ionic product of water to calculate $[H^+]$ from $[OH^-]$ and vice versa

▲ estimate the pH of a solution from given $[H^+]$ which may lead to pH values which are not whole numbers, negative or above 14

▲ explain why aqueous primary amine solutions are weakly alkaline

▲ explain, in terms of a shift in the position of the water equilibrium, why some salt solutions are acidic, some are neutral and some are alkaline.

PROBLEM SOLVING EXERCISES

1. The relative electrical conductivities of ethanoic and hydrochloric acids were measured at three different concentrations. The results are given in the table.

Concentration of acid /mol l^{-1}	Relative conductivity	
	$CH_3COOH(aq)$	HCl(aq)
0.1	5	390
0.01	16	410
0.001	48	420

What can you deduce from these observations about the effect of concentration on the dissociation of each acid?

(PS skill 8)

2. The acid hydrogen fluoride, HF, is 10 per cent dissociated in a 0.1 mol l^{-1} aqueous solution. What is the pH of 0.1 mol l^{-1} HF(aq)?

(PS skill 4)

3. Sodium carbonate, Na_2CO_3, is manufactured as a feedstock for glass production. It is made from cheap and readily available raw materials, salt and limestone.

$$2NaCl + CaCO_3 \rightleftharpoons CaCl_2 + Na_2CO_3$$

In the above reaction, the reverse reaction is favoured so a one-stage process is not possible. Instead the following sequence of reactions is required.

(i) $NaCl + NH_3 + H_2O \rightleftharpoons NH_4Cl + NaHCO_3$
(ii) $NaHCO_3 \xrightarrow{heat} Na_2CO_3 + CO_2 + H_2O$
(iii) $CaCO_3 \longrightarrow CaO + CO_2$
(iv) $CaO + H_2O \longrightarrow Ca(OH)_2$
(v) $Ca(OH)_2 + 2NH_4Cl \rightleftharpoons CaCl_2 + 2NH_3 + 2H_2O$

Outline the process by means of a simple flow diagram.

(PS skill 2)

4. Bromothymol blue, the acid–base indicator, is a weak acid whose ionisation can be represented as follows:

$$\underset{\text{yellow}}{HBtb(aq)} \rightleftharpoons H^+(aq) + \underset{\text{blue}}{Btb^-}$$

Explain why bromothymol blue has:
(a) a yellow colour in acid solution (below pH 6)
(b) a blue colour in alkaline solution (above pH 8).

(PS skill 9)

5. At a temperature of 700 K, hydrogen iodide is 20 per cent dissociated into its elements as shown below.

$$HI \rightleftharpoons \tfrac{1}{2}H_2 + \tfrac{1}{2}I_2$$

A 0.5 mol sample of HI is placed in a closed container at 700 K. Calculate the amount (in moles) of each component in the equilibrium mixture which is formed.

(PS skill 4)

6. The pH of an aqueous solution of potassium tartrate is greater than 7. What conclusion can you draw about the strength of tartaric acid?

(PS skill 8)

7. When copper carbonate is heated in air it decomposes as shown below.

$$\underset{\text{green}}{CuCO_3} \rightleftharpoons \underset{\text{black}}{CuO + CO_2}$$

Describe, with the aid of a labelled diagram, an experiment you could carry out to show that the presence of CO_2 reduces or even prevents the decomposition of $CuCO_3$. Indicate the observations you would make.

(PS skill 5)

8. Synthesis gas is converted over a copper-based catalyst to form methanol.

$$CO + 2H_2 \rightleftharpoons CH_3OH \qquad \Delta H = -91 \text{ kJ mol}^{-1}$$

If the relative rate of methanol production is plotted against temperature, the graph obtained is as shown below.

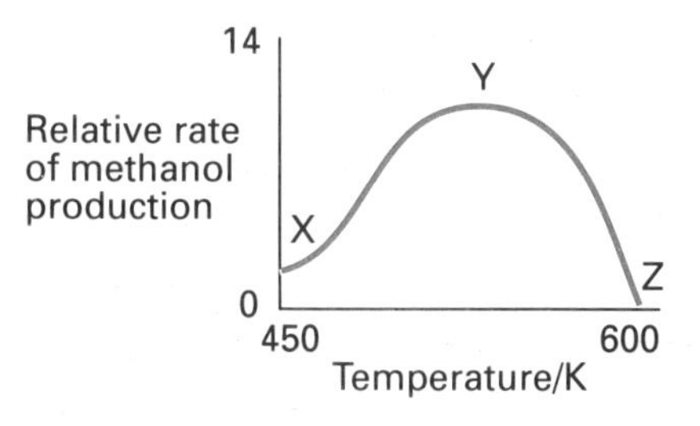

Figure 7.38

Explain (a) the increasing rate of CH_3OH production from X to Y, (b) the decreasing rate of CH_3OH production from Y to Z.

(PS skill 9)

PROBLEM SOLVING EXERCISES

9. 100 cm^3 of HCl(aq) in a flask were titrated with 1 mol l^{-1} NaOH(aq) added from a burette. The pH of the mixture in the flask was recorded and the results plotted on the graph below.

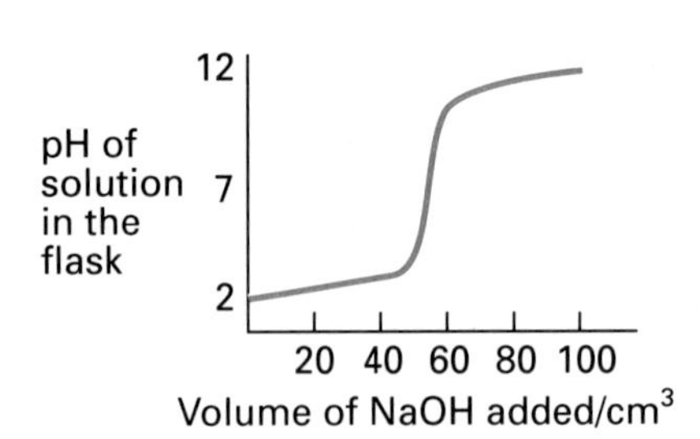

Figure 7.39

Estimate the concentration of the HCl acid solution.

(PS skill 8)

10. The same mass of powdered chalk is added to two cylinders, A and B, containing equal volumes of:

(a) 2 mol l^{-1} CH_3COOH acid
(b) 2 mol l^{-1} CH_3COOH saturated with dissolved $CH_3COO^-Na^+$.

The evolution of CO_2 is faster in cylinder A than in cylinder B. How can you explain this difference in rate?

(PS skill 9)

11. A researcher was asked to compare the effects of F, Cl, Br and I atoms on the degree of ionisation of carboxylic acids. Which four molecules from the grid below could be used to provide a fair comparison?

A	B	C	D
I_3COOH	$F_2CHCOOH$	$I_2CHCOOH$	$ClCH_2COOH$
E	F	G	H
$BrCF_2COOH$	$Cl_2CHCOOH$	FCH_2COOH	$Br_2CHCOOH$

(PS skill 7)

12. The pH values of two 0.1 mol l^{-1} solutions are listed below.

Solution	pH
$NaHSO_4$(aq)	1.4
$NaHCO_3$(aq)	8.4

By considering the reaction of the negative ion in each case with water, explain the difference in pH values.

(PS skill 9)

13. You are given three identical sealed tubes containing nitrogen dioxide in equilibrium with dinitrogen tetroxide

$$2NO_2(g) \rightleftharpoons N_2O_4(g)$$

dark brown colour — pale yellow colour

How could you demonstrate experimentally that the position of equilibrium shifts to the left at higher temperatures and to the right at lower temperatures? Outline the observations you would make.

(PS skill 5)

14. The key reaction in the manufacture of sulphuric acid is the conversion of sulphur dioxide to sulphur trioxide over vanadium oxide.

$$SO_2(g) + \tfrac{1}{2}O_2(g) \rightleftharpoons SO_3(g) \qquad \Delta H = -99 \text{ kJ mol}^{-1}$$

This reaction takes place in a converter over four separate beds of catalyst. The heat evolved during each catalysed stage is removed by a heat exchanger.

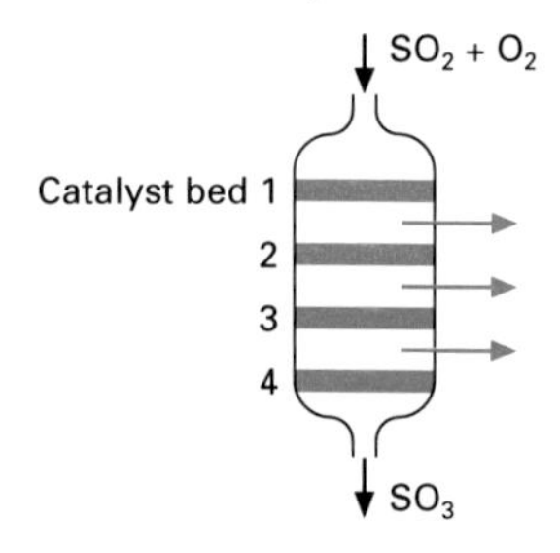

Figure 7.40

As the reaction mixture progresses down through the converter, it experiences changes in temperature and composition (figure 7.41).
Explain the shape of each graph.

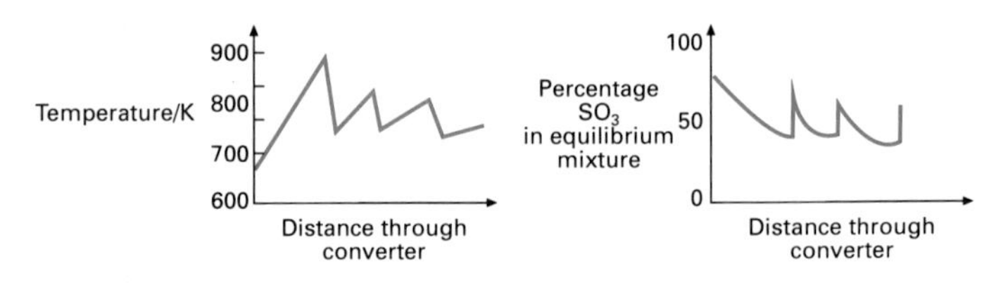

Figure 7.41

(PS skill 9)

Unit 8
Radioisotopes

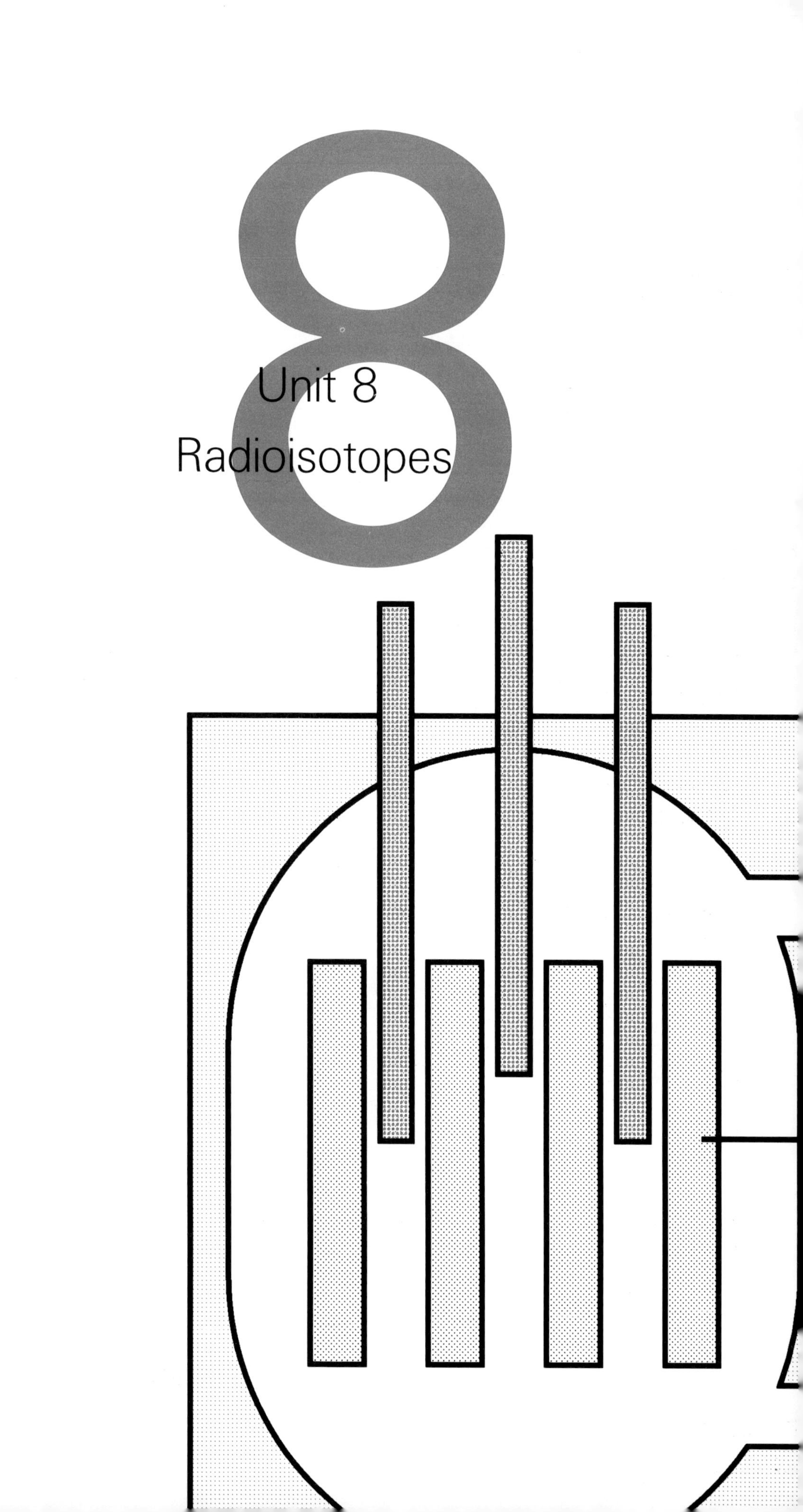

ASSUMED KNOWLEDGE AND UNDERSTANDING

Before starting on Unit 8 you should know and understand:

the composition of an atom in terms of protons, neutrons and electrons

atomic number and mass number

the representation of a nuclide as $^{\text{mass number}}_{\text{atomaic number}}\text{symbol}$

the nature of isotopes

the relative atomic mass of an element which contains isotopes.

OUTLINE OF THE CORE AND EXTENSION MATERIAL

As you progress through Unit 8 you should, at least, try to grasp the fundamental content listed under **core** and, if possible, pick up the extra and/or more difficult content listed under **extension**.

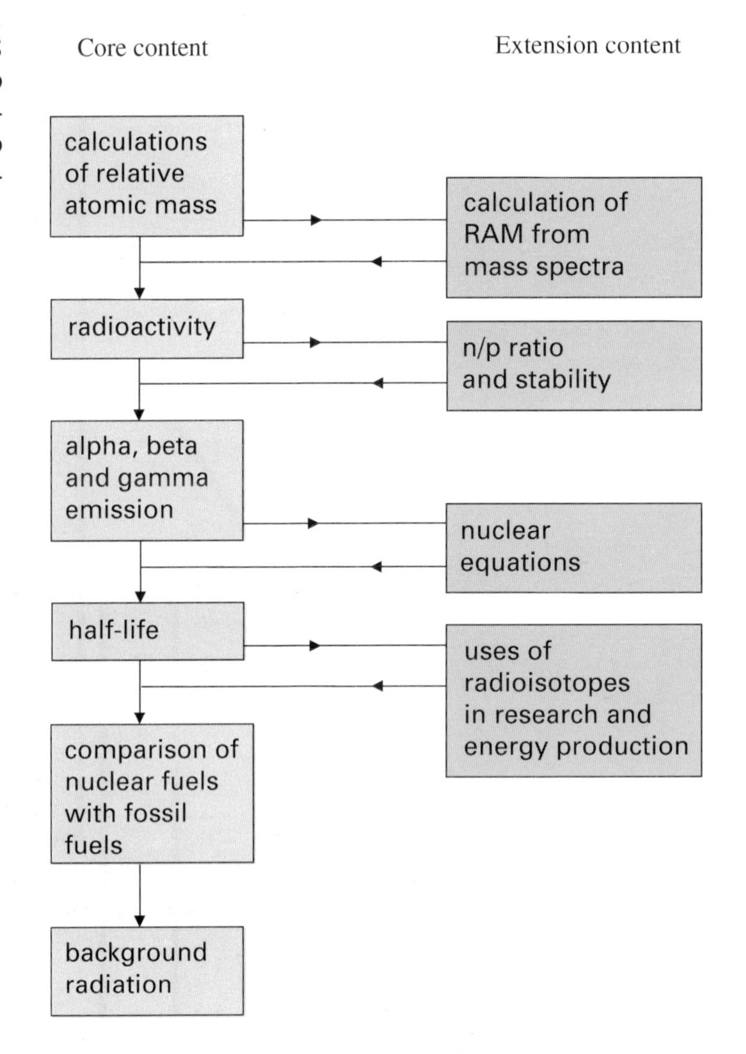

8 RADIOISOTOPES

8.1 The mass spectrometer

Atoms are too small for any one to be weighed directly. However, it is possible to compare the masses of different atoms. The invention, in 1919, of the **mass spectrometer** by Francis Aston gave chemists an accurate and reliable method of comparing the atomic masses of different elements. Any mass spectrometer does three things:

(a) it produces ions
(b) it separates them according to their masses
(c) it records the relative intensities of the ions of different mass, to form a mass spectrum.

Figure 8.1 outlines the function of a simple mass spectrometer.

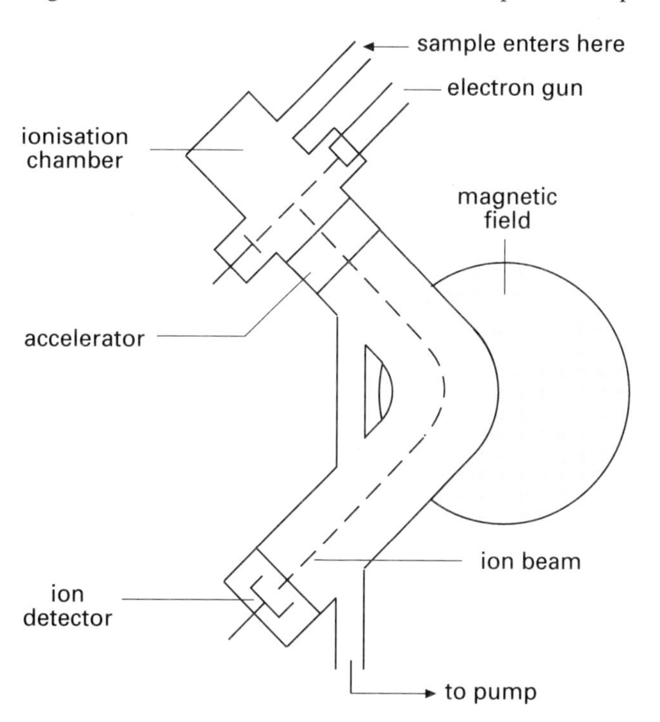

Figure 8.1 The layout of a mass spectrometer

The element whose atomic mass is to be determined is vaporised and then bombarded by high-energy electrons from an electron gun. These collide with atoms or molecules of the element and knock electrons out of them, causing positive ions to be formed. In most cases one electron per atom or molecule will be ejected and 1+ ions formed in the ionisation chamber.

$$E(g) \longrightarrow E^{+}(g) + e^{-}$$

The positively charged ions are focused into a narrow beam and accelerated by a high voltage, towards a magnetic field. The magnetic field deflects the ion beam out of its straight-line path and towards the detector. For ions of the **same charge**, say 1+, the extent of deflection depends only on the mass of the ion.

The heavier the ion, the less it will be deflected. The lighter the ion, the more it will be deflected.

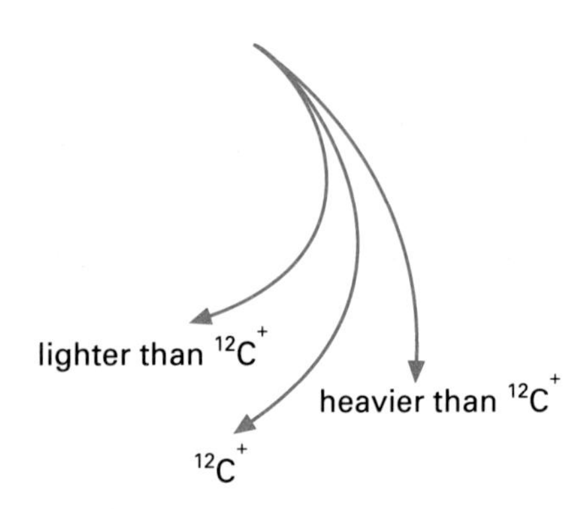

Figure 8.2 Deflection varies with mass of ion

The atom to which all other atoms or molecules are compared is the atom of the carbon-12 isotope, ^{12}C. At particular values of accelerating voltage and magnetic field strength, a $^{12}C^+$ ion will be deflected so as to arrive precisely at the detector. Under the same conditions ions lighter than $^{12}C^+$ would be over-deflected and ions heavier than $^{12}C^+$ would be under-deflected.

By adjusting the accelerating voltage it is possible to direct ions, whatever their mass, towards the detector. At a given voltage V, all ions of the same mass will pass through the slit in front of the detector (all other ions, lighter or heavier, will strike the walls of the deflecting tube). All ions of identical mass are therefore focused as closely as possible into a single sharp line. By comparing the accelerating voltage required to focus carbon-12 ions, V_c, with that required to focus ions of the sample element, V_E, we can calculate the relative atomic mass of the sample ions.

$$\frac{V_C}{V_E} = \frac{\text{mass of (sample ion)}^+}{\text{mass of } ^{12}\text{C}^+}$$

The accelerating voltage required to collect fluorine ions is 0.63 times that required to collect carbon-12 ions.

$$\text{Therefore } \frac{V_C}{V_E} = \frac{1}{0.63} = 1.58 = \frac{\text{mass of F}^+ \text{ ion}}{\text{mass of } ^{12}\text{C}^+ \text{ ion}}$$

The mass of the carbon-12 atom has already been given a value of 12.000 atomic mass units (u).

Therefore mass of F^+ ions $= 1.58 \times 12$ u
$= 19$ u

8.2 Atomic masses of elements

Fluorine is one of a minority of elements which contain atoms of which 100 per cent have the same mass. Most elements contain two or more **isotopes**, i.e. atoms of slightly different mass. When copper gas, for example, is analysed in a mass spectrometer, two lines (or peaks) are observed on the mass spectrum obtained (Figure 8.3). The height of each line indicates the relative intensity of each ion beam and thus the relative proportion (abundance) of each isotope.

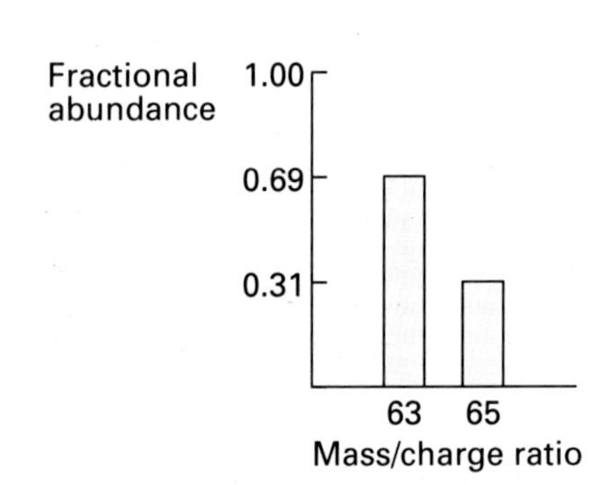

Figure 8.3 Mass spectrum of copper

The mass spectrum indicates that there are two isotopes of copper (two lines). The lighter isotope is clearly more abundant than the heavier isotope. In 100 atoms of copper, 69 of the atoms will be of the lighter type and 31 of the atoms will be of the heavier variety. The contribution of each isotope to the overall mass of copper is in proportion to its abundance.

The amount of mass that each isotope contributes to the average atomic mass of the element is equal to the product of its mass and its fractional abundance. The relative atomic mass of copper is therefore the average of the masses of its isotopes **weighted so as to take into account their different proportions**.

$$\text{RAM of copper} = \begin{bmatrix} \text{mass of first isotope} \\ \times \\ \text{its fractional abundance} \end{bmatrix} + \begin{bmatrix} \text{mass of second isotope} \\ \times \\ \text{its fractional abundance} \end{bmatrix}$$
$$= 63 \times 0.69 + 65 \times 0.31$$
$$= 43.47 + 20.15$$
$$= 63.62$$

The relative atomic mass of copper is therefore 63.62. No copper atom has a mass of 63.62 u but 100 copper atoms would have a total mass of 6362 u.

Mass spectrometry is a well established technique in chemistry. It is more often concerned with the ions derived from molecules and molecular fragments than simply with ions derived from atoms.

8.2.1 Isotope mass and mass number

On the relative atomic mass scale (mass of $^{12}C = 12$ u) the mass of one proton is 1.0074 u and the mass of one neutron is 1.0089 u. Since these relative masses of proton and neutron are both **extremely close to 1**, and since the mass of an electron is negligible, it follows that all isotope masses will be extremely close to being whole numbers. For most purposes we can happily assume that the relative atomic mass of an isotope and its mass number (number of protons + neutrons) have the same numerical value. For example, if we consider aluminium,

$^{27}_{13}Al$ has 13 protons and 14 neutrons.
Its mass number is 27.
Its atomic mass is 27 u.
Its relative atomic mass is 27.

8.3 Stable and unstable isotopes

An atom of an element with a **specific mass number** is called an **isotope** or **nuclide**. 287 naturally occurring isotopes or nuclides are known to exist. Of these, 269 are stable and the other 18 are unstable and radioactive.

What causes an isotope to be unstable? The stability of a nucleus can be related to its neutron-to-proton ratio. If the number of neutrons is plotted against the number of protons (atomic number) for all the naturally occurring **stable** isotopes, a chart such as the one shown in Figure 8.4 (overleaf) is obtained.

The stable nuclei of the lighter elements contain approximately equal numbers of neutrons and protons, i.e. their n/p ratio is equal to 1. As the atomic number increases and more and more protons have to be packed into the tiny nucleus, the forces of repulsion between these positive protons rise sharply. A larger and larger excess of neutrons is required to moderate these repulsions between protons. The n/p ratio increases until it is roughly 1.5/1 for the heavy nuclei.

The stable isotopes form an area in the graph known as the stability belt. Nuclei whose n/p ratios lie **outside** this belt spontaneously undergo changes (transformations) that tend to bring their n/p ratios into or closer to the stability belt.

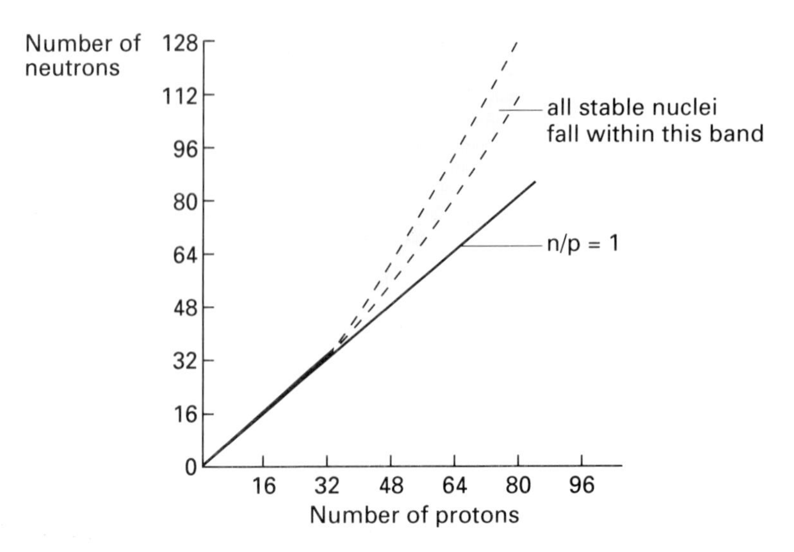

Figure 8.4 n/p ratio for stable isotopes (nuclides)

Conversion from an unstable n/p ratio to a more stable n/p ratio brings about emission of radiation. The **unstable** nucleus or atom is said to be radioactive. Stable nuclei are non-radioactive.

8.4 Radioactivity

An unstable atom (nucleus) is transformed into a more stable atom (nucleus) by the emission of radiation. The radiation may be in the form of particles or waves. The emitted particle or wave carries off the excess energy released by the change, which is described as **radioactive decay**.

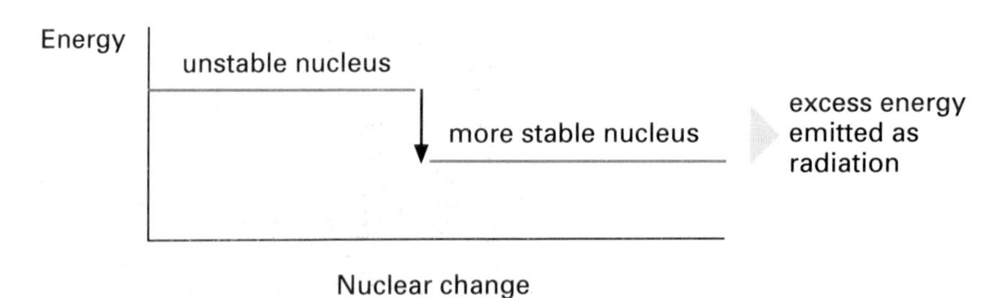

Figure 8.5 Nuclear decay

The phenomenon of radioactivity was first discovered by a Frenchman, Henri Becquerel, in 1896. Covered photographic plates accidentally left close to a uranium salt were found to be exposed. An invisible radiation must have passed through the covering. Two of Becquerel's colleagues, Pierre and Marie Curie, successfully isolated two other radioactive elements, polonium and radium. For their pioneering work all three were awarded the Nobel prize for physics in 1903.

There are three distinct types of radiation emitted:

(a) alpha (α) particles
(b) beta (β) particles
(c) gamma (γ) rays.

8.4.1 Alpha emission

An alpha particle is made up of two protons and two neutrons and is therefore identical to a helium nucleus. For that reason it has the symbol $^{4}_{2}He$.

Alpha emission is common for **heavy** radioactive nuclides which have too many protons for stability. The largest stable, non-radioactive nucleus is $^{209}_{83}Bi$ which has 83 protons. Nuclei with more than 83 protons are unstable.

> ***The ejection of an alpha particle reduces the atomic number by 2 and the mass number by 4.***

A new element is therefore formed with a more favourable nuclear size and n/p ratio. An example is the emission of alpha particles from **americium-241** to form neptunium-237.

$$^{241}_{95}Am \longrightarrow {}^{237}_{93}Np + {}^{4}_{2}He + \text{ energy}$$

Note that in the above nuclear equation both mass number (superscripts) and atomic number (subscripts) are conserved.

$$241 = 237 + 4$$
$$95 = 93 + 2$$

Such equations are written to indicate changes which take place only in the nucleus. The orbital electrons outside the nucleus are ignored.

Alpha particles, being relatively heavy, can ionise atoms and fracture molecules which are in their path. The fast-moving alpha particles released by the radionuclide americium-241 are put to good use in smoke detectors. In a smoke detector a very tiny current from a battery passes through a small column of air ionised by alpha-radiation from americium-241. When smoke enters the device, the electrical conductivity of the ionised air is changed and this activates the alarm.

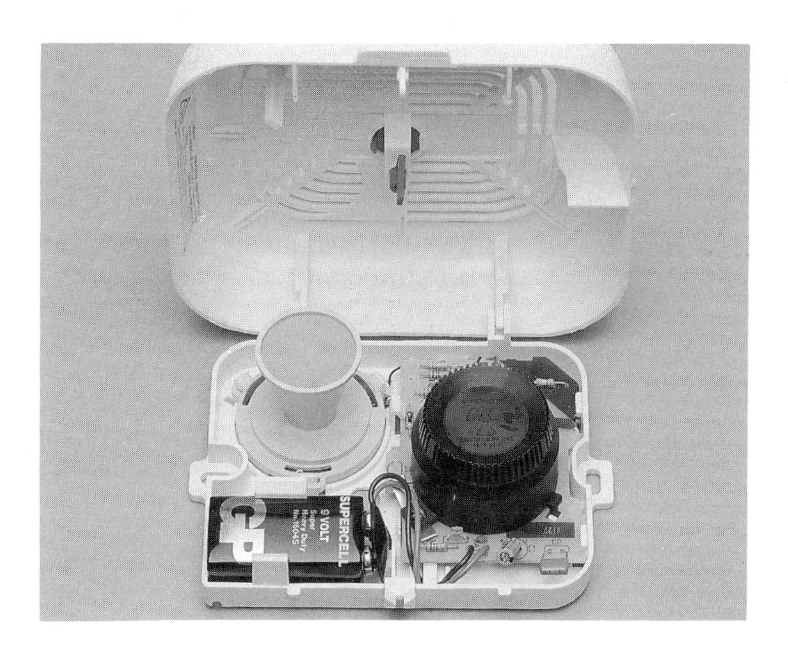

Figure 8.6 The components of a smoke alarm: the radioactive source is contained in the dark-coloured component on the right

As alpha particles go through materials they continually collide with atoms and molecules of the material. The energy of the alpha particles is rapidly scattered. The range of travel of alpha particles is therefore short. They travel only a few centimetres in air and can be stopped by paper or a thin layer of skin. Alpha-emitting radioisotopes (radionuclides) are therefore harmless **outside** the body. If accidentally swallowed or inhaled, they may be deposited in body tissue and cause some problems.

A tiny pellet of another alpha-emitter, the isotope **plutonium-239**, is used to power heart pacemakers for sufferers from some heart-blocking defects. A steady heart rate is maintained by the pacemaker.

8.4.2 Beta emission

A nuclide which has too high an n/p ratio to be stable can improve matters by converting an unwanted neutron into a proton plus an electron.

$$\underset{\text{neutron}}{{}^{1}_{0}\text{n}} \longrightarrow \underset{\text{proton}}{{}^{1}_{1}\text{H}} + \underset{\text{electron}}{{}^{0}_{-1}\text{e}}$$

The proton stays in the nucleus but the electron is ejected with high speed and is known as a **beta particle**. Beta particles are therefore simply electrons produced in and shot out from the nucleus of an atom. A new element is formed as the product nucleus contains **one more proton** than before. The mass number remains unchanged.

> ***The ejection of a beta particle raises the atomic number by 1 but causes no change in the mass number.***

Iodine-131 is a neutron-rich isotope which decays to xenon-131 by firing out a beta particle.

$${}^{131}_{53}\text{I} \longrightarrow {}^{131}_{54}\text{Xe} + {}^{0}_{-1}\text{e}$$

The equation is balanced by ensuring that the total mass (upper numbers) and charge (lower numbers) after the transformation are the same as they were before the change.

Beta emission is a very common form of radioactive disintegration. Radioactive isotopes (radioisotopes), beta emitters included, are easily detected and measured because radiation ionises surrounding gases and also affects a photographic emulsion. When gas molecules are ionised their electrical conductivity rises; a property used in the **Geiger-Müller** counter for detecting radiation. Radioisotopes can be used as **tracers** to follow a substance through physical, chemical or biological processes without interfering with the processes themselves.

In medicine, radioisotopes are used for diagnosis, treatment and research. They can be introduced into the body as 'beacons'. Useful information can be gained by measuring the rate of uptake by an organ of a particular element or compound. Iodine, for example, accumulates in a gland in the neck called the **thyroid** gland. A normal thyroid contains 8 mg iodine of which 0.05 mg is replaced each day. The functioning of the gland can be assessed by giving the patient a drink containing a tiny amount of radioactive iodine-131. The rate of uptake is then measured through a Geiger counter held externally.

The beta particles which are ejected from an unstable nucleus travel up to 30 cm in air before the buffeting by air molecules brings them to rest. Being so small, beta particles, unlike alpha particles, can actually pass through a sheet of paper but are absorbed by sheets of metal.

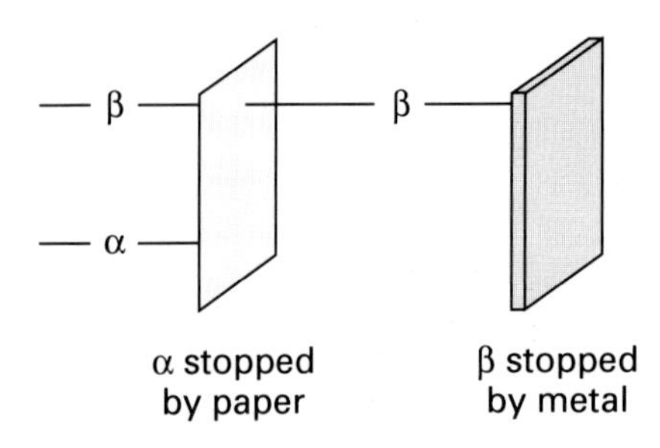

Figure 8.7 Penetration by alpha and beta particles

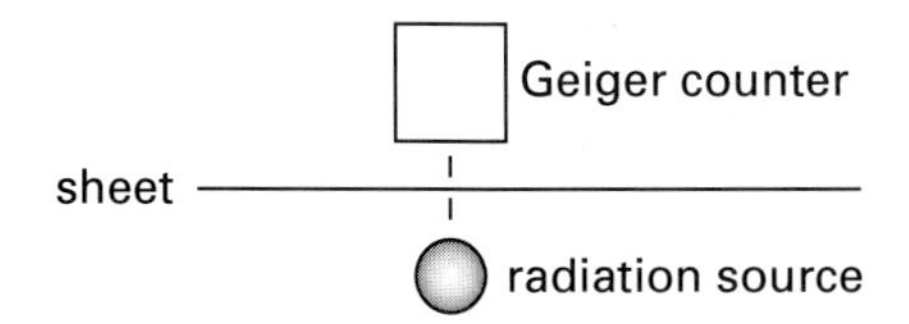

Figure 8.8 Monitoring thickness by radiation

When paper or plastic sheet is being manufactured, its thickness can be checked by placing the sheet between a beta-emitting source and a Geiger country. By measuring the fraction of radiation passing through the sheet, its thickness can be estimated quite accurately and controlled on a production line.

8.4.3 Gamma emission

Gamma rays are electromagnetic waves of very short wavelengths (even shorter than X-rays).

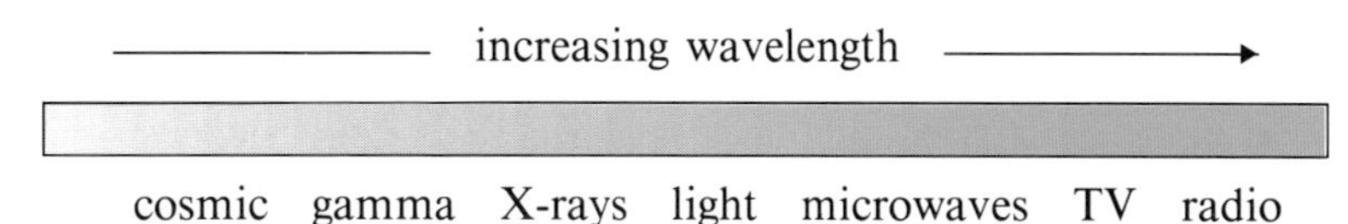

When an unstable nucleus decays to an even slightly more stable arrangement the excess energy must be released. Alpha and beta particles carry off some of this energy in the form of kinetic (movement) energy, but energy in the form of gamma waves may also be given off. Gamma emission therefore **accompanies** other radioactive emissions and is a means by which a high energy (excited) nucleus can lose energy and become somewhat more stable and placid.

Cobalt-60m is an artificially produced radioisotope which is said to be meta-stable (m). It is over-energetic and emits gamma rays to form a less excited cobalt-60 isotope. This in turn decays by beta emission to form the stable isotope, nickel-60.

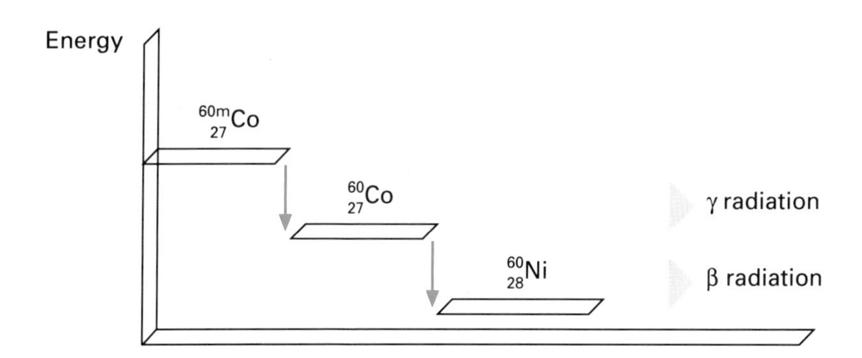

Figure 8.9 Radioactive decay of cobalt-60m

Unlike alpha and beta particles, gamma rays have **neither mass nor charge**. They are undeflected by electric or magnetic fields, travel long distances in air and are only absorbed by thick sections of dense materials such as lead or concrete.

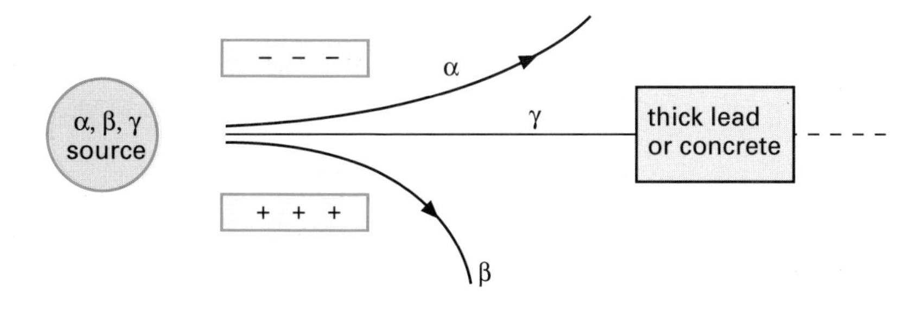

Figure 8.10 Deflection of alpha, beta and gamma radiation

It is a paradox that, while radiation in small amounts may be instrumental in initiating cancer, in very large doses it can destroy cancer. Cancer cells are more easily destroyed by radiation than normal cells. If gamma radiation can be precisely concentrated or focused on the cancerous tissue itself, the cancer can be eliminated with only slight damage to surrounding normal tissue. A brain tumour, for instance, can be destroyed by irradiating it from various angles with gamma rays from a **cobalt-60** source.

8.4.4 Successive radioactive emissions

Some radioactive nuclei, particularly the heavy ones, cannot gain stability by a single emission. Instead a **series** of successive emissions occurs until eventually a stable nucleus is formed.

Thorium-232, for example, is a naturally occurring radioactive isotope found in the earth. It decays to lead-208 (over a considerable length of time) in a sequence involving the emission of six alpha and four beta particles.

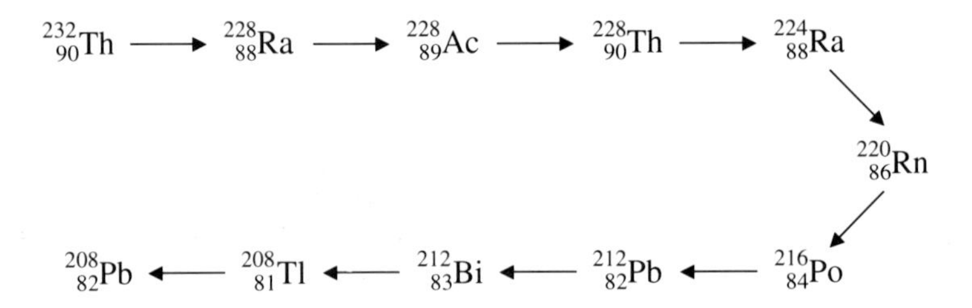

Each step in the process provides some decrease in energy and some increase in stability.

Sample exercise

The naturally occurring radioisotope **uranium-238** emits a total of eight alpha and six beta particles in a sequence of disintegrations. What is the mass number, atomic number and identity of the final product nuclide?

Method

The first step is to find the atomic number of uranium from the periodic table and to write a nuclear equation which represents the overall change. Let the symbol $^{m}_{a}E$ represent the product nuclide.

$$^{238}_{92}U \longrightarrow 8\,^{4}_{2}He + 6\,^{0}_{-1}e + ^{m}_{a}E$$

Applying conservation of **mass** to the equation,

$$238 = 32 + 0 + m$$
$$m = 206 \text{ (mass number of product is 206).}$$

Applying conservation of **charge**,

$$92 = 16 - 6 + a$$
$$a = 82 \text{ (atomic number of product is 82).}$$

Element of atomic number 82 is lead.
The final product nuclide is $^{206}_{82}Pb$.

8.5 Rate of radioactive decay

Unstable nuclei can emit different forms of radiation and form different products. They also decay or disintegrate at **different rates**. The rate of decay of a radioactive nucleus is measured in terms of a period of time known as the **half-life** ($t_{\frac{1}{2}}$). Each radioactive isotope (nuclide) has its own characteristic half-life.

> ***Half-life is the time taken for any amount of a radioisotope (radionuclide) to fall to half its previous amount.***

The term 'amount' may refer to the **number** of radioactive nuclei (N), or the **mass** of these radioactive nuclei (M). Neither of these can be measured directly. Half-life is thus often calculated from measurements of the **intensity** (I) of emitted radiation, as recorded on some kind of detector. Intensity of emission is directly proportional to the amount of the isotope in the sample.

$$N \xrightarrow{t_{\frac{1}{2}}} \tfrac{1}{2}N$$

$$M \xrightarrow{t_{\frac{1}{2}}} \tfrac{1}{2}M$$

$$I \xrightarrow{t_{\frac{1}{2}}} \tfrac{1}{2}I$$

Sodium-24 is an artificially produced radionuclide which decays, by beta emission, to magnesium-24.

$${}^{24}_{11}\text{Na} \longrightarrow {}^{24}_{12}\text{Mg} + {}^{0}_{-1}\text{e}$$

The half life of sodium-24 is 15 hours. This means that after 15 hours, one half of the original ^{24}Na atoms will have changed into ^{24}Mg atoms. After 30 hours the fraction of Na atoms left will be ¼, after 45 hours ⅛ remains, and so on. The gradual change in composition is illustrated in Figure 8.11.

Using a Geiger counter, we can record the decrease in the intensity of emitted radiation over a period of time. A plot of intensity (counts per minute) against time for sodium-24 produces a graph (Figure 8.12) from which the half-life of Na-24 can be calculated (15 hours).

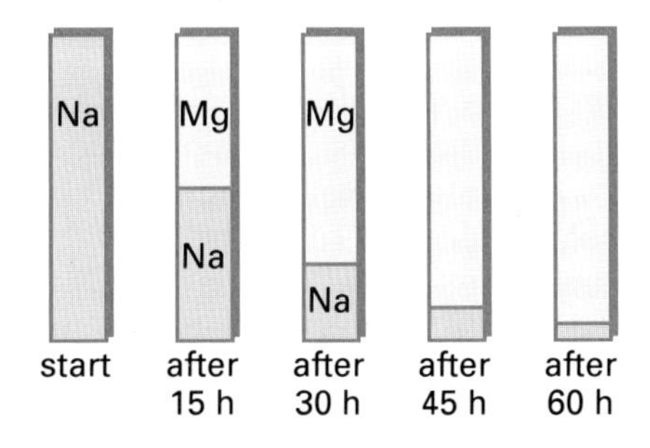

Figure 8.11 Rate of decay of sodium-24

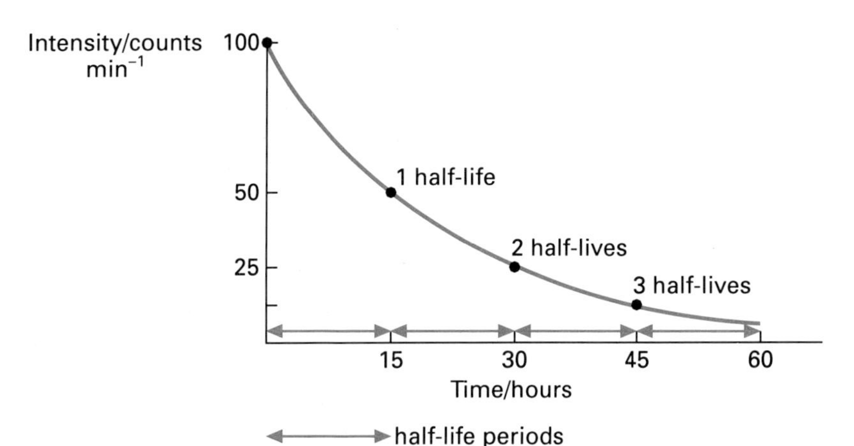

Figure 8.12 Decay graph for sodium-24

From this type of graph we can estimate how long it will take for the radioactivity to fall to any selected intensity. Similarly, we are able to predict the intensity of radiation at any given time. As you can see from Figure 8.12, the intensity of radiation from sodium-24 is relatively low after 60 hours. The sample would be harmless within one week and gone within two weeks.

Trace amounts of a radioactive sample inserted into the blood supply can be used to detect circulatory disorders. Salt (NaCl) solution containing a small amount of sodium-24 may be injected into the bloodstream. By measuring the build-up of radiation at various places in the body, doctors can readily ascertain whether circulation of blood to these areas is normal.

Sample exercise

A hospital obtains a sample of the radioisotope iodine-131 for use in the diagnosis of a thyroid disorder. The half-life of the isotope is eight days. What fraction of iodine-131 remains after eight weeks?

Method

8 weeks = 56 days = 7×8 days = 7 half-lives

This shows that the time interval is exactly seven whole half-lives and the fraction left can be calculated without having to compile a graph of amount versus time.

$$1 \longrightarrow \frac{1}{2} \longrightarrow \frac{1}{4} \longrightarrow \frac{1}{8} \longrightarrow \frac{1}{16} \longrightarrow \frac{1}{32} \longrightarrow \frac{1}{64} \longrightarrow \frac{1}{128}$$

So the fraction of iodine-131 which remains after eight weeks is $\frac{1}{128}$.

The unstable atoms of a radioactive element disintegrate and eject alpha and/or beta particles plus gamma radiation. This happens in a spontaneous and random fashion. We cannot predict which atoms will disintegrate or when. We can only say that, within a particular interval of time (the half-life), half of the radioactive atoms will break down.

The half-life of a radioisotope is constant. It depends solely on the composition of the nucleus and cannot be altered by external conditions such as temperature, pressure, concentration and what chemical compound the element is part of. **Neon**, for instance, has three stable isotopes and three unstable (radioactive) isotopes, all of them artificially made.

Table 8.1 Stable and unstable isotopes of neon

Isotope	Description
^{20}Ne	stable
^{21}Ne	stable
^{22}Ne	stable
^{23}Ne	β emitter: $t_{\frac{1}{2}} = 37.2$ s
^{24}Ne	β emitter: $t_{\frac{1}{2}} = 3.38$ min
^{25}Ne	β emitter: $t_{\frac{1}{2}} = 0.61$ s

The more unstable the nucleus, the faster the rate of decay and the shorter its half-life. Conversely, the more a nucleus approaches some sort of stability, the longer its half-life. A stable nucleus can be thought of as having an infinitely long half-life: it is non-radioactive.

8.6 Radiation from outer space

A sizeable proportion of the natural background radiation received by an individual at sea level is from **cosmic radiation**, while the rest comes from radioactive materials in the earth's crust. At one time it was believed that all background radiation came entirely from radioactive substances in the earth. In 1910, Albert Gockel of the USA sent a radiation detector attached to a balloon three miles up into the atmosphere. He expected radiation to diminish the higher it went but, to his surprise, the opposite happened. It was eventually concluded that the observed radiation must be entering the earth's atmosphere from outer space. This was the first recognition of cosmic rays.

The actual origin of cosmic rays is still obscure. The particles arriving from outer space are mainly protons, some alpha particles and a few nuclei of other elements. These particles collide with atoms of nitrogen, oxygen and other elements in the outer atmosphere and generate a host of new particles plus high energy gamma rays. Cosmic radiation is most intense at an altitude of 10–15 miles and gradually decreases as one gets closer to the ground.

Figure 8.13 Airliners fly through regions of relatively intense cosmic radiation

8.6.1 Carbon dating

One naturally occurring radioisotope produced by the action of cosmic rays is **carbon-14**. Neutrons in the cosmic radiation crash into nitrogen atoms and create protons and carbon-14 atoms.

$$^{1}_{0}n + ^{14}_{7}N \longrightarrow ^{1}_{1}H + ^{14}_{6}C$$

The carbon-14 continuously formed in the upper atmosphere becomes incorporated into the CO_2 of the air. It reaches a steady concentration of about one carbon-14 atom for every 10^{12} atoms of carbon-12. This radioactive CO_2 becomes readily incorporated into biological organisms. Our bodies, for example, have the same fraction of carbon-14 as exists in the air. This is a very small amount and emits radiation whose intensity is only about **16 disintegrations per minute** for each gram of carbon.

Although the carbon-14 is continuously disintegrating, it is also being replaced at a steady rate in living plants through photosynthesis: the amount of carbon-14 in plants and trees is therefore constant. When a plant dies, the intake of carbon-14 abruptly stops but the radioactive decay of the carbon-14 continues as before.

In very old wood, paper or cloth the decay of carbon-14 over thousands of years reduces the ratio of ^{14}C:^{12}C. By measuring the radioactivity due to carbon-14 in a sample of once-living material, we can estimate the time at which its life ceased. This method is used to determine the age of relics or archaeological finds and is known as **radiocarbon dating**. The method is not sensitive beyond 50 000 years because by that time almost all the carbon-14 will have decayed.

$$^{14}_{6}C \longrightarrow ^{14}_{7}N + ^{0}_{-1}e \qquad t_{\frac{1}{2}} = 5.7 \times 10^3 \text{ years}$$

Estimating the age of a wooden object by comparing its carbon-14 radioactivity with that of a growing tree depends on a critical assumption: that the ratio ^{14}C:^{12}C in the past was the same as it is today. That is a dubious assumption.

Sample exercise

The radioactivity due to carbon-14 in a fragment of 'bog oak' found buried on a peat moor is measured as 3.5 counts/min/g carbon. If the half-life of carbon-14 is 5.7×10^3 years and the activity of a living plant is 16 counts/min/g carbon, how old is the wood?

Method

The fraction of radioactivity remaining is not equal to $1/2^n$ where n is a whole number, i.e. not $\frac{1}{2}, \frac{1}{4}, \frac{1}{8}, \frac{1}{16}$, etc. As a result, the time which has elapsed since the oak was a living tree is not a whole number of half-lives. A graph of activity versus time will therefore have to be constructed.

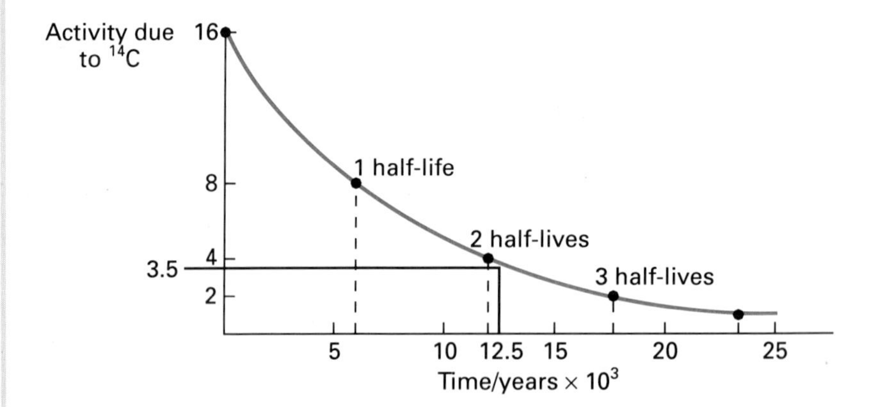

Figure 8.14

From the graph one would predict that an activity of 3.5 units would be reached after a period of time equal to 12.5×10^3 years. The bog oak is therefore approximately 12 500 years old.

8.7 Radiation from the earth

Naturally occurring radioactive isotopes fall into two groups:
(a) marginally unstable isotopes which were created when the earth was formed and which have sufficiently long half-lives (millions of years) for a certain amount of them still to exist.
(b) unstable isotopes which are themselves the products of decay of long-lived isotopes from group (a).

The **long-lived** isotopes consist mainly of thorium-232, uranium-238 and uranium-235. Each of these radionuclides decays in a chain of disintegrations which finally stops when a stable isotope of lead is formed (Table 8.2).

Table 8.2 Decay of long-lived isotopes

Radionuclide	Half-life/years	Final product nuclide
^{232}Th	1.41×10^{10}	^{205}Pb
^{238}U	4.51×10^{9}	^{206}Pb
^{235}U	7.04×10^{8}	^{207}Pb

8.7.1 Radon

Different rocks and oils contain varying amounts of uranium and thorium. Inevitably, building materials made of stone, sand, gravel, etc., will contain natural radioactive substances. More than 99 per cent of natural uranium is in the form of uranium-238. The decay of ^{238}U to ^{206}Pb proceeds by way of 14 different radioactive species. One intermediate product is **radon-222** which, because of its noble element electron structure, is a gas. ^{222}Ra has a short half-life of 3.8 days and rapidly decays away. However, it is always being resupplied by the slow decay of isotopes preceding it in the decay chain from ^{232}U.

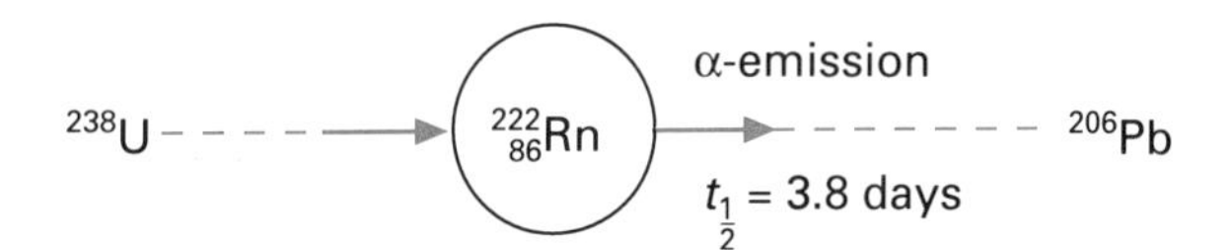

Figure 8.15 Formation and decay of radon-222

Another alpha-emitting radon radioisotope, **radon-220**, is generated in the decay chain stemming from thorium-232. It is, therefore, often called **thoron** and has a half-life of 55 seconds.

Radon, being a gas, can seep out of rocks which are cracked and diffuse to the surface. Granite rocks in the south-west of England are extensively split and more inclined to allow radon to escape in this way.

Figure 8.16 Weathered granite 'tors' on Dartmoor, an area where atmospheric radon levels are relatively high

Since radon is much denser than air, it does not diffuse away into the atmosphere but hangs about inside poorly ventilated houses and buildings, which it has infiltrated through cracks in the floor, etc. We are not so much concerned with radon itself, since it is short-lived, but with its 'daughter' decay products, which are all metallic solids and can lodge in the lungs.

Five to ten per cent of lung cancer cases in people who live in industrialised countries is believed to be caused by radon. Clearly there is a need to produce more radon-proof housing in danger areas above uranium and thorium bearing rock. A transparent plastic strip called **Tastrak** can be used to detect the alpha particles emitted by radon and this offers a simple method of detecting the presence of the gas.

8.7.2 Radioactive potassium and rubidium

Another important natural radioisotope is **potassium-40**. It makes up 0.012 per cent of all potassium, and potassium is the seventh most common element on earth. ^{40}K has a half-life of 1.3×10^9 years and so probably dates back to the formation of the earth. In fact the constant rate of the change $^{40}K \rightarrow {}^{40}Ar$ allows us to measure the K/Ar ratio in rocks and use it to determine the age of the rock.

Potassium is an essential ingredient in the growth of plants and thus K is absorbed into the **food chain.** Added to that, the fertiliser industry helps to spread potassium compounds over the land and this ensures a uniform distribution of ^{40}K. The level of radiation from potassium-40 is, however, very low and has been with us since the beginning of agriculture.

Rubidium, as an element, is inconspicuous since it has no major uses. However, it is the sixteenth most common element, commoner than zinc, copper or nickel. Nearly 28 per cent of natural rubidium is the isotope **rubidium-87**, which is a weak beta-emitter with a long half-life of 4.9×10^{10} years.

Naturally occurring radioactive materials in the earth generally end up dissolved in water and are taken up by plants, animals and humans. Everything we eat or drink is therefore likely to be very slightly radioactive.

Figure 8.17 A helicopter spraying decontaminants on to the wrecked Chernobyl nuclear plant after the catastrophic explosion; the ruined reactor has now been sheathed in concrete

8.8 Artificially produced radiation and radioisotopes

Radioactive nuclides made artificially by humans account for less than a quarter of the radiation to which we are exposed. These radioisotopes or radionuclides are generally produced, under strictly controlled conditions, inside **nuclear reactors**. However, the explosion of nuclear weapons or accidents in nuclear reactors can release radioactive material into the atmosphere. This **radioactive fallout** is carried by the wind and falls to the ground in rain. There it may be absorbed by plants and taken in by animals. The disastrous explosion of the Russian nuclear reactor at Chernobyl in 1986 was an extreme but fortunately isolated example of the release of large amounts of radioactive material into the environment.

One of the commoner isotopes produced in nuclear explosions is **strontium-90**. It is a weak beta-emitter with a half-life of 28 years but decays to yttrium-90 which emits strong beta radiation. Strontium gets easily into the food chain. It behaves in body chemistry much like calcium which is a major component of bones and teeth. Milk is one of the best sources of calcium for humans, and milk is the commonest means by which humans absorb strontium-90.

Caesium-137 follows much the same pattern as strontium-90. It has a similar half-life, 30 years, but chemically speaking it copies potassium rather than calcium. ^{137}Cs is absorbed by plants and ends up in human muscle.

8.9 Nuclear fission

Caesium and strontium are common products of **nuclear fission**. Here a very heavy nucleus is broken up to give two lighter nuclei and a large output of energy. Very few heavy nuclei can be split apart in this way. The only **fissionable** isotopes of importance are **plutonium-239** and **uranium-235**. Their splitting is fairly random in the sense that the two nuclear fragments formed may be any two nuclei from perhaps 200 isotopes of 35 different elements.

Uranium-235, when bombarded by neutrons, cracks into two unequal fragments plus two or three neutrons.

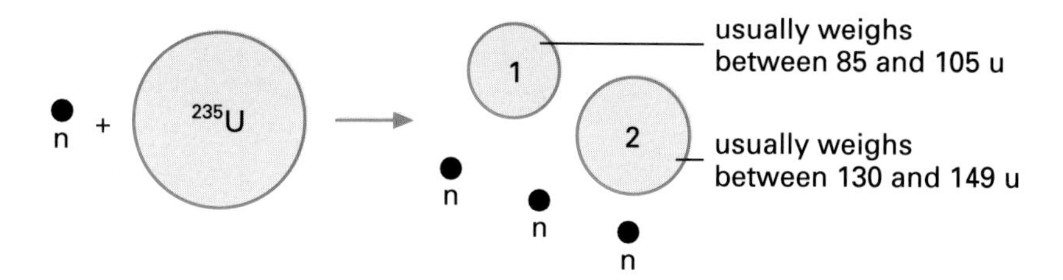

Figure 8.18 Fission of uranium-235

Each neutron-induced fission event produces two or three more neutrons. Every neutron produced is capable of creating additional fissions and releasing yet more neutrons. A **chain reaction** is started and enormous amounts of energy can be released in a very short time. This is what happens in a nuclear fission (atomic) bomb.

In a nuclear power station the extra neutrons have somehow to be removed so that they cannot promote the multiple fissions which produce an 'avalanche' effect and a runaway reaction. The ideal situation is the continuous production of energy **at a fixed rate** (Figure 8.19).

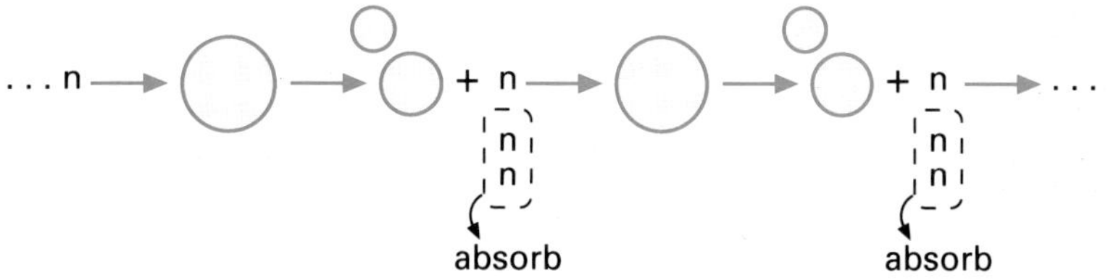

... n → ○ → ○ + n → ○ → ○ + n → ...

absorb absorb

Figure 8.19 Controlled fission

In a nuclear reactor, the rate of fissioning (splitting) is maintained at a reasonable level by the use of **control rods**, made from boron steel or cadmium. These metals have the capacity to **absorb** large numbers of neutrons and thus remove them from circulation. The reaction, once started by a source of neutrons, can be slowed down or speeded up, simply by moving the control rods in or out of the reactor core where the nuclear fission is taking place.

The neutrons emitted in a nuclear fission have excess energy. They are more effective in causing further fissions if they can be slowed down. In all but 'fast' reactors the core contains a material, such as graphite or water, which will **moderate** the neutrons by absorbing some of their kinetic energy and thus slow them down.

In addition to control rods and perhaps moderator, a nuclear reactor requires a **coolant**. The coolant, either water, liquid sodium or CO_2 gas, circulates through the reactor core in order to carry off the large amount of heat generated by the fission of the nuclei.

There are many different designs of nuclear reactor. Figure 8.20 outlines the components of a typical nuclear reactor core which is about twice the size of a double-decker bus.

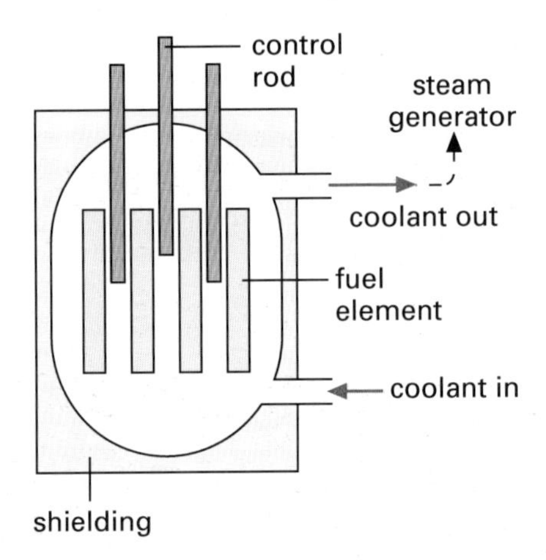

Figure 8.20 Components of a nuclear reactor

Finally we come to the **fuel** for the reactor. British Nuclear Fuels produce the fuel for all the British nuclear power stations at the Springfields plant near Preston. Some reactors use natural uranium fuel which contains 0.7 per cent uranium-235. Other reactors require enriched uranium fuel in which the uranium-235 content has been artificially raised to 3 per cent.

The uranium-235 generates heat energy by undergoing fission when it captures neutrons. Two typical fission reactions are shown below.

$$ {}^{1}_{0}n + {}^{235}_{92}U \longrightarrow {}^{97}_{40}Zr + {}^{137}_{52}Te + 2{}^{1}_{0}n + \boxed{\text{energy}} $$

$$ {}^{1}_{0}n + {}^{235}_{92}U \longrightarrow {}^{91}_{36}Kr + {}^{142}_{56}Ba + 3{}^{1}_{0}n + \boxed{\text{energy}} $$

After several years in a thermal reactor the fuel becomes less efficient and needs to be replaced. The **spent fuel** is a mixture of unused uranium fuel, plutonium and waste fission products.

The **plutonium** is produced through combination of the uranium-238 nuclei with slow neutrons.

$$ {}^{238}_{92}U + {}^{1}_{0}n \longrightarrow {}^{239}_{93}Pu + {}^{0}_{-1}e $$

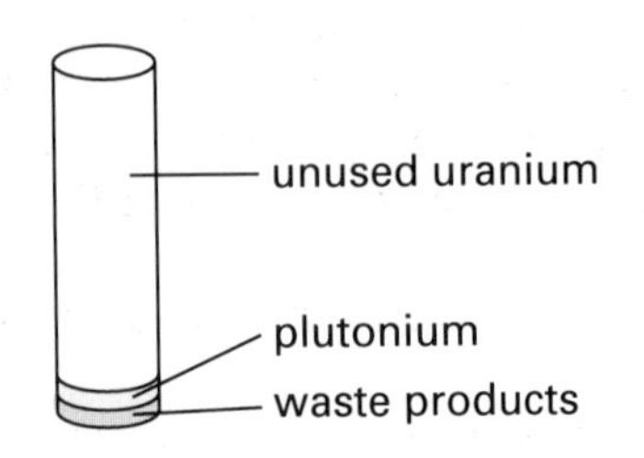

Figure 8.21 Typical composition of spent fuel

Plutonium does not occur naturally but is capable of fission and is therefore an alternative nuclear fuel. It can be split by **fast** travelling neutrons and so does not require a moderator.

The spent fuel waste products contain both short-lived and long-lived isotopes. They are stored on site in cooling ponds to allow the radioactivity and heat generated from the short-lived isotopes to die away. When this has been achieved the spent fuel may be transported to Sellafield, the nuclear fuel reprocessing plant in Cumbria. Here the uranium and plutonium can be recovered. The highly active waste containing long-lived isotopes is converted into solid glass blocks. These blocks will ultimately be deposited hundreds of metres underground in stable rock formations, and left to decay safely at their own slow rate. This type of radioactive waste attracts public concern, not because of the volume produced annually (which is in fact very small) but because of the long time it will take for the radiation to die away.

The heat produced from nuclear reactors turns water into steam and drives the blades of the turbo generators which create electricity. The electricity produced by Scottish Nuclear's power stations at Hunterston and Torness is sold to Scottish Power and Scottish Hydro-Electric.

Figure 8.22 The Torness nuclear power complex

8.10 Nuclear fusion

The reverse of nuclear fission is nuclear **fusion**. In this process two light nuclei are combined or fused together to form a heavier nucleus. In so doing, a huge amount of energy is released.

The energy which we unceasingly receive from the sun is produced by nuclear fusion. The sun consists mainly of hydrogen and helium plus a small proportion of other light elements. The continuous fusion of light isotopes to form heavier ones enables the sun to generate the energy radiated to us on earth. Approximately 6×10^{18} kJ of energy are received by the earth each day.

As a potential commercial source of energy, fusion has several advantages over fission.

(a) Much larger amounts of energy per unit mass of fuel are produced.
(b) The product nuclei are generally stable and not radioactive. Disposal of waste is therefore not a problem.

(c) The light nuclei suitable for fusion, e.g. isotopes of hydrogen, are far more abundant than the heavy isotopes needed for fission.

The major obstacle to achieving controlled fusion power is the enormous **activation energy** required to get positive nuclei to overcome the mutual repulsion between them and collide. Temperatures around two hundred million kelvins (2×10^8 K) have to be achieved in order to bring about the combination of nuclei. Perhaps the most promising fusion reaction is that between the heavy isotopes of hydrogen, namely deuterium, ^2_1H, and tritium, ^3_1H. Deuterium is abundant in seawater while the other reactant, tritium, can be made from lithium, a readily available metal.

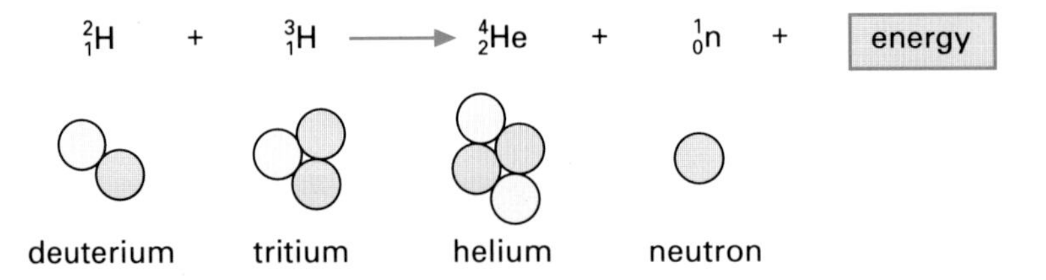

Figure 8.23 Fusion reaction

At present scientists from various countries are working together to solve the enormous practical problems which have to be overcome before useful energy can be obtained from nuclear fusion reactions. To bring about the reaction described, they have to find a way to heat a mixture of deuterium and tritium gases to temperatures around 2×10^8 K and also contain the resulting plasma (electrons and bare nuclei) while fusion occurs. Since no material can withstand these huge temperatures, prototype fusion reactors rely on the use of strong magnetic fields to confine the charged particles of the hot plasma.

If fusion reactors did become a commercial reality, they could be the solution to the world's energy needs. We would then have our own 'mini-suns' on earth.

8.11 Nuclear and fossil fuels

The ever-increasing world demand for energy has placed a tremendous strain on the fuels which can provide that energy. These fuels include nuclear fuels such as uranium and plutonium, and fossil fuels such as oil, natural gas and coal. Any policy on fuel must take into account both the size of the fuel reserves and the hazards associated with the use of the fuel.

8.11.1 Size of fuel reserves

Fast reactors, like the one pioneered at Dounreay in Caithness, use mixed uranium–plutonium fuel, obtained as a by-product from ordinary thermal reactors. There is enough plutonium present for the fissioning process to be sustained by fast, rather than slow (moderated), neutrons. The reactor core can be surrounded by a blanket of uranium-238 which captures some neutrons and forms plutonium-239. It is possible to alter the relative amounts of splitting and capturing, so that more fresh fuel is generated (bred) than existing fuel is consumed. The great advantage of these **fast breeder reactors** is that they convert the relatively plentiful but non-fissionable uranium-238 into fissionable but not naturally occurring plutonium fuel.

Figure 8.24 compares reserves of uranium with the major fossil fuels.

coal
oil
gas
^{235}U
^{238}U
World reserves

Figure 8.24 Comparison of fuel reserves

8.11.2 Hazards

Fuels which give rise to pollutants are a hazard to the environment and all living things on the planet. The fission of nuclear fuels such as uranium or plutonium produces products which may be described as **pollutants**. The products are radioactive isotopes, some short-lived and some with longer half-lives.

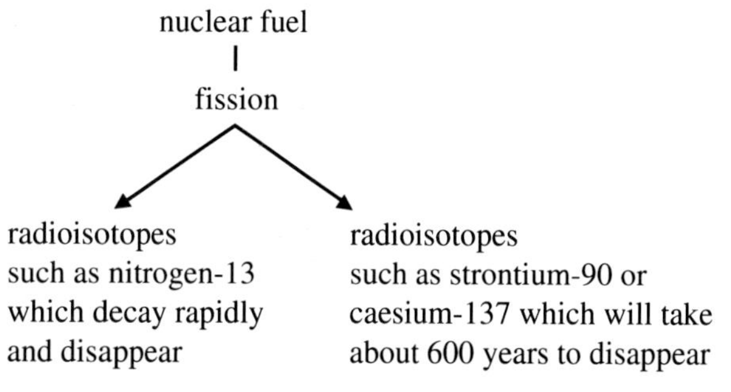

The main hazard is the longer-lived group of isotopes produced in nuclear reactors. These cannot be 'neutralised' in any way but can only be confined out of harm's way and left to decay at their own rate.

Fossil fuels also produce pollutants but some of these are released to the environment when the fuel is burned.

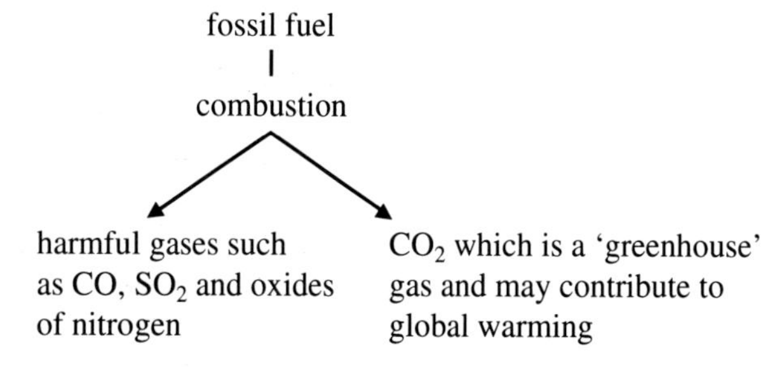

Much effort is made to remove sulphur from fuels, before burning, and to convert other pollutant chemically, after burning. The emission of carbon dioxide is now a problem for all users of fossil fuels as its over-production may bring about damaging changes in climate.

8.12 Radiation in our daily lives

Since 1895, when X-rays were first discovered, the use of radiation has brought enormous benefit to people. Nuclear reactors have produced power for generating electricity on a grand scale. They have also been the means of producing radioisotopes in sufficient quantities to allow their widespread use at relatively low cost. Radioisotopes have proved invaluable in chemical research, industry and medicine.

8.12.1 Chemical research

Radioisotopes are often used to trace the path of an element as it passes through the various steps from reactant to product. Biochemists, for example, have learnt a lot about the mechanisms of complex reactions by using carbon-14 as a tracer or **radioisotope label** for molecules. One such reaction is photosynthesis.

$$6CO_2 + 6H_2O \longrightarrow C_6H_{12}O_6 + 6O_2$$

The reaction proceeds through a series of steps. To study the pathway, plants were exposed to $^{14}CO_2$, i.e. CO_2 containing ^{14}C. At various times the plants were analysed to determine which compounds contained ^{14}C and were therefore products of photosynthesis. For his research in this area, Melvin Calvin of California received the Nobel Prize for Chemistry in 1961.

8.12.2 Industry

Radioactive isotopes find widespread use as **tracers** in every area of industry. The frictional wear of engine metals, for instance, can be checked by making the metal slightly radioactive. The measured activity of the isotope in the lubricating oil indicates the extent of surface erosion of the metal.

A beta-emitting radioisotope can be employed in a whole range of applications, one of which is to control the **automatic filling** of containers with powder or liquid (Figure 8.25).

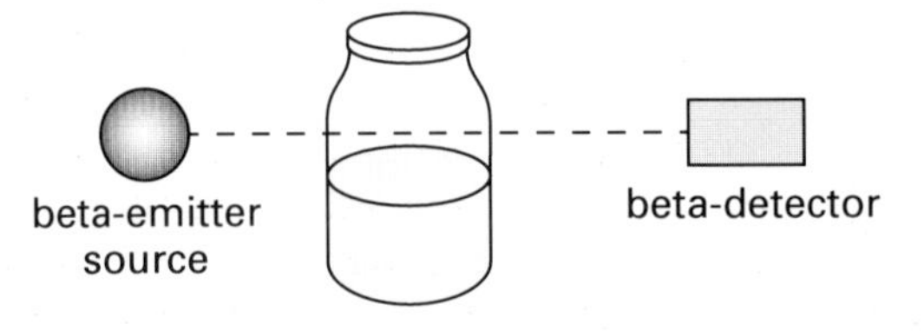

Figure 8.25 Automatic filling

The detector is connected, electrically, to the circuit which controls the filling operation. When the contents rise to the maximum level (just above the beta beam), the beta radiation will be absorbed by the contents and reception cut off. This triggers the electrical mechanism which stops the filling process.

8.12.3 Medicine

By far the largest exposure of the public to radioisotopes is in the field of medical diagnosis, treatment and research. Where the radioisotope is chemically attached to a pharmaceutical or drug for tracer purposes, the radioactive substance formed is called a **radiopharmaceutical**. The radiopharmaceutical is composed of an appropriate pharmaceutical compound and a radioisotope which effectively labels the compound and allows it to be traced anywhere in the body.

The first use of the radiopharmaceutical tracer technique was by Moore in 1948. To investigate brain disorders, Moore synthesised difluorescein containing ^{131}I. He injected this into the patient's vein and measured the radioactive emission over regions of the head using a Geiger counter. Nowadays the detecting device is a **gamma camera** which records gamma radiation emitted by the radioisotope and is linked to a data processor. The radioisotope used is commonly **technetium-99m**. The 'm' indicates that it is the **metastable** form of Tc. This excited radionuclide is ideal for the purpose as it decays quickly (half-life = 6 hours) and emits gamma rays suitable for scanning purposes.

Depending on which part of the body is under investigation, an appropriate pharmaceutical compound is selected. The technetium radioisotope reacts with this carrier compound and becomes chemically attached to it.

[^{99m}Tc]BAT represents a new radiopharmaceutical for tracing the flow of blood in the brain. The radioactive technetium is incorporated into the middle of the molecular structure of the pharmaceutical bisaminothiol (BAT), as shown in Figure 8.26. Technetium is an intriguing element. It does not exist naturally but can be specially produced in a nuclear reactor by bombardment of natural molybdenum with neutrons to form molybdenum-99. This isotope emits beta particles to form technetium-99m.

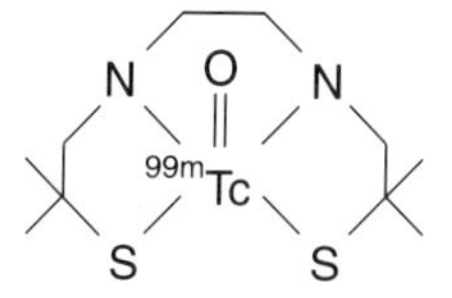

Figure 8.26 [^{99m}Tc]BAT

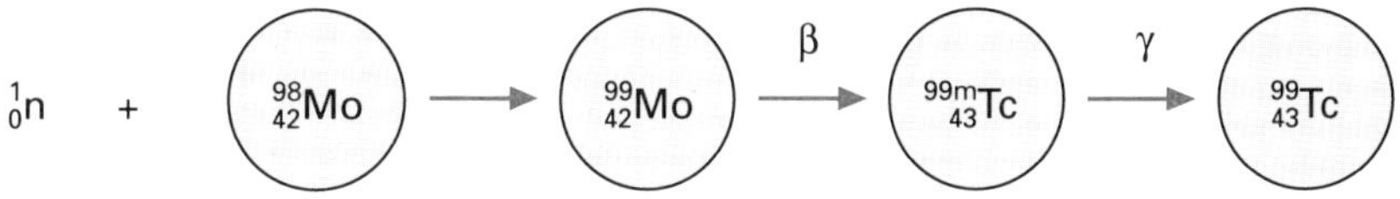

$$^{1}_{0}n + {}^{98}_{42}Mo \longrightarrow {}^{99}_{42}Mo \xrightarrow{\beta} {}^{99m}_{43}Tc \xrightarrow{\gamma} {}^{99}_{43}Tc$$

Figure 8.27

Once technetium is separated chemically as pure ^{99}Tc, it decays quickly and has completely disappeared after about three days. This means that any hospital wishing to use the radioisotope would have to acquire it and use it immediately. To overcome this snag, the hospital generally orders the molybdenum- 99 isotope; this acts as a 'cow' from which the technetium-99m can be 'milked' at the hospital site where it is needed.

The molybdenum isotope, made in the nuclear reactor, is loaded into a glass tube, packed with aluminium oxide powder. This ^{99}Mo has a half-life of 66 hours and decays to ^{99m}Tc. The molybdenum cow or technetium generator, whatever you want to call it, is delivered to the hospital. The technetium can be easily extracted when required by passing salt solution through the column. Molybdenum does not dissolve in salt solution but technetium does and can be collected.

Thanks to its ready availability, suitable half-life and favourable gamma emission energy, technetium-99m has established itself as the key radioisotope in **nuclear medicine**. It can be chemically bound to a range of pharmaceutical 'carriers' and used to trace the movement of biochemicals in order to diagnose the function of various organs of the body.

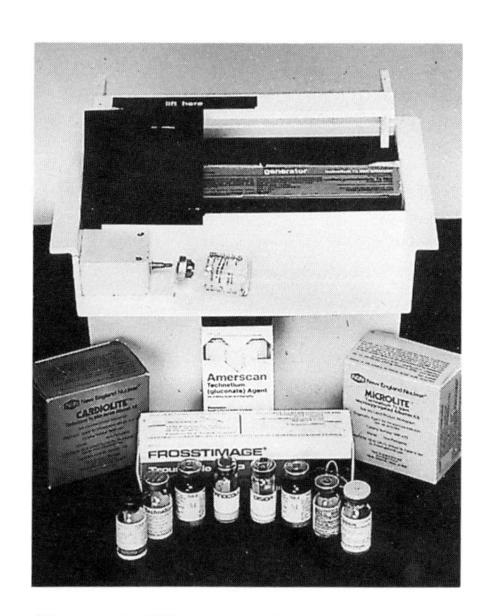

Figure 8.28 A technetium generator and radio-pharmaceutical kit

Summary of Unit 8

Having read and understood the information and ideas given in this unit, you should now be able to:

calculate the relative atomic mass of an element from the masses and relative proportions of its isotopes

explain radioactivity, in terms of unstable isotopes rearranging their nuclei and emitting radiation energy to form more stable compositions

describe the nature and properties of alpha, beta and gamma radiation

state that the spontaneous disintegration of individual nuclei in a radioactive sample is random and not affected by the chemical or physical state of the sample

state that each radioisotope has a characteristic half-life

calculate the value of either amount of radioisotope, time elapsed or half-life, given the values of the other two variables

give examples of radioisotopes being used to treat cancer, as tracers to follow movement of chemicals, and as measuring or controlling devices in industry

state how heat energy can be produced by the fission of radioisotopes of uranium and plutonium

outline common sources of background radiation such as cosmic rays, radon and artificially produced radioisotopes

compare nuclear and fossil fuels in terms of the worldwide reserves of raw materials and of the possible hazards to the environment brought on by their use

▲ use information from an element's mass spectrum to calculate its relative atomic mass

▲ explain the instability of a nucleus in terms of its neutron-to-proton ratio

▲ write a balanced nuclear equation which may include any nuclide, neutrons, protons, alpha and beta particles and gamma rays

▲ give examples of radioisotopes being used as a means of dating material, labelling molecules and producing energy by nuclear fusion.

PROBLEM SOLVING EXERCISES

1. The graph shows the rate of decay of a sample of bismuth-212.

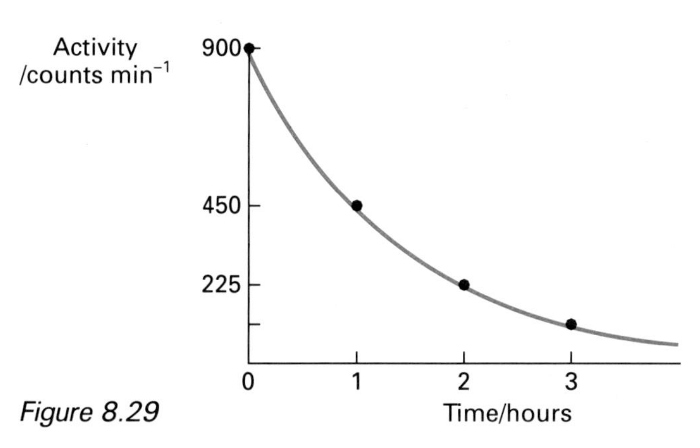

Figure 8.29

Draw a graph showing the curve that would be obtained if the sample of bismuth-212 used was half the above mass.

(PS skill 6)

2. Radioactive salt solution, $^{24}NaCl$, was injected into a patient's bloodstream and radiation detectors placed at four sites on the patient's body. The detectors were connected electrically to separate meters which measured the radioactivity at each point.

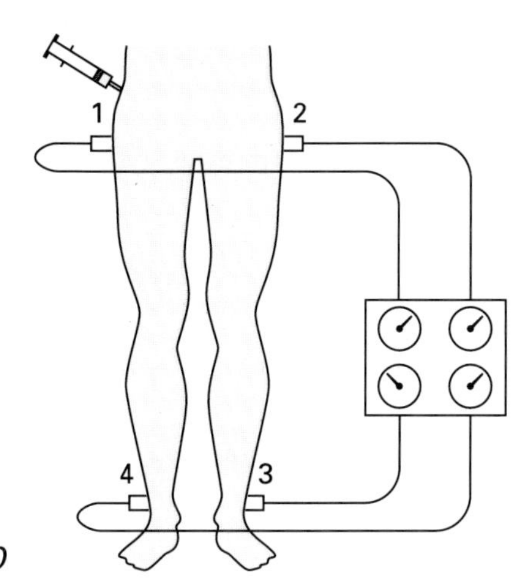

Figure 8.30

A low reading was obtained on meter 3.

What can you infer about the circulation of blood through the patient's blood vessels?

(PS skill 8)

3. A solution of $AgNO_3(aq)$ in a flask is tested with the stepwise addition of NaI(aq) to which a trace of radioactive iodine ($^*I$) has been added. Silver iodide is precipitated.

$$AgNO_3(aq) + Na^*I(aq) \rightarrow NaNO_3(aq) + Ag^*I(s)$$

The **solution** registers no radioactivity until all the $Ag^+(aq)$ has been used up.

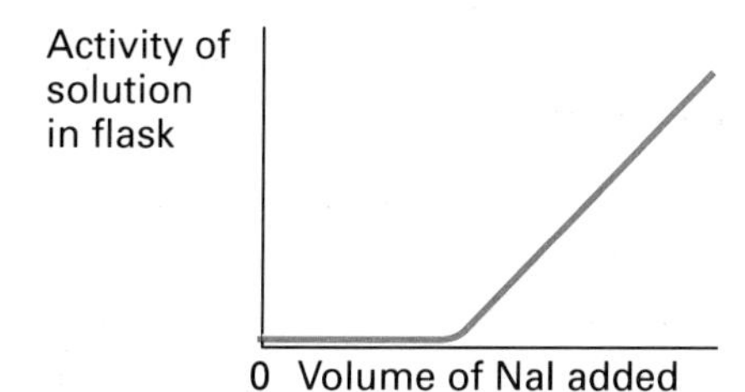

Figure 8.31

In a second experiment, $AgNO_3(aq)$ is added stepwise to a flask containing NaI(aq) to which a trace of *I has been added.

Sketch the graph you would expect when the activity of the solution in the flask is plotted against the volume of $AgNO_3$ added.

(PS skill 10)

4. The following is a description of the uranium fuel cycle. Uranium ore is first mined and then fed into an extraction plant. Here uranium, containing 0.7 per cent ^{235}U, is isolated and then transferred to an enrichment plant where the proportion of ^{235}U is raised to about 3 per cent. This enriched uranium is then fabricated into fuel elements which are used in nuclear reactors to provide energy. Spent fuel is moved to a fuel reprocessing plant where separation of unused fuel from fission products takes place. Unconverted uranium fuel is recycled to the enrichment plant. Gaseous waste products are removed and liquid waste products transferred to a storage tank to cool down.

Summarise this information by means of a flow diagram.

(PS skill 2)

5. The table overleaf lists the half-lives of alpha-emitting uranium isotopes and the energies of alpha particles emitted.

PROBLEM SOLVING EXERCISES

Uranium isotope	Half-life	Energy of alpha particle emitted/MeV
226	0.5 second	7.43
227	1.1 minutes	6.87
228	9.1 minutes	6.68
230	20.8 days	5.88
232	68.9 years	5.32
233	1.59×10^5 years	4.82
234	2.45×10^5 years	4.78
235	7.04×10^8 years	4.40
236	2.34×10^7 years	4.49
238	4.46×10^9 years	4.20

From these results, what can you say (infer) about the relationship between the stability of a nucleus and the energy carried off by the alpha particle which it emits?

(PS skill 8)

6. Thermochemical equations for the nuclear fusion of hydrogen isotopes and the combustion of hydrogen gas are given below.

$$\Delta H/\text{kJ mol}^{-1}$$

$$^{2}_{1}H + ^{3}_{1}H \longrightarrow ^{4}_{2}He + ^{1}_{0}n \qquad -16.9 \times 10^8$$

$$H_2 + \tfrac{1}{2}O_2 \longrightarrow H_2O \qquad -286$$

What amount of hydrogen gas (in moles) would need to be burned in order to produce the same amount of heat energy as that released by the fusion of 1 mol hydrogen-2 nuclei and 1 mol hydrogen-3 nuclei?

(PS skill 4)

7. Four methods of presenting information are listed in the grid.

A line graph	B pie chart
C flow diagram	D bar graph

Which method(s) could be used to display the following information?

(a) The neutron:proton ratio in carbon-14.

(b) The rate of disintegration of carbon-14.

(c) The process by which carbon-14 enters living material.

(PS skill 3)

8. Germanium is used in semiconductors, alloys and special glasses. Naturally occurring germanium consists of the isotopes listed below.

Relative atomic mass of isotope	Relative abundance/per cent
70	20.5
72	27.4
73	7.8
74	36.5
75	7.8

Sketch the mass spectrum that would be obtained from a sample of naturally occurring germanium (assume all the ions produced are Ge^+).

(PS skill 2)

9. Strontium-90 is a 'neutron-rich' radioisotope and decays by beta-emission.

Calculate the n/p ratios of strontium-90 and the isotope formed as a result of the beta-emission.

(PS skill 4)

10. In an ester-forming reaction, methanol was radioactively tagged with oxygen-18. The ester rather than the water was found to be radioactive.

$$CH_3\text{–}\overset{\overset{\displaystyle O}{\|}}{C}\text{–}O\text{–}H + CH_3\text{–}^{18}O\text{–}H \longrightarrow CH_3\text{–}\overset{\overset{\displaystyle O}{\|}}{C}\text{–}^{18}O\text{–}CH_3 + H_2O$$

In a second experiment, on ester hydrolysis, the water is tagged with ^{18}O.

Complete the equation below to show whether the radioactive oxygen ends up in the acid or alcohol molecule.

$$CH_3\text{–}\overset{\overset{\displaystyle O}{\|}}{C}\text{–}O\text{–}CH_3 + H\text{–}^{18}O\text{–}H \longrightarrow$$

(PS skill 10)

11. Strontium-90 is a beta-emitter with a half-life of 28 years.

Calculate the number of beta particles emitted from 0.02 g of ^{90}Sr in 56 years.

(PS skill 4)

ANSWERS TO PROBLEM SOLVING EXERCISES

Unit 1

1.

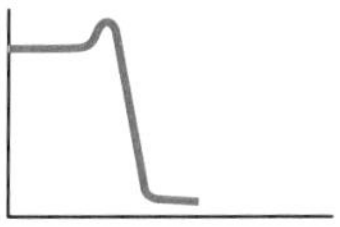

2. The sulphur is poisoning the nickel catalyst. When all the active sites on the nickel surface have been filled the reaction will stop and the rate will be zero.

3. Change in concentration = 0.000 81 × 210 = 0.17 mol l^{-1}.
Concentration of sucrose after 210 min = 0.32 − 0.17 = 0.15 mol l^{-1}

4. Generally the rate of reaction increases with increasing surface area.

5.

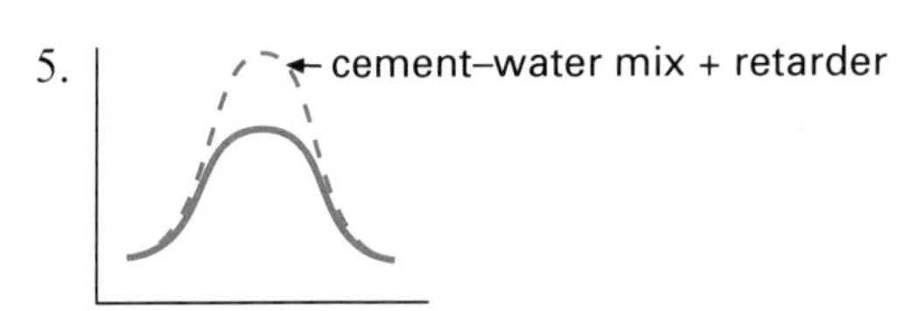

6. (a) A and D (b) B and D (c) B and F

7. Rate = 0.25 s^{-1}; time taken = 1/r = 4 s

8. Apparatus should contain reaction flask or tube, stoppered and with a delivery tube to either a gas syringe or a measuring cylinder inverted over water.

9. Mn^{2+}

10. C

Unit 2

1. (a) A and E (b) B and F (c) A and C

2.

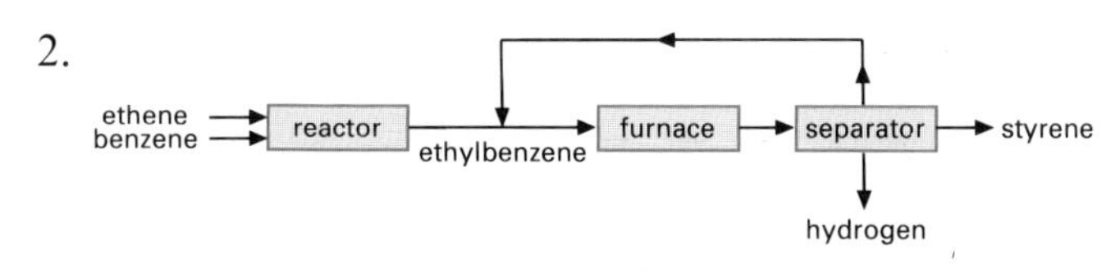

3. (a) CO, H_2, NH_3, O_2, CH_4 (b) $CH_2{=}CHCN$
(c) HCN, CH_3CN, CH_3CH_2CN

4. (a) Recycle the unreacted hydrocarbons.
(b) Fractional distillation.

5. A possible subdivision sequence:

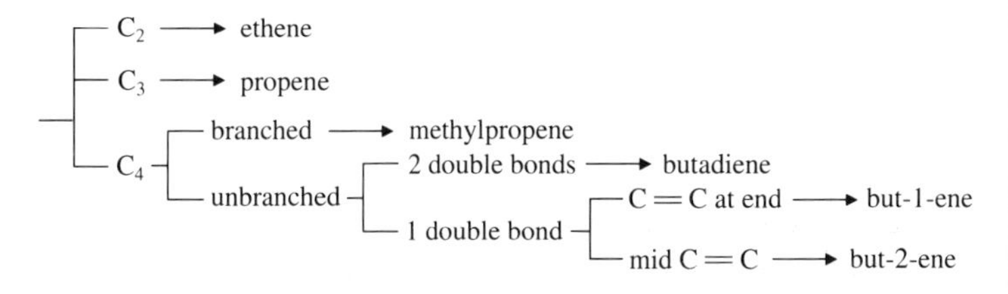

6.

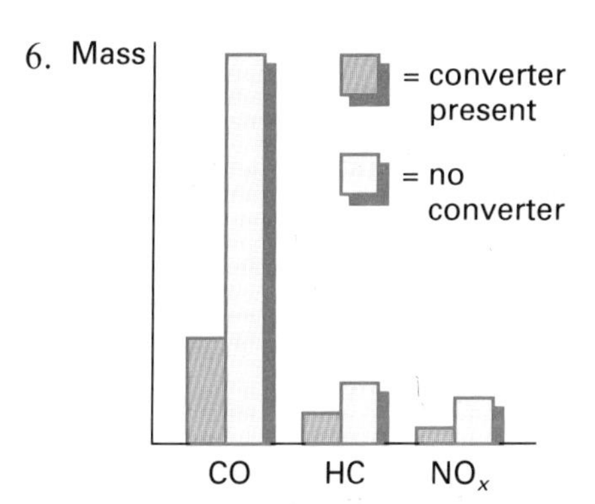

7. A contains a C=C bond and will rapidly decolorise bromine.

8. 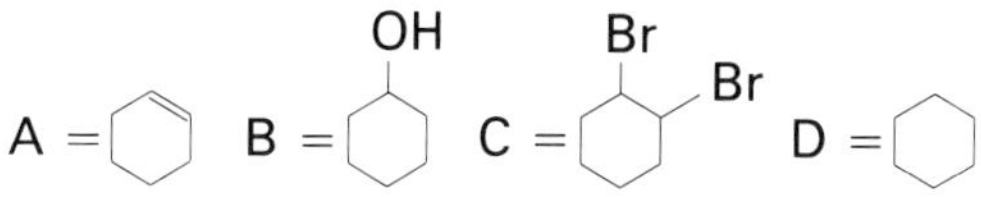 A = B = C = D =
OH
Br
Br

9. Supplying oxygen (oxidation) would be likely to convert aldehyde RCHO to carboxylic acid, RCOOH.

10. $CH_3CH_2OCH_2CH(CH_3)CH_3$

11. $CH_3CH_2CH{=}CHCH(CH_3)CH_2CH_3$

12. The alkene may be either hept-2-ene or hept-3-ene.

ANSWERS TO PROBLEM SOLVING EXERCISES

Unit 3

1. C
2. $2CO:3H_2$
3. Current = 16×10^3A
4. 15.3 g air
5. Au^{3+}
6. A and D
7. C_3H_6
8. 0.73 litre H_2
9. 1 vol. CH_4:3 vol. Cl_2
10. The apparatus should contain a power supply and two electrodes, perhaps fixed through the base of the container. The gases may be collected in test-tubes filled with water at the start of the experiment and inverted over the electrodes.
11. The bars for CO_2 in experiments 2, 3 and 4 should each be 1 mol in size.

Unit 4

1. C and A
2. ...—O—C(CH$_3$)(H)—CH$_2$—C(=O)—O—C(CH$_3$)(H)—CH$_2$—C(=O)—O—C(CH$_3$)(H)—CH$_2$—C(=O)—...
3. The tripeptide is composed of lysine, methionine and a third amino acid which is not used as reference in this experiment.
4. Energy released = 39.5 J
5. $CH_3(CH_2)_6$ C(=O)—O—CH(CH$_3$)—$(CH_2)_5$—CH_3
6. Titrate the fat or oil with an alkali of known concentration.
7.

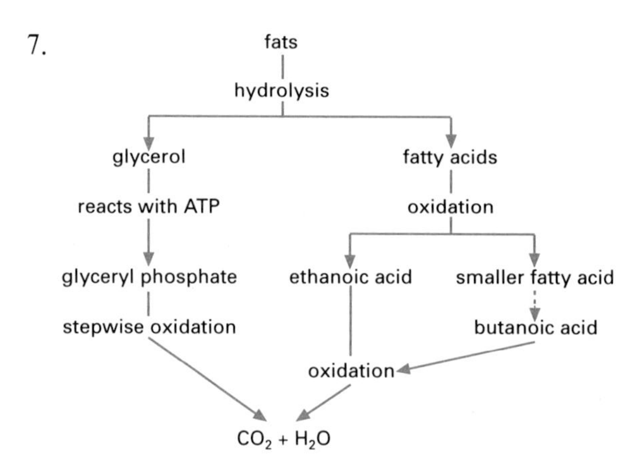

8. Methionine is an essential amino acid and must be included in the diet. Glycine is not essential and can be synthesised internally by chicks.
9. (a) B (b) C (c) A or B
10. Enzymes offer lower activation energy routes than non-enzyme catalysts.
11. B and F are unsaturated, whereas the other four acids are saturated.

Unit 5

1. (a) The fluoride has an ionic lattice structure. (b) The bromide has a molecular structure.
2. A and D
3. C—Si bond length = $0.5 \times 154 + 0.5 \times 232$ = 193 pm

ANSWERS TO PROBLEM SOLVING EXERCISES

4. The number of occupied electron shells is reduced from three to two in changing from Mg to Mg^{2+}.

5. Solubility in water, melting point range, conductivity of melt, etc.

6. linear: H—Be—H; bent: O—N—O; flat: BF_3 (F, F, F around B); pyramidal (3-D): NCl_3 (Cl, Cl, Cl around N)

7. F...H...F The stability is due to hydrogen bonding, with H acting as a bridge between the two F atoms.

8. C

9. A possible method would be to fill a gas syringe with the gas and pass it through water, collecting the undissolved gas in an empty gas syringe.

10. $SiBr_4$ and $SiCl_4$.

11. Example: Group 1, Li ⟶ Cs As covalent radius increases, mp decreases.

12. Maleic acid has internal hydrogen bonding *within* the molecule, as the COOH groups are adjacent. In fumaric acid the COOH groups are on opposite sides of the molecule: hydrogen bonding is between neighbouring molecules (intermolecular). This produces extra attraction between fumaric acid molecules and results in a higher mp.

Unit 6

1. Predicted ΔH = about -8077 kJ mol^{-1}

2. The heat required to sustain the endothermic decomposition of limestone is supplied from that released by the exothermic reaction between coke and air.

3. IF

4. Some of the bonds in the molecules are different and the energy inputs for bond breaking will therefore differ.

5. Estimated bond enthalpy = 325 kJ mol^{-1}

6. ΔT values (descending) are 5 K, 5 K, 10 K.

7. The higher the charges on the ions, the stronger the interionic attraction. More energy is needed to break up the lattice.

8. 386 kJ mol^{-1}

9. There is no obvious correlation. Oxides of Mg and Ti, for example are both very stable (large negative ΔH values) but Mg oxidises very slowly while Ti oxidises relatively rapidly.

10. Somewhere between $+14.2$ and $+2.9$ kJ mol^{-1}.

11. As the proportion of air rises, (1) hydrocarbons burn more completely to CO_2 and less CO is formed, (2) more N_2 and O_2 combine, raising the level of NO_x.

Unit 7

1. The degree of dissociation increases with increasing dilution. This trend is less marked in the case of a strong acid which is almost fully ionised to start with.

2. pH = 2

3.

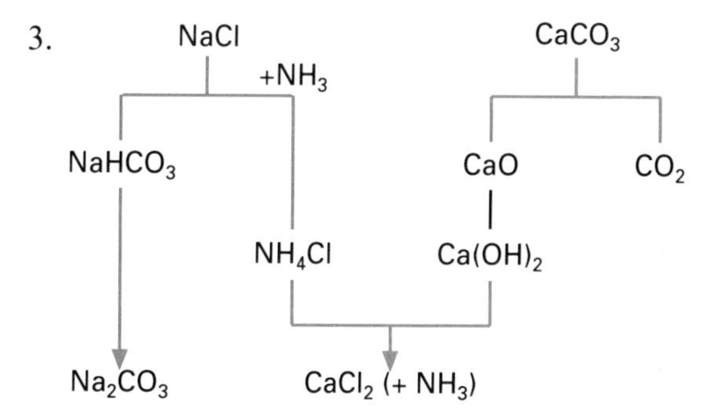

Note: the NH_3 formed in stage (v) can be shown recycled to stage (i).

4. (a) In acid $[H^+]$ is raised and the position of equilibrium shifts to the left, favouring the yellow molecules.
(b) In alkaline solution $[H^+]$ is reduced by neutralisation and the equilibrium position shifts to the right, favouring the blue ions.

ANSWERS TO PROBLEM SOLVING EXERCISES

5. Mixture contains 0.4 mol HI + 0.05 mol H_2 + 0.05 mol I_2.

6. Tartaric acid is a weak acid.

7. Apparatus might consist of two horizontal combustion tubes containing samples of $CuCO_3$. One sample is heated in a stream of air while the other is heated in a stream of CO_2. The first turns black readily but the second sample barely changes colour.

8 (a) At the low temperature, formation of CH_3OH is controlled by the rate factor: as T rises, rate rises and methanol is produced more rapidly.
(b) At the higher temperatures, equilibrium is achieved and the formation of CH_3OH is controlled by the equilibrium position. Being exothermic, the yield of methanol decreases as T rises: the amount produced in a given time diminishes.

9. 0.5 mol l^{-1}

10. Ethanoic acid is weak:

$$CH_3COOH \rightleftharpoons CH_3COO^- + H^+$$

Addition of CH_3COO^- ions shifts equilibrium in favour of the undissociated molecules. $[H^+]$ falls and the reaction rate with chalk decreases.

$$2H^+ + CO_3^{2-} \longrightarrow H_2O + CO_2$$

11. B, C, F and H

12. $NaHSO_4$ is acidic as its negative ion donates an H^- ion to water.

$$HSO_4^- + H_2O \longrightarrow H_3O^+ + SO_4^{2-}$$

H^+ (H^+)

$NaHCO_3$ is basic as its negative ion removes an H^+ ion from water, leaving OH^-

$$HCO_3^- + H_2O \longrightarrow H_2O + CO_2 + OH^-$$

H^+

13. Put tube 1 in a freezing mixture, tube 2 in boiling water and leave tube 3 at room temperature. The colours to be observed are, respectively, yellow, dark brown and somewhere in between yellow and dark brown.

14. (a) As the reaction in each catalyst bed is exothermic, the temperature rises gradually but drops sharply as heat is removed between beds by a heat exchanger.
(b) As the formation of SO_3 is exothermic, rising temperature shifts the equilibrium to the left and lowers the proportion of SO_3 in the mixture. With the fall in temperature between beds the yield of SO_3 rises sharply each time.

Unit 8

1. The graph should show the activity at the start as 450 counts min^{-1}, 225 after one half-life (approximately one hour), and so on.

2. Circulation is restricted somewhere en route to position 3.

3.

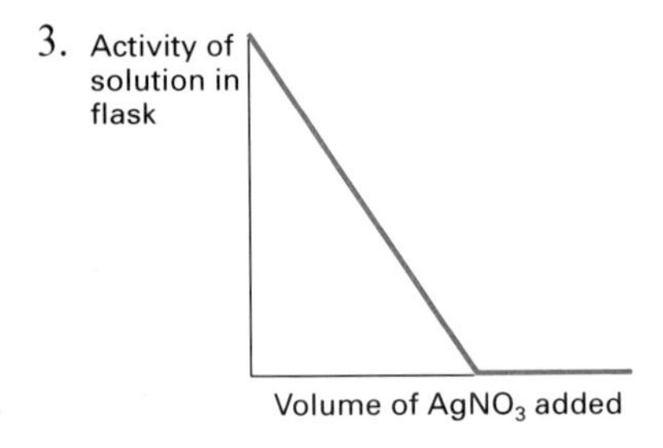

4.

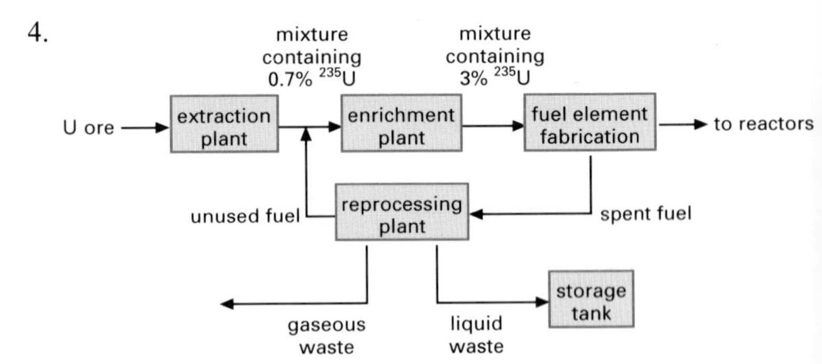

5. The more stable a nucleus, the longer is its half-life. From this data it appears that the longer the half-life, the lower the energy of the emitted alpha particles. We can deduce that low energy alpha radiation is emitted from the more stable nuclei and high energy radiation from those that are less stable.

6. Amount of H_2 gas required = 6×10^6 mol.

ANSWERS TO PROBLEM SOLVING EXERCISES

7. (a) B or D (b) A (c) C

8.

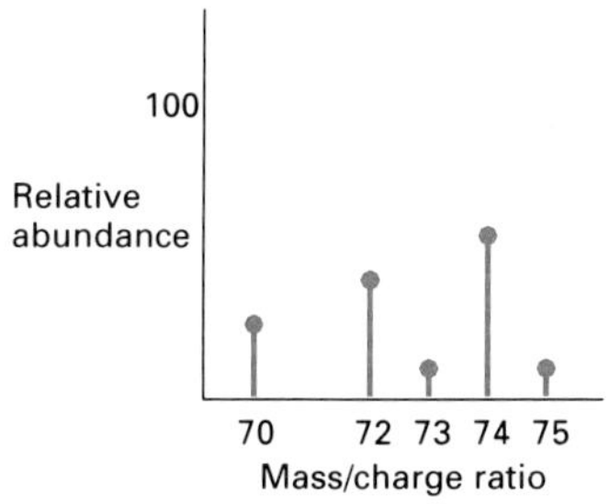

9. n/p ratio for $^{90}_{38}Sr$ = 1.37; n/p ratio for $^{90}_{39}Y$ = 1.31

10. $$CH_3\overset{\overset{\displaystyle O}{\|}}{C}{-}O{-}CH_3 + H{-}{}^{18}O{-}H \longrightarrow CH_3{-}\overset{\overset{\displaystyle O}{\|}}{C}{-}{}^{18}O{-}H + CH_3OH$$

11. Number of beta particles = 1×10^{20}

INDEX

C

W

X

Y